Undergraduate Topics in Computer Science

D0939440

'Undergraduate Topics in Computer Science' (UTiCS) delivers high-quality instructional content for undergraduates studying in all areas of computing and information science. From core foundational and theoretical material to final-year topics and applications, UTiCS books take a fresh, concise, and modern approach and are ideal for self-study or for a one- or two-semester course. The texts are all authored by established experts in their fields, reviewed by an international advisory board, and contain numerous examples and problems, many of which include fully worked solutions.

The UTiCS concept relies on high-quality, concise books in softback format, andgenerally a maximum of 275–300 pages. For undergraduate textbooks that arelikely to be longer, more expository, Springer continues to offer the highly regardedTexts in Computer Science series, to which we refer potential authors.

More information about this series at http://www.springer.com/series/7592

Gerard O'Regan

Mathematics in Computing

An Accessible Guide to Historical, Foundational and Application Contexts

Second Edition

 Springer

Gerard O'Regan
SQC Consulting
Mallow, Cork, Ireland

ISSN 1863-7310   ISSN 2197-1781   (electronic)
Undergraduate Topics in Computer Science
ISBN 978-3-030-34208-1   ISBN 978-3-030-34209-8   (eBook)
https://doi.org/10.1007/978-3-030-34209-8

This Springer imprint is published by the registered company Springer Nature Switzerland AG
The registered company address is: Gewerbestrasse 11, 6330 Cham, Switzerland

To

*My dear aunts Mrs. Rita Lowry and
Mrs. Kitty Butler*

and

In memory of my late uncle Moses Fenton.

Preface

Overview

The objective of this book is to give the reader a flavour of mathematics used in the computing field. We emphasize the applicability of mathematics rather than the study of mathematics for its own sake, and the goal is that the reader will appreciate the rich applications of mathematics to the computing field. This includes applications to the foundations of computing; to error detection and correcting codes with finite field theory; to the field of cryptography with the results of number theory; to the modelling of telecommunication networks with graph theory; to the application of discrete mathematics and proof techniques to the software correctness field (especially safety-critical systems using formal methods and model checking); to language theory and semantics; and to computability and decidability.

Organization and Features

The first chapter introduces analog and digital computers, and the von Neumann architecture which is the fundamental architecture underlying a digital computer. Chapter 2 discusses the foundations of computing, and we describe the binary number system and the step reckoner calculating machine that were invented by Leibniz. Babbage designed the difference engine as a machine to evaluate polynomials, and his analytic engine provided the vision of a modern computer. Boole was an English mathematician who made important contributions to mathematics and logic, and his symbolic logic is the foundation for digital computing.

Chapter 3 provides an introduction to fundamental building blocks in mathematics including sets, relations and functions. A set is a collection of well-defined objects and it may be finite or infinite. A relation between two sets A and B indicates a relationship between members of the two sets and is a subset of the Cartesian product of the two sets. A function is a special type of relation such that for each element in A there is at most one element in the co-domain B. Functions may be partial or total and injective, surjective or bijective.

Chapter 4 presents a short introduction to algorithms, where an algorithm is a well-defined procedure for solving a problem. It consists of a sequence of steps that takes a set of values as input and produces a set of values as output. An algorithm is an exact specification of how to solve the problem, and it explicitly defines the procedure so that a computer program may implement the solution in some programming language.

Chapter 5 presents the fundamentals of number theory and discusses prime number theory and the greatest common divisor and least common multiple of two numbers.

Chapter 6 discusses algebra and we discuss simple and simultaneous equations, including the method of elimination and the method of substitution to solve simultaneous equations. We show how quadratic equations may be solved by factorization, completing the square or using the quadratic formula. We present the laws of logarithms and indices. We discuss various structures in abstract algebra, including monoids, groups, rings, integral domains, fields and vector spaces.

Chapter 7 discusses sequences and series and permutations and combinations. Arithmetic and geometric sequences and series are discussed, and we discuss applications of geometric sequences and series to the calculation of compound interest and annuities.

Chapter 8 discusses mathematical induction and recursion. Induction is a common proof technique in mathematics, and there are two parts to a proof by induction (the base case and the inductive step). We discuss strong and weak induction, and we discuss how recursion is used to define sets, sequences and functions. This leads us to structural induction, which is used to prove properties of recursively defined structures.

Chapter 9 discusses graph theory where a graph $G = (V, E)$ consists of vertices and edges. It is a practical branch of mathematics that deals with the arrangements of vertices and edges between them, and it has been applied to practical problems such as the modelling of computer networks, determining the shortest driving route between two cities and the travelling salesman problem.

Chapter 10 discusses cryptography, which is an important application of number theory. The codebreaking work done at Bletchley Park in England during the Second World War is discussed, and the fundamentals of cryptography, including private- and public-key cryptosystems, are discussed.

Chapter 11 presents coding theory and is concerned with error detection and error correction codes. The underlying mathematics includes abstract mathematics such as group theory, rings, fields and vector spaces.

Chapter 12 discusses language theory and includes a discussion on grammars, parse trees and derivations from a grammar. The important area of programming language semantics is discussed, including axiomatic, denotational and operational semantics.

Chapter 13 discusses computability and decidability. The Church–Turing thesis states that anything that is computable is computable by a Turing machine. Church and Turing showed that mathematics is not decidable. In other words, there is no mechanical procedure (i.e. algorithm) to determine whether an arbitrary mathematical proposition is true or false, and so the only way is to determine the truth or falsity of a statement is try to solve the problem.

Chapter 14 discusses matrices including 2×2 and general $n \times m$ matrices. Various operations such as the addition and multiplication of matrices are considered, and the determinant and inverse of a square matrix are discussed. The application of matrices to solving a set of linear equations using Gaussian elimination is considered.

Chapter 15 presents a short history of logic, and we discuss Greek contributions to syllogistic logic, stoic logic, fallacies and paradoxes. Boole's symbolic logic and its application to digital computing are discussed, and we consider Frege's work on predicate logic.

Chapter 16 provides an introduction to propositional and predicate logic. Propositional logic may be used to encode simple arguments that are expressed in natural language and to determine their validity. The nature of mathematical proof is discussed, and we present proof by truth tables, semantic tableaux and natural deduction. Predicate logic allows complex facts about the world to be represented, and new facts may be determined via deductive reasoning. Predicate calculus includes predicates, variables and quantifiers, and a predicate is a characteristic or property that the subject of a statement can have.

Chapter 17 presents some advanced topics in logic including fuzzy logic, temporal logic, intuitionistic logic, undefined values, theorem provers and the applications of logic to AI. Fuzzy logic is an extension of classical logic that acts as a mathematical model for vagueness. Temporal logic is concerned with the expression of properties that have time dependencies, and it allows temporal properties about the past, present and future to be expressed. Intuitionism was a controversial theory on the foundations of mathematics based on a rejection of the law of the excluded middle, and an insistence on constructive existence. We discuss three approaches to deal with undefined values, including the logic of partial functions; Dijkstra's approach with his cand and cor operators; and Parnas's approach which preserves a classical two-valued logic.

Chapter 18 discusses the nature of proof and theorem proving, and we discuss automated and interactive theorem provers. We discuss the nature of mathematical proof and formal mathematical proof. Chapter 19 discusses software engineering and the mathematics to support software engineering.

Chapter 20 discusses software reliability and dependability, and covers topics such as software reliability and software reliability models; the Cleanroom methodology, system availability, safety and security-critical systems; and dependability engineering.

Chapter 21 discusses formal methods, which consist of a set of mathematical techniques to rigorously specify and derive a program from its specification. Formal methods may be employed to rigorously state the requirements of the proposed

system; they may be employed to derive a program from its mathematical specification; and they may provide a rigorous proof that the implemented program
satisfies its specification. They have been mainly applied to the safety-critical field.

Chapter 22 presents the Z specification language, which is one of the most
widely used formal methods. It was developed at Oxford University in the U.K.

Chapter 23 discusses automata theory, including finite-state machines, pushdown automata and Turing machines. Finite-state machines are abstract machines
that are in only one state at a time, and the input symbol causes a transition from the
current state to the next state. Pushdown automata have greater computational
power, and they contain extra memory in the form of a stack from which symbols
may be pushed or popped. The Turing machine is the most powerful model for
computation, and this theoretical machine is equivalent to an actual computer in the
sense that it can compute exactly the same set of functions.

Chapter 24 discusses model checking which is an automated technique such that
given a finite-state model of a system and a formal property, then it systematically
checks whether the property is true or false in a given state in the model. It is an
effective technique to identify potential design errors, and it increases the confidence in the correctness of the system design.

Chapter 25 discusses probability and statistics and includes a discussion on
discrete and continuous random variables, probability distributions, sample spaces,
sampling, the abuse of statistics, variance and standard deviation and hypothesis
testing. The application of probability to the software reliability field is discussed.

Chapter 26 discusses complex numbers and quaternions. Complex numbers are
of the form $a + bi$ where a and b are real numbers, and $i^2 = -1$. Quaternions are a
generalization of complex numbers to quadruples that satisfy the quaternion formula $i^2 = j^2 = k^2 = -1$.

Chapter 27 provides a very short introduction to calculus and provides a
high-level overview of limits, continuity, differentiation, integration, numerical
analysis, Fourier series, Laplace transforms and differential equations.

Chapter 28 is the concluding chapter in which we summarize the journey that we
have travelled in this book.

Audience

The audience of this book includes computer science students who wish to obtain
an overview of mathematics used in computing, and mathematicians who wish to
get an overview of how mathematics is applied in the computing field. The book
will also be of interest to the motivated general reader.

Acknowledgements

I am deeply indebted to friends and family who supported my efforts in this endeavour. My thanks to the team at Springer for suggesting this new edition and for their professional work. A special thanks to my aunts (Mrs. Rita Lowry and Mrs. Kitty Butler) who are always a pleasure to visit in Co.Tipperary and Co.Cork, and who have clearly shown that it is possible to be over 90 and yet to have the energy and sense of fun of teenagers.

Cork, Ireland   Gerard O'Regan

Contents

List of Figures

What Is a Computer? 1

Key Topics

Analog computers
Digital computers
Vacuum tubes
Transistors
Integrated circuits
von Neumann architecture
Generation of computers
Hardware
Software

1.1 Introduction

It is difficult to think of western society today without modern technology. We have witnessed in recent decades a proliferation of high-tech computers, mobile phones, text messaging, the Internet, the World Wide Web and social media. Software is pervasive, and it is an integral part of automobiles, airplanes, televisions and mobile communication. The pace of change is relentless, and communication today is instantaneous with technologies such as Skype, Twitter and WhatsApp.

Today, people may book flights over the World Wide Web as well as keeping in contact with friends and family members around the world. In previous generations, communication involved writing letters that often took months to reach the

© Springer Nature Switzerland AG 2020   1
G. O'Regan, *Mathematics in Computing*, Undergraduate Topics
in Computer Science, https://doi.org/10.1007/978-3-030-34209-8_1

recipient. However, today's technology has transformed the modern world into a global village, and the modern citizen may make video calls over the Internet or post pictures and videos on social media sites such as Facebook and Twitter. The World Wide Web allows business to compete in a global market.

A computer is a programmable electronic device that can process, store and retrieve data. It processes data according to a set of instructions or program. All computers consist of two basic parts, namely, *hardware* and *software*. The hardware is the physical part of the machine, and the components of a digital computer include memory for short-term storage of data or instructions; an arithmetic/logic unit for carrying out arithmetic and logical operations; a control unit responsible for the execution of computer instructions in memory; and peripherals that handle the input and output operations. Software is a set of instructions that tells the computer what to do.

The original meaning of the word '*computer*' referred to someone who carried out calculations rather than an actual machine. The early digital computers built in the 1940s and 1950s were enormous machines consisting of thousands of vacuum tubes. They typically filled a large room but their computational power was a fraction of the personal computers and mobile devices used today.

There are two distinct families of computing devices, namely, *digital computers* and the historical *analog computer*. The earliest computers were analog not digital, and these two types of computer operate on quite different principles.

The computation in a digital computer is based on binary digits, i.e. '0' and '1'. Electronic circuits are used to represent binary numbers, with the state of an electrical switch (i.e. 'on' or 'off') representing a binary digit internally within a computer.

A digital computer is a sequential device that generally operates on data one step at a time, and the earliest digital computers were developed in the 1940s. The data are represented in binary format, and a single transistor (initially bulky vacuum tubes) is used to represent a binary digit. Several transistors are required to store larger numbers.

An *analog computer* operates in a completely different way to a digital computer. The representation of data in an analog computer reflects the properties of the data that are being modelled. For example, data and numbers may be represented by physical quantities such as electric voltage, whereas a stream of binary digits is used to represent them in a digital computer.

1.2   Analog Computers

James Thompson (who was the brother of the physicist Lord Kelvin) did early foundational work on analog computation in the nineteenth century. He invented a wheel-and-disc integrator, which was used in mechanical analog devices, and he worked with Kelvin to construct a device to perform the integration of a product of two functions. Kelvin later described a general-purpose analog machine (he did not build it) for integrating linear differential equations. He built a tide predicting analog computer that remained in use at the Port of Liverpool up to the 1960s.

The operations in an analog computer are performed in parallel, and they are useful in simulating dynamic systems. They have been applied to flight simulation, nuclear power plants and industrial chemical processes.

Vannevar Bush developed the first large-scale general-purpose mechanical analog computer at the Massachusetts Institute of Technology. Bush's differential analyser (Fig. 1.1) was a mechanical analog computer designed to solve sixth-order differential equations by integration, using wheel-and-disc mechanisms to perform the integration. The mechanization allowed integration and differential equations problems to be solved more rapidly. The machine took up the space of a large table in a room and weighed about 100 tonnes.

It contained wheels, discs, shafts and gears to perform the calculations. It required a considerable setup time by technicians to solve an equation. It contained 150 motors and miles of wires connecting relays and vacuum tubes.

Data representation in an analog computer is compact, but it may be subject to corruption with noise. A single capacitor can represent one continuous variable in an analog computer. Analog computers were replaced by digital computers shortly after the Second World War.

Fig. 1.1 Vannevar Bush with the differential analyser

1.3 Digital Computers

Early digital computers used vacuum tubes to store binary information, and a vacuum tube may represent the binary value '0' or '1'. These tubes were large and bulky and generated a significant amount of heat. Air conditioning was required to cool the machine, and there were problems with the reliability of the tubes.

Shockley and others invented the transistor in the late 1940s, and it replaced vacuum tubes from the late 1950s onwards. Transistors are small and consume very little power, and the resulting machines were smaller, faster and more reliable.

Integrated circuits were introduced in the early 1960s, and a massive amount of computational power could now be placed on a very small chip. Integrated circuits are small and consume very little power, and may be mass-produced to a very high-quality standard. However, integrated circuits are difficult to modify or repair, and are nearly always replaced on failure.

The fundamental architecture of a computer has remained basically the same since von Neumann and others proposed it in the mid-1940s. It includes a central processing unit which includes the control unit and the arithmetic unit, an input and output unit, and memory.

1.3.1 Vacuum Tubes

A vacuum tube is a device that relies on the flow of an electric current through a vacuum. Vacuum tubes (*thermionic valves*) were widely used in electronic devices such as televisions, radios and computers until the invention of the transistor.

The basic idea of a vacuum tube is that the current passes through the filament, which then heats it up so that it gives off electrons. The electrons are negatively charged and are attracted to the small positive plate (or anode) within the tube. A unidirectional flow is thus established between the filament and the plate. Thomas Edison had observed this while investigating the reason for breakage of lamp filaments. He noted an uneven blackening (darkest near one terminal of the filament) of the bulbs in his incandescent lamps and noted that current flows from the lamp's filament and a plate within the vacuum.

The first generation of computers used several thousand bulky vacuum tubes, with several racks of vacuum tubes taking up the space of a large room. The vacuum tube used in the early computers was a three-terminal device, and it consisted of a cathode, a grid and a plate. It was used to represent one of two binary states, i.e. the binary value '0' or '1'.

The filament of a vacuum tube becomes unstable over time. In addition, if air leaks into the tube then oxygen will react with the hot filament and damage it. The size and unreliability of vacuum tubes motivated research into more compact and reliable technologies. This led to the invention of the transistor in the late 1940s.

The first generation of digital computers all used vacuum tubes, e.g. the Atanasoff–Berry computer (ABC) developed at the University of Iowa in 1942;

Colossus developed at Bletchley Park, England in 1944; and ENIAC developed in the United States in the mid-1940s.

1.3.2 Transistors

The transistor is a fundamental building block in modern electronic systems, and its invention revolutionized the field of electronics. It was smaller, cheaper and more reliable than the existing vacuum tubes.

The transistor is a three-terminal, solid-state electronic device. It can control electric current or voltage between two of the terminals by applying an electric current or voltage to the third terminal. The three-terminal transistor enables an electric switch to be made which can be controlled by another electrical switch. Complicated logic circuits may be built up by cascading these switches (switches that control switches that control switches, and so on.).

These logic circuits may be built very compactly on a silicon chip with a density of over a million transistors per square centimetre. The switches may be turned on and off very rapidly (e.g. every 0.000000001 s). These electronic chips are at the heart of modern electronic devices.

The transistor (Fig. 1.2) was developed at Bell Labs after the Second World War. The goal of the research was to find a solid-state alternative to vacuum tubes, as this technology was too bulky and unreliable. Three Bell Labs inventors (Shockley, Bardeen and Brattain) were awarded the Nobel Prize in physics in 1956 in recognition of their invention of the transistor.

William Shockley was involved in radar research and anti-submarine operations research during the Second World War, and after the war he led the Bell Labs research group (that included Bardeen and Brattain) that aimed to find a solid-state alternative to the glass-based vacuum tubes.

Bardeen and Brattain succeeded in creating a point-contact transistor in 1947 independently of Shockley who was working on a junction-based transistor. Shockley believed that the points contact transistor would not be commercially viable, and his junction point transistor was announced in 1951.

Shockley formed Shockley Semiconductor Inc. (part of Beckman Instruments) in 1955. The second generation of computers used transistors instead of vacuum tubes. The University of Manchester's experimental Transistor Computer was one of the earliest transistor computers. The prototype machine appeared in 1953, and the full-size version was commissioned in 1955. The invention of the transistor is discussed in more detail in (O'Regan 2018).

1.3.3 Integrated Circuits

Jack Kilby of Texas Instruments invented the integrated circuit in 1958. His invention used a wafer of germanium, and Robert Noyce of Fairchild Semiconductors did subsequent work on silicon-based integrated circuits. The integrated

Fig. 1.2 Replica of transistor. Public domain

circuit was an effective solution to the problem of building a circuit with many components, and the Nobel Prize in Physics was awarded to Kirby in 2000 for his contributions to its invention.

An integrated circuit consists of a set of electronic circuits on a small chip of semiconductor material, and it is much smaller than a circuit made from independent components. Integrated circuits today are extremely compact and may contain billions of transistors and other electronic components in a tiny area. The width of each conducting line has got smaller and smaller over the years due to advances in technology, and it is now measured in tens of nanometers.

The number of transistors per unit area has been doubling (roughly) every 1–2 years over the last 30 years. This amazing progress in circuit fabrication is known as Moore's law after Gordon Moore (one of the founders of Intel) who formulated the law in the mid-1960s (O'Regan 2018).

Kilby was designing micromodules for the military, and this involved connecting many germanium[1] wafers of discrete components together by stacking each wafer on top of one another. The connections were made by running wires up the sides of the wafers.

Kilby saw this process as unnecessarily complicated and realized that if a piece of germanium was engineered properly that it could act as many components simultaneously. That is, instead of making transistors one-by-one several transistors

---

[1]Germanium is an important semiconductor material used in transistors and other electronic devices.

could be made at the same time on the same piece of semiconductor. In other words, transistors and other electric components such as resistors, capacitors and diodes can be made by the same process with the same materials.

This idea led to the birth of the first integrated circuit and its development involved miniaturizing transistors and placing them on silicon chips called semi-conductors. The use of semiconductors led to third-generation computers, with a major increase in speed and efficiency.

Users interacted with third-generation computers through keyboards and monitors and interfaced with an operating system, which allowed the device to run several applications at one time with a central program that monitored the memory. Computers became accessible to a wider audience, as they were smaller and cheaper than their predecessors.

1.3.4 Microprocessors

The Intel P4004 microprocessor was the world's first microprocessor, and it was released in 1971. It was the first semiconductor device that provided, at the chip level, the functions of a computer.

The invention of the microprocessor happened by accident rather than design. Busicom, a Japanese company, requested Intel to design a set of integrated circuits for its new family of high-performance programmable calculators. Ted Hoff, an Intel engineer, studied Busicom's design and rejected it as unwieldy. He proposed a more elegant solution requiring just four integrated circuits (Busicom required twelve integrated circuits), and his design included a chip that was a general-purpose logic device that derived its application instructions from the semiconductor memory. This was the Intel 4004 microprocessor.

It provided the basic building blocks that are used in today's microcomputers, including the arithmetic and logic unit and the control unit. The 4-bit Intel 4004 ran at a clock speed of 108 kHz and contained 2,300 transistors. It processed data in 4 bits, but its instructions were 8-bit long. It could address up to 1 Kb of program memory and up to 4 Kb of data memory.

Gary Kildall of Digital Research was one of the early people to recognize the potential of a microprocessor as a computer. He worked as a consultant with Intel, and he began writing experimental programs for the Intel 4004 microprocessor. He later developed the CP/M operating system for the Intel 8080 chip, and he set up Digital Research to commercialize the operating system.

The development of the microprocessor led to the fourth generation of computers with thousands of integrated circuits placed onto a single silicon chip. A single chip could now contain all the components of a computer from the CPU and memory to input and output controls. It could fit in the palm of the hand, whereas first generation of computers filled an entire room.

Table 1.1 von Neumann architecture

Component	Description
Arithmetic unit	The arithmetic unit can perform basic arithmetic operations
Control unit	The program counter contains the address of the next instruction to be executed. This instruction is fetched from memory and executed. This is the basic fetch-and-execute cycle (Fig. 1.4) The control unit contains a built-in set of machine instructions
Input–output unit	The input and output unit allows the computer to interact with the outside world
Memory	The one-dimensional memory stores all program instructions and data. These are usually kept in different areas of memory The memory may be written to or read from, i.e. it is random access memory (RAM) The program instructions are binary values, and the control unit decodes the binary value to determine which instruction to execute

1.4 von Neumann Architecture

The earliest computers were fixed programs machines that were designed to do a specific task. This proved to be a major limitation as it meant that a complex manual rewiring process was required to enable the machine to solve a different problem.

The computers used today are general-purpose machines designed to allow a variety of programs to be run on the machine. von Neumann and others (von Neumann 1945) described the fundamental architecture underlying the computers used today in the late 1940s. It is known as von Neumann architecture (Fig. 1.3).

The von Neumann architecture arose on work done by von Neumann, Eckert, Mauchly and others on the design of the EDVAC computer (which was the successor to ENIAC computer). von Neumann's draft report on EDVAC (von Neumann 1945) described the new architecture[2] (Table 1.1).

The architecture led to the birth of stored-program computers, where a single store is used for both machine instructions and data. Its key components are as follows:

The key approach to building a general-purpose device according to von Neumann was in its ability to store not only its data and the intermediate results of computation, but also to store the instructions or commands for the computation. The computer instructions can be part of the hardware for specialized machines, but for general-purpose machines the computer instructions must be as changeable as

---

[2]Eckert and Mauchly were working with him on this concept during their work on ENIAC and EDVAC, but their names were removed from the final report due to their resignation from the University of Pennsylvania to form their own computer company. von Neumann architecture includes a central processing unit which includes the control unit and the arithmetic unit, an input and output unit, and memory.

Fig. 1.3 von Neumann
architecture

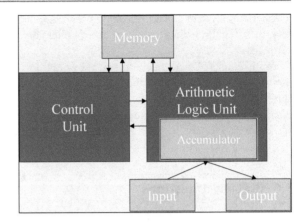

Fig. 1.4 Fetch/execute cycle

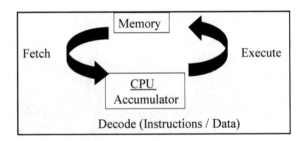

the data that is acted upon by the instructions. *His insight was to recognize that both the machine instructions and data could be stored in the same memory.*

The key advantage of the von Neumann architecture over the existing approach was that it was much simpler to reconfigure a computer to perform a different task. All that was required was to enter new machine instructions in computer memory rather than physically rewiring a machine as was required with ENIAC. The limitations of von Neumann architecture include that it is limited to sequential processing and not very suitable for parallel processing.

1.5   Hardware and Software

Hardware is the physical part of the machine. It is tangible and may be seen or touched, and includes punched cards, vacuum tubes, transistors and circuit boards, integrated circuits and microprocessors. The hardware of a personal computer includes a keyboard, network cards, a mouse, a DVD drive, hard disc drive, printers and scanners and so on.

Software is intangible and consists of a set of instructions that tells the computer what to do. It is an intellectual creation of a programmer (or a team of programmers).

Operating system software manages the computer hardware and resources and acts as an intermediary between the application programs and the computer hardware.

Application software refers to software programs that provide functionality for users to exploit the power of the computer to perform useful tasks such as business applications including spreadsheets and accountancy packages, financial applications, editors, compilers for programming languages, computer games, social media and so on.

1.6 Review Questions

1. Explain the difference between analog and digital computers.
2. Explain the difference between hardware and software.
3. What is a microprocessor?
4. Explain the difference between vacuum tubes, transistors and integrated circuits.
5. Explain the von Neumann architecture.
6. What are the advantages and limitations of the von Neumann architecture?
7. Explain the difference between a fixed program machine and a stored-program machine.

1.7 Summary

A computer is a programmable electronic device that can process, store and retrieve data. It processes data according to a set of instructions or program. All computers consist of two basic parts, namely, the hardware and software. The hardware is the physical part of the machine, whereas software is intangible and is the set of instructions that tells the computer what to do.

There are two distinct families of computing devices, namely, digital computers and the historical analog computer. These two types of computer operate on quite different principles. The earliest digital computers were built in the 1940s, and these were large machines consisting of thousands of vacuum tubes. However, their computational power was a fraction of what is available today.

A digital computer is a sequential device that generally operates on data one step at a time. The data are represented in binary format, and a single transistor in a digital computer can store only two states, i.e. on and off. Several transistors are required to store larger numbers.

The representation of data in an analog computer reflects the properties of the data that is being modelled. Data and numbers may be represented by physical quantities such as electric voltage, whereas a stream of binary digits represents the data in a digital computer.

von Neumann architecture is the fundamental architecture used on digital computers, and a single store is used for both machine instructions and data. Its introduction made it much easier to reconfigure a computer to perform a different task. All that was required was to enter new machine instructions in computer memory rather than physically rewiring a machine as was required with ENIAC.

References

O'Regan G (2018) World of computing. Springer
O'Regan G (2018) Giants of computing. Springer
von Neumann J (1945) First draft of a report on the EDVAC. University of Pennsylvania

Foundations of Computing

2

2.1 Introduction

This chapter considers important foundational work done by Wilhelm Leibniz, Charles Babbage, George Boole, Lady Ada Lovelace and Claude Shannon. Leibniz was a seventeenth-century German mathematician, philosopher and inventor, and he is recognized (with Isaac Newton) as the inventor of Calculus. He developed the Step Reckoner calculating machine that could perform all four basic arithmetic operations (i.e. addition, subtraction, multiplication and division), and he also invented the binary number system (which is used extensively in the computer field).

© Springer Nature Switzerland AG 2020   13
G. O'Regan, *Mathematics in Computing*, Undergraduate Topics
in Computer Science, https://doi.org/10.1007/978-3-030-34209-8_2

Boole and Babbage are considered grandfathers of the computing field, with Babbage's Analytic Engine providing a vision of a mechanical computer, and Boole's logic providing the foundation for modern digital computers.

Babbage was a nineteenth-century scientist and inventor who did pioneering work on calculating machines. He invented the Difference Engine (a sophisticated calculator that could be used to produce mathematical tables), and he also designed the Analytic Engine (the world's first mechanical computer). The design of the Analytic Engine included a processor, memory and a way to input information and output results.

Lady Ada Lovelace was introduced into Babbage's ideas on the analytic engine at a dinner party. She was fascinated and predicted that such a machine could be used to compose music, produce graphics, as well as solving mathematical and scientific problems. She explained how the Analytic Engine could be programmed, and she wrote what is considered the first computer program.

Boole was a nineteenth-century English mathematician who made important contributions to mathematics, probability theory and logic. Boole's logic provides the foundation for digital computers.

Shannon was the first person to apply Boole's logic to switching theory, and he showed that this could simplify the design of circuits and telephone routing switches. It provides the perfect mathematical model for switching theory and for the subsequent design of digital circuits and computers.

2.2 Step Reckoner Calculating Machine

Leibniz (Fig. 27.6) was a German philosopher, mathematician and inventor in the field of mechanical calculators. He developed the binary number system used in digital computers, and he invented the Calculus independently of Sir Isaac Newton. He became familiar with Pascal's calculating machine, the *Pascaline*, while in Paris in the early 1670s. He recognized its limitations as the machine could perform addition and subtraction operations only.

He designed and developed a calculating machine that could perform addition, subtraction, multiplication, division and the extraction of roots. He commenced work on the machine in 1672, and the machine was completed in 1694. It was the first calculator that could perform all four arithmetic operations, and it was called the *Step Reckoner* (Fig. 2.2). It allowed the common arithmetic operations to be carried out mechanically (Fig. 2.1).

The operating mechanism used in his calculating machine was based on a counting device called the stepped cylinder or '*Leibniz wheel*'. This mechanism allows a gear to represent a single decimal digit from zero to nine in just one revolution, and this remained the dominant approach to the design of calculating machines for the next 200 years. It was essentially a counting device consisting of a set of wheels that were used in calculation. The Step Reckoner consisted of an accumulator which could hold 16 decimal digits and an 8-digit input section. The

Fig. 2.1 Replica of step reckoner at Technische Sammlungen Museum, Dresden

eight dials at the front of the machine set the operand number, which was then employed in the calculation.

The machine performed multiplication by repeated addition and division by repeated subtraction. The basic operation is to add or subtract the operand from the accumulator as many times as desired. The machine could add or subtract an 8-digit number to the 16-digit accumulator to form a 16-digit result. It could multiply two 8-digit numbers to give a 16-digit result, and it could divide a 16-bit number by an 8-digit number. Addition and subtraction are performed in a single step, with the operating crank turned in the opposite direction for subtraction. The result is stored in the accumulator.

2.3 Binary Numbers

Arithmetic has traditionally been done using the decimal notation,[1] and Leibniz was one of the first to recognize the potential of the binary number system. This system uses just two digits, namely, '0' and '1', with the number two represented by 10, the

---

[1]The sexagesimal (or base-60) system was employed by the Babylonians c. 2000 BC. Indian and Arabic mathematicians developed the decimal system between 800 and 900 AD.

Table 2.1 Binary number system

Binary	Dec.	Binary	Dec.	Binary	Dec.	Binary	Dec.
0000	0	0100	4	1000	8	1100	12
0001	1	0101	5	1001	9	1101	13
0010	2	0110	6	1010	10	1110	14
0011	3	0111	7	1011	11	1111	15

number four by 100 and so on. Leibniz described the binary system in *Explication de l'Arithmétique Binaire* (Leibniz 1703), which was published in 1703. A table of values for the first 15 binary numbers is given in Table 2.1.

Leibniz's (1703) paper describes how binary numbers may be added, subtracted, multiplied and divided, and he was an advocate of their use. The key advantage of the use of binary notation is in digital computers, where a binary digit may be implemented by an *on/off* switch, with the digit 1 representing that the switch is on, and the digit 0 representing that the switch is off.

The use of binary arithmetic allows more complex mathematical operations to be performed by relay circuits, and Boole's Logic (described in a later section) is the perfect model for simplifying such circuits and is the foundation underlying digital computing.

The binary number system (base 2) is a positional number system, which uses two binary digits 0 and 1, and an example binary number is 1001.01_2 which represents $1 \times 2^3 + 0 \times 2^2 + 0 \times 2^1 + 1 \times 2^0 + 0 \times 2^{-1} + 1 \times 2^{-2} = 1 \times 2^3 + 1 \times 2^0 + 1 \times 2^{-2} = 8 + 1 + 0.25 = 9.25$.

The decimal system (base 10) is more familiar for everyday use, and there are algorithms to convert numbers from decimal to binary and vice versa. For example, to convert the decimal number 25 to its binary representation 11001_2 we proceed as follows (Fig. 2.2):

The base 2 is written on the left, and the number to be converted to binary is placed in the first column. At each stage in the conversion, the number in the first column is divided by 2 to form the quotient and remainder, which are then placed on the next row. For the first step, the quotient when 25 is divided by 2 is 12 and the remainder is 1. The process continues until the quotient is 0, and the binary representation result is then obtained by reading the second column from the bottom up. Thus, we see that the binary representation of 25 is 11001_2.

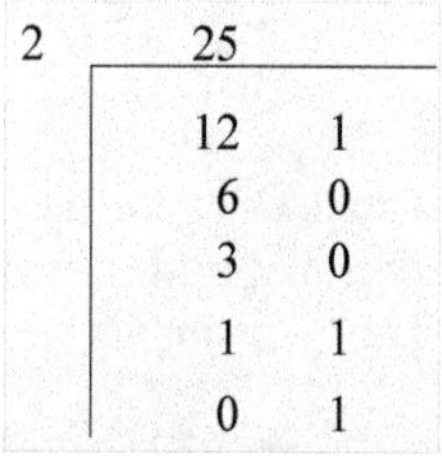

```
2      25
       ───────
       12    1
        6    0
        3    0
        1    1
        0    1
```

Fig. 2.2 Decimal to binary conversion

Similarly, there are algorithms to convert decimal fractions to binary representation (to a defined number of binary digits as the representation may not terminate), and the conversion of a number that contains an integer part and a fractional part involves converting each part separately and then combining them.

The octal (base 8) and hexadecimal (base 16) are often used in computing, as the bases 2, 8 and 16 are related bases and easy to convert between, as to convert between binary and octal involves grouping the bits into groups of three on either side of the point. Each set of 3 bits corresponds to one digit in the octal representation. Similarly, the conversion between binary and hexadecimal involves grouping into sets of 4 digits on either side of the point. The conversion from octal to binary or hexadecimal to binary is equally simple and involves replacing the octal (or hexadecimal) digit with the 3-bit (or 4-bit) binary representation.

Numbers are represented in a digital computer as sequences of bits of fixed length (e.g. 16 bits, 32 bits). There is a difference in the way in which integers and real numbers are represented, with the representation of real numbers being more complicated.

An integer number is represented by a sequence of (usually 2 or 4) bytes where each byte is 8 bits. For example, a 2-byte integer has 16 bits with the first bit used as the sign bit (the sign is 1 for negative numbers and 0 for positive integers), and the remaining 15 bits represent the number. This means that 2 bytes may be used to represent all integer numbers between −32768 and 32767. A positive number is represented by the normal binary representation discussed earlier, whereas a negative number is represented using 2's complement of the original number (i.e. 0 changes to 1 and 1 changes to 0 and the sign bit is 1). All the standard arithmetic operations may then be carried out (using modulo-2 arithmetic).

The representation of floating-point real numbers is more complicated, and a real number is represented to a fixed number of significant digits (the significand) and scaled using an exponent in some base (usually 2). That is, the number is represented (approximated) as

$$\text{significand} \times \text{base}^{\text{exponent}}$$

The significand (also called mantissa) and exponent have a sign bit. For example, in simple floating-point representation (4 bytes), the mantissa is generally 24 bits and the exponent 8 bits, whereas for double precision (8 bytes) the mantissa is generally 53 bits and the exponent 11 bits. There is an IEEE standard for floating-point numbers (IEEE 754).

2.4 The Difference Engine

Babbage (Fig. 2.3) is considered (along with Boole) to be one of the grandfathers of the computing field. He contributed to several areas including mathematics, statistics, astronomy, calculating machines, philosophy, railways and lighthouses. He founded the British Statistical Society and the Royal Astronomical Society.

Fig. 2.3 Charles Babbage

Babbage was interested in accurate mathematical tables for scientific work. However, there was a high error rate in the existing tables due to human error introduced during calculation. He became interested in finding a mechanical method to perform calculation to eliminate the errors introduced by humans. He planned to develop a more advanced machine than the Pascaline or the Step Reckoner, and his goal was to develop a machine that could compute polynomial functions.

He designed the Difference Engine (No. 1) in 1821 for the production of mathematical tables. This was essentially a mechanical calculator (analogous to modern electronic calculators), and it was designed to compute polynomial functions of degree 4. It could also compute logarithmic and trigonometric functions such as sine or cosine (as these may be approximated by polynomials).[2]

The accurate approximation of trigonometric, exponential and logarithmic functions by polynomials depends on the degree of the polynomials, the number of decimal digits that it is being approximated to, and on the error function. A higher degree polynomial is generally able to approximate the function more accurately.

Babbage produced prototypes for parts of the Difference Engine, but he never actually completed the machine. The Swedish engineers, Georg and Edvard Scheutz, built the first working Difference Engine (based on Babbage's design) in 1853 with funding from the Swedish government. Their machine could compute

---

[2]The power series expansion of the Sine function is given by $Sin(x) = x - x^3/3! + x^5/5! - x^7/7! + \cdots$. The power series expansion for the Cosine function is given by $Cos(x) = 1 - x^2/2! + x^4/4! - x^6/6! + \cdots$. Functions may be approximated by interpolation and the approximation of a function by a polynomial of degree n requires $n + 1$ points on the curve for the interpolation. That is, the curve formed by the polynomial of degree n that passes through the $n + 1$ points of the function to be approximated is an approximation to the function. The error function also needs to be considered.

Table 2.2 Analytic engine

Part	Function
Store	This contains the variables to be operated upon as well as all those quantities, which have arisen from the result of intermediate operations
Mill	The mill is essentially the processor of the machine into which the quantities about to be operated upon are brought

polynomials of degree 4- on 15-digit numbers, and the 3rd Scheutz Difference Engine is on display at the Science Museum in London.

It was the first machine to compute and print mathematical tables mechanically. The machine was accurate, and it showed the potential of mechanical machines as a tool for scientists and engineers.

The machine is unable to perform multiplication or division directly. Once the initial value of the polynomial and its derivative are calculated for some value of x, the difference engine may calculate any number of nearby values using the numerical method of finite differences. This method replaces computational intensive tasks involving multiplication or division, by an equivalent computation that just involves addition or subtraction.

The British government cancelled Babbage's project in 1842. He designed an improved difference engine No.2 (Fig. 2.4) in 1849. It could operate on seventh-order differences (i.e. polynomials of order 7) and 31-digit numbers. The machine consisted of 8 columns with each column consisting of 31 wheels. However, it was over 150 years later before it was built (in 1991) to mark the two hundredth anniversary of his birth. The Science Museum in London also built the printer that Babbage designed, and both the machine and the printer worked correctly according to Babbage's design (after a little debugging).

2.5 The Analytic Engine—Vision of a Computer

The Difference Engine was designed to produce mathematical tables, but it required human intervention to perform the calculations. Babbage recognized its limitations, and he proposed a revolutionary solution by outlining his vision of a mechanical computer. His plan was to construct a new machine that would be capable of executing all tasks that may be expressed in algebraic notation. His vision of such a computer (Analytic Engine) consisted of two parts (Table 2.2).

Babbage intended that the operation of the Analytic Engine would be analogous to the operation of the *Jacquard loom*.[3] The latter is capable of weaving (i.e. executing on the loom) a design pattern that has been prepared by a team of skilled

---

[3]The Jacquard loom was invented by Joseph Jacquard in 1801. It is a mechanical loom which used the holes in punch cards to control the weaving of patterns in a fabric. The use of punched cards allowed complex designs to be woven from the pattern defined on the punched cards. Each punched card corresponds to one row of the design, and the cards were appropriately ordered. It

Fig. 2.4 Difference engine no. 2. Photo public domain

artists. The design pattern is represented by a set of cards with punched holes,
where each card represents a row in the design. The cards are then ordered, placed
in the loom and the loom produces the exact pattern.

The use of the punched cards in the Analytic Engine allowed the formulae to be
manipulated in a manner dictated by the programmer. The cards commanded the
analytic engine to perform various operations and to return a result. Babbage distinguished between two types of punched cards:

–   *Operation Cards*
–   *Variable Cards.*

Operation cards are used to define the operations to be performed, whereas the
variable cards define the variables or data that the operations are performed upon.
His planned use of punched cards to store programs in the Analytic Engine is
similar to the idea of a stored computer program in Von Neumann architecture.
However, Babbage's idea of using punched cards to represent machine instructions

was very easy to change the pattern of the fabric being weaved on the loom, as this simply
involved changing cards.

Fig. 2.5 Lady Ada Lovelace

and data was over 100 years before digital computers. *Babbage's Analytic Engine is therefore an important milestone in the history of computing.*

Babbage intended that the program be stored on read-only memory using punch cards and that the input and output would be carried out using punch cards. He intended that the machine would be able to store numbers and intermediate results in memory that could then be processed. There would be several punch card readers in the machine for programs and data. He envisioned that the machine would be able to perform conditional jumps as well as parallel processing where several calculations could be performed at once.

The Analytic Engine was designed in 1834 as the world's first mechanical computer (Babbage and Menabrea 1842). It included a processor, memory and a way to input information and output results. However, the machine was never built, as Babbage was unable to secure funding from the British Government.

2.5.1   Applications of Analytic Engine

Lady Augusta Ada Lovelace (nee Byron)[4] (Fig. 2.5) was a mathematician who collaborated with Babbage on applications for the analytic engine. She is considered the world's first programmer, and the Ada programming language is named in her honour.

She was introduced to Babbage at a dinner party in 1833, and she visited Babbage's studio in London, where the prototype Difference Engine was on

---

[4]Lady Ada Lovelace was the daughter of the poet, Lord Byron.

display. She recognized the beauty of its invention, and she was fascinated by the idea of the analytic engine. She communicated regularly with Babbage with ideas on its applications.

Lovelace produced an annotated translation of Menabrea's '*Notions sur la machine analytique de Charles Babbage*' (Babbage and Menabrea 1842). She added copious notes to the translation,[5] which were about three times the length of the original memoir, and considered many of the difficult and abstract questions connected with the subject. These notes are regarded as a description of a computer and software.

She explained in the notes how the Analytic Engine could be programmed and wrote what is considered to be the first computer program. This program detailed a plan be written for how the engine would calculate *Bernoulli numbers*. Lady Ada Lovelace is therefore considered to be the first computer programmer, and Babbage called her the '*enchantress of numbers*'.

She saw the potential of the analytic engine to fields other than mathematics. She predicted that the machine could be used to compose music, produce graphics, as well as solving mathematical and scientific problems. She speculated that the machine might act on other things apart from numbers, and be able to manipulate symbols according to rules. In this way, a number could represent an entity other than a quantity.

2.6 Boole's Symbolic Logic

George Boole (Fig. 2.6) was born in Lincoln, England in 1815. His father (a cobbler who was interested in mathematics and optical instruments) taught him mathematics and showed him how to make optical instruments. Boole inherited his father's interest in knowledge, and he was self-taught in mathematics and Greek. He taught at various schools near Lincoln, and he developed his mathematical knowledge by working his way through Newton's Principia, as well as applying himself to the work of mathematicians such as Laplace and Lagrange.

He published regular papers from his early 20s, and these included contributions to probability theory, differential equations and finite differences. He developed Boolean algebra, which is the foundation for modern computing, and he is considered (along with Babbage) to be one of the grandfathers of computing. His work was theoretical, and he never actually built a computer or calculating machine. *However, Boole's symbolic logic was the perfect mathematical model for switching theory and for the design of digital circuits.*

Boole became interested in formulating a calculus of reasoning, and he published 'Mathematical Analysis of Logic' in 1847 (Boole 1848). This work developed novel ideas on a logical method, and he argued that logic should be considered

---

[5]There is some controversy as to whether this was entirely her own work or a joint effort by Lovelace and Babbage.

Fig. 2.6 George Boole

as a separate branch of mathematics, rather than as a part of philosophy. He argued that there are mathematical laws to express the operation of reasoning in the human mind, and he showed how Aristotle's syllogistic logic could be reduced to a set of algebraic equations. He corresponded regularly on logic with Augustus De Morgan.[6]

His paper on logic introduced two quantities '0' and '1'. He used the quantity 1 to represent the universe of thinkable objects (i.e. the universal set), and the quantity 0 represents the absence of any objects (i.e. the empty set). He then employed symbols, such as x, y, z, etc., to represent collections or classes of objects given by the meaning attached to adjectives and nouns. Next, he introduced three operators $(+, -$ and $\times)$ that combined classes of objects.

The expression xy (i.e. x multiplied by y or $x \times y$) combines the two classes x, y to form the new class xy (i.e. the class whose objects satisfy the two meanings represented by class x *and* class y). Similarly, the expression $x + y$ combines the two classes x, y to form the new class $x + y$ (that satisfies either the meaning represented by class x *or* class y). The expression $x - y$ combines the two classes x, y to form the new class $x - y$. This represents the class (that satisfies the meaning represented by class x but not class y). The expression $(1 - x)$ represents objects that do not have the attribute that represents class x.

Thus, if $x =$ black and $y =$ sheep, then xy represents the class of black sheep. Similarly, $(1 - x)$ would represent the class obtained by the operation of selecting all things in the world except black things; $x (1 - y)$ represents the class of all things that are black but not sheep; and $(1 - x) (1 - y)$ would give us all things that are neither sheep nor black.

---

[6]De Morgan was a nineteenth-century British mathematician based at University College London. De Morgan's laws in Set Theory and Logic state that $(A \cup B)^c = A^c \cap B^c$ and $\neg (A \vee B) \equiv \neg A \wedge \neg B$.

He showed that these symbols obeyed a rich collection of algebraic laws and could be added, multiplied, etc., in a manner that is like real numbers. These symbols may be used to reduce propositions to equations, and algebraic rules may be employed to solve the equations. The rules include the following:

1.	$x + 0 = x$	(Additive identity)
2.	$x + (y + z) = (x + y) + z$	(Associative)
3.	$x + y = y + x$	(Commutative)
4.	$x + (1 - x) = 1$	
5.	$\times \cdot 1 = x$	(Multiplicative identity)
6.	$x \cdot 0 = 0$	
7.	$x + 1 = 1$	
8.	$xy = yx$	(Commutative)
9.	$x(yz) = (xy)z$	(Associative)
10.	$x(y + z) = xy + xz$	(Distributive)
11.	$x(y - z) = xy - xz$	(Distributive)
12.	$x^2 = x$	(Idempotent)

These operations are similar to the modern laws of set theory with the set union operation represented by '+', and the set intersection operation is represented by multiplication. The universal set is represented by '1' and the empty by '0'. The associative and distributive laws hold. Finally, the set complement operation is given by $(1 - x)$.

He applied the symbols to encode Aristotle's Syllogistic Logic, and he showed how the syllogisms could be reduced to equations. This allowed conclusions to be derived from premises by eliminating the middle term in the syllogism. He refined his ideas on logic further in 'An Investigation of the Laws of Thought' which was published in 1854 (Boole 1958). This book aimed to identify the fundamental laws underlying reasoning in the human mind and to give expression to these laws in the symbolic language of a calculus.

He considered the equation $x^2 = x$ to be a fundamental law of thought. It allows the principle of contradiction to be expressed as (i.e. for an entity to possess an attribute and at the same time not to possess it)

$$x^2 = x$$
$$\Rightarrow x - x^2 = 0$$
$$\Rightarrow x(1 - x) = 0$$

For example, if x represents the class of horses then $(1 - x)$ represents the class of 'not-horses'. The product of two classes represents a class whose members are common to both classes. Hence, $x (1 - x)$ represents the class whose members are at once both horses and 'not-horses', and the equation $x (1 - x) = 0$ expresses that fact that there is no such class. That is, it is the empty set.

Fig. 2.7 Binary AND
operation

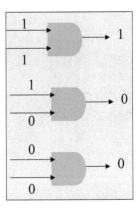

Boole contributed to other areas in mathematics including differential equations, finite differences[7] and to the development of probability theory. Des McHale has written an interesting biography of Boole and Des McHale (1985). Boole's logic appeared to have no practical use, but this changed with Claude Shannon's (1937) Master's Thesis, which showed its applicability to switching theory and to the design of digital circuits.

2.6.1 Switching Circuits and Boolean Algebra

Claude Shannon's Master's Thesis showed that Boole's algebra provided the perfect mathematical model for switching theory and for the design of digital circuits. It may be employed to optimize the design of systems of electromechanical relays, and circuits with relays solve Boolean algebra problems. The use of the properties of electrical switches to process logic is the basic concept that underlies all modern electronic digital computers. Digital computers use the binary digits 0 and 1, and Boolean logical operations may be implemented by electronic AND, OR and NOT gates. More complex circuits (e.g. arithmetic) may be designed from these fundamental building blocks.

Modern electronic computers use billions of transistors that act as switches and can change state rapidly. A high voltage represents the binary value 1 with low voltage representing the binary value 0. A silicon chip may contain billions of tiny electronic switches arranged into logical gates. The basic logic gates are AND, OR and NOT. These gates may be combined in various ways to allow the computer to perform more complex tasks such as binary arithmetic. Each gate has binary value inputs and outputs.

The example in Fig. 2.7 is that of an 'AND' gate which produces the binary value 1 as output only if both inputs are 1. Otherwise, the result will be the binary

---

[7]Finite Differences are a numerical method used in solving differential equations.

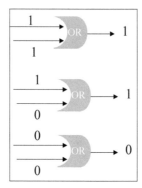

Fig. 2.8 Binary OR operation

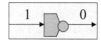

Fig. 2.9 NOT operation

value 0. Figure 2.8 shows an 'OR' gate which produces the binary value 1 as output if any of its inputs is 1. Otherwise, it will produce the binary value 0.

Finally, a NOT gate (Fig. 2.9) accepts only a single input which it inverts. That is, if the input is '1' the value '0' is produced and vice versa.

The logic gates may be combined to form more complex circuits. The example in Fig. 2.10 is that of a half adder of 1 + 0. The inputs to the top OR gate are 1 and 0 which yields the result of 1. The inputs to the bottom AND gate are 1 and 0 which yields the result 0, which is then inverted through the NOT gate to yield binary 1. Finally, the last AND gate receives two 1's as input and the binary value 1 is the result of the addition. The half adder computes the addition of two arbitrary binary digits, but it does not calculate the carry. It may be extended to a full adder that provides a carry for addition.

Fig. 2.10 Half adder

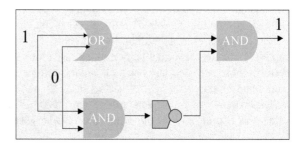

Fig. 2.11 Claude Shannon

2.7   Application of Symbolic Logic to Digital Computing

Claude Shannon (Fig. 2.11) was the first person[8] to see the applicability of Boole's algebra to simplify the design of circuits and telephone routing switches. He showed that Boole's symbolic logic was the perfect mathematical model for switching theory and for the subsequent design of digital circuits and computers.

His influential *Master's Thesis is a key milestone in computing*, and it shows how to lay out circuits according to Boolean principles. It provides the theoretical foundation of switching circuits, and *his insight of using the properties of electrical switches to do Boolean logic is the basic concept that underlies all electronic digital computers.*

Shannon realized that you could combine switches in circuits in such a manner as to carry out symbolic logic operations. This allowed binary arithmetic and more complex mathematical operations to be performed by relay circuits. He designed a circuit, which could add binary numbers, and he later designed circuits that could make comparisons and thus capable of performing a conditional statement. *This was the birth of digital logic and the digital computing age.*

---

[8]Victor Shestakov at Moscow State University also proposed a theory of electric switches based on Boolean algebra (published in Russian in 1941 whereas Shannon's were published in 1937).

Vannevar Bush (Regan 2013) was Shannon's supervisor at MIT, and Shannon's initial work was to improve Bush's mechanical computing device known as the Differential Analyser. This machine had a complicated control circuit that was composed of 100 switches that could be automatically opened and closed by an electromagnet. Shannon's insight was his realization that an electronic circuit is similar to Boolean algebra, and he showed how Boolean algebra could be employed to optimize the design of systems of electromechanical relays used in the analog computer. He also realized that circuits with relays could solve Boolean algebra problems.

His Master's thesis '*A Symbolic Analysis of Relay and Switching Circuits*' (Shannon 1937) showed that the binary digits (i.e. 0 and 1) can be represented by electrical switches. This allowed binary arithmetic and more complex mathematical operations to be performed by relay circuits, and provided electronics engineers with the mathematical tool that they needed to design digital electronic circuits and provided the foundation for the field.

The design of circuits and telephone routing switches could be simplified with Boole's symbolic algebra. Shannon showed how to lay out circuits according to Boolean principles, and his Master's thesis became the foundation for the practical design of digital circuits. These circuits are fundamental to the operation of modern computers and telecommunication systems, and his insight of using the properties of electrical switches to do Boolean logic is the basic concept that underlies all electronic digital computers.

2.8   Review Questions

1. Explain the significance of binary numbers in the computing field.
2. Explain the importance of Shannon's Master Thesis.
3. Explain the significance of the Analytic Engine.
4. Explain why Ada Lovelace is considered the world's first programmer.
5. Explain the significance of Boole to the computing field.
6. Explain the significance of Babbage to the computing field.
7. Explain the significance of Leibniz to the computing field.

2.9   Summary

This chapter considered foundational work done by Leibniz, Babbage, Boole, Ada Lovelace and Shannon. Leibniz developed a calculating machine (the Step Reckoner) that could perform the four basic arithmetic operations. He also invented the binary number system, which is used extensively in the computer field.

Babbage did pioneering work on calculating machines. He designed the Difference Engine (a sophisticated calculator that could be used to produce mathematical tables), and he also designed the Analytic Engine (the world's first mechanical computer).

Lady Ada Lovelace was introduced to Babbage's ideas on the analytic engine, and she predicted that such a machine could be used to compose music, produce graphics, as well as solving mathematical and scientific problems.

Boole was a nineteenth-century English mathematician who made important contributions to mathematics, and his symbolic logic provides the foundation for digital computers.

Shannon was a twentieth-century American mathematician and engineer, and he showed that Boole's symbolic logic provided the perfect mathematical model for switching theory and for the subsequent design of digital circuits and computer.

References

Babbage C, Menabrea LF (1842) Sketch of the analytic engine invented by charles babbage. Bibliothèque Universelle de Genève, October, No. 82 Translated by Ada, Augusta, Countess of Lovelace
Boole G (1848) The calculus of logic. Camb Dublin Math J III:183–98
Boole G (1958) An investigation into the laws of thought. Dover Publications (First published in 1854)
Leibniz WG (1703) Explication de l'Arithmétique Binaire. Memoires de l'Academie Royale des Sciences
McHale D (1985) George Boole, His Life and Work. Cork University Press
O'Regan G (2013) Giants of computing. Springer
Shannon C (1937) A symbolic analysis of relay and switching circuits. Masters thesis, Massachusetts Institute of Technology

Overview of Mathematics in Computing

3

Key Topics

Sets
Set operations
Russell's paradox
Computer representation of sets
Relations
Composition of relations
Functions
Partial and total functions
Functional programming
Number theory
Automata theory
Graph theory

3.1 Introduction

This chapter introduces essential mathematics for computing and discusses fundamental concept such as sets, relations and functions. Sets are collections of well-defined objects; relations indicate relationships between members of two sets

© Springer Nature Switzerland AG 2020
G. O'Regan, *Mathematics in Computing*, Undergraduate Topics
in Computer Science, https://doi.org/10.1007/978-3-030-34209-8_3

A and B; and functions are a special type of relation where there is exactly (or at most)[1] one relationship for each element $a \in A$ with an element in B.

A set is a collection of well-defined objects that contains no duplicates. The term '*well defined*' means that for a given value it is possible to determine whether or not it is a member of the set. There are many examples of sets such as the set of natural numbers $\mathbb{N}$, the set of integer numbers $\mathbb{Z}$ and the set of rational numbers $\mathbb{Q}$. The natural number $\mathbb{N}$ is an infinite set consisting of the numbers $\{1, 2, \ldots\}$. Venn diagrams may be used to represent sets pictorially.

A binary relation R (A, B) where A and B are sets is a subset of the Cartesian product $(A \times B)$ of A and B. The domain of the relation is A and the co-domain of the relation is B. The notation aRb signifies that there is a relation between a and b and that $(a, b) \in R$. An n-ary relation R $(A_1, A_2, \ldots A_n)$ is a subset of $(A_1 \times A_2 \times \cdots \times A_n)$. However, an n-ary relation may also be regarded as a binary relation $R(A, B)$ with $A = A_1 \times A_2 \times \cdots \times A_{n-1}$ and $B = A_n$.

Functions may be total or partial. A total function $f : A \to B$ is a special relation such that for each element $a \in A$ there is exactly one element $b \in B$. This is written as $f(a) = b$. A partial function differs from a total function in that the function may be undefined for one or more values of A. The domain of a function (denoted by **dom** f) is the set of values in A for which the partial function is defined. The domain of the function is A if f is a total function. The co-domain of the function is B.

We introduce topics such as number theory, automata theory and graph theory. Number theory is the branch of mathematics that is concerned with the mathematical properties of the natural numbers and integers. Automata Theory is the branch of computer science that is concerned with the study of abstract machines and automata. These include finite-state machines, pushdown automata and Turing machines. Graph theory is a practical branch of mathematics that deals with the arrangements of certain objects known as vertices (or nodes) and the relationships between them. We briefly discuss computability and decidability, and the succeeding chapters contain more detailed information on the mathematics in computing.

3.2 Set Theory

A set is a fundamental building block in mathematics, and it is defined as a collection of well-defined objects. The elements in a set are of the same kind, and they are distinct with no repetition of the same element in the set.[2] Most sets encountered in computer science are finite, as computers can only deal with finite entities. Venn

---

[1]We distinguish between total and partial functions. A total function $f : A \to B$ is defined for every element in A, whereas a partial function may be undefined for one or more values in A.
[2]There are mathematical objects known as *multi-sets* or *bags* that allow duplication of elements. For example, a bag of marbles may contain three green marbles, two blue and one red marble.

diagrams[3] are often employed to give a pictorial representation of a set, and to illustrate various set operations such as set union, intersection and set difference.

There are many well-known examples of sets including the set of natural numbers denoted by $\mathbb{N}$, the set of integers denoted by $\mathbb{Z}$, the set of rational numbers denoted by $\mathbb{Q}$, the set of real numbers denoted by $\mathbb{R}$ and the set of complex numbers denoted by $\mathbb{C}$.

Example 3.1 The following are examples of sets:

–   The books on the shelves in a library,
–   The books that are currently overdue from the library,
–   The customers of a bank,
–   The bank accounts in a bank,
–   The set of Natural Numbers $\mathbb{N} = \{1, 2, 3, \ldots\}$,
–   The Integer Numbers $\mathbb{Z} = \{\ldots, -3, -2, -1, 0, 1, 2, 3, \ldots\}$,
–   The non-negative integers $\mathbb{Z}^+ = \{0, 1, 2, 3, \ldots\}$ and
–   The set of Prime Numbers $= \{2, 3, 5, 7, 11, 13, 17, \ldots\}$.
–   The Rational Numbers is the set of quotients of integers

$$\mathbb{Q} = \{p/q : p, q \in \mathbb{Z} \text{ and } q \neq 0\}$$

A finite set may be defined by listing all its elements. For example, the set $A = \{2, 4, 6, 8, 10\}$ is the set of all even natural numbers less than or equal to 10. The order in which the elements are listed is not relevant, i.e. the set $\{2, 4, 6, 8, 10\}$ is the same as the set $\{8, 4, 2, 10, 6\}$.

Sets may be defined by using a predicate to constrain set membership. For example, the set $S = \{n : \mathbb{N} : n \leq 10 \wedge n \bmod 2 = 0\}$ also represents the set $\{2, 4, 6, 8, 10\}$. That is, the use of a predicate allows a new set to be created from an existing set by using the predicate to restrict membership of the set. The set of even natural numbers may be defined by a predicate over the set of natural numbers that restricts membership to the even numbers. It is defined by

$$\text{Evens} = \{x | x \in \mathbb{N} \wedge even(x)\}.$$

---

[3]The British logician, John Venn, invented the Venn diagram. It provides a visual representation of a set and the various set-theoretical operations. Their use is limited to the representation of two or three sets as they become cumbersome with a larger number of sets.

In this example, *even(x)* is a predicate that is true if x is even and false otherwise. In general, $A = \{x \in E \mid P(x)\}$ denotes a set A formed from a set E using the predicate P to restrict membership of A to those elements of E for which the predicate is true.

The elements of a finite set S are denoted by $\{x_1, x_2, \ldots x_n\}$. The expression $x \in S$ denotes that the element x is a member of the set S, whereas the expression $x \notin S$ indicates that x is not a member of the set S.

A set S is a subset of a set T (denoted $S \subseteq T$) if whenever $s \in S$ then $s \in T$, and in this case the set T is said to be a superset of S (denoted $T \supseteq S$). Two sets S and T are said to be equal if they contain identical elements, i.e. $S = T$ if and only if $S \subseteq T$ and $T \subseteq S$. A set S is a proper subset of a set T (denoted $S \subset T$) if $S \subseteq T$ and $S \neq T$. That is, every element of S is an element of T and there is at least one element in T that is not an element of S. In this case, T is a proper superset of S (denoted $T \supset S$).

The empty set (denoted by $\emptyset$ or $\{\}$) represents the set that has no elements. Clearly, $\emptyset$ is a subset of every set. The singleton set containing just one element x is denoted by $\{x\}$, and clearly $x \in \{x\}$ and $x \neq \{x\}$. Clearly, $y \in \{x\}$ if and only if $x = y$.

Example 3.2

(i) $\{1, 2\} \subseteq \{1, 2, 3\}$.
(ii) $\emptyset \subset \mathbb{N} \subset \mathbb{Z} \subset \mathbb{Q} \subset \mathbb{R} \subset \mathbb{C}$.

The cardinality (or size) of a finite set S defines the number of elements present in the set. It is denoted by $|S|$. The cardinality of an infinite[4] set S is written as $|S| = \infty$.

Example 3.3

(i)   Given $A = \{2, 4, 5, 8, 10\}$ then $|A| = 5$.
(ii) Given $A = \{x \in \mathbb{Z} : x^2 = 9\}$ then $|A| = 2$.
(iii) Given $A = \{x \in \mathbb{Z} : x^2 = -9\}$ then $|A| = 0$.

---

[4]The natural numbers, integers and rational numbers are countable sets (i.e. they may be put into a one-to-one correspondence with the Natural numbers), whereas the real and complex numbers are uncountable sets.

3.2.1 Set-Theoretical Operations

Several set-theoretical operations are considered in this section. These include the Cartesian product operation, the power set of a set, the set union operation, the set intersection operation, the set difference operation and the symmetric difference operation.

Cartesian Product

The Cartesian product allows a new set to be created from existing sets. The Cartesian[5] product of two sets S and T (denoted $S \times T$) is the set of ordered pairs $\{(s, t) \mid s \in S, t \in T\}$. Clearly, $S \times T \neq T \times S$ and so the Cartesian product of two sets is not commutative. Two ordered pairs (s_1, t_1) and (s_2, t_2) are considered equal if and only if $s_1 = s_2$ and $t_1 = t_2$.

The Cartesian product may be extended to that of n sets $S_1, S_2, \ldots, S_n$. The Cartesian product $S_1 \times S_2 \times \cdots \times S_n$ is the set of ordered n-tuples $\{(s_1, s_2, \ldots, s_n) \mid s_1 \in S_1, s_2 \in S_2, \ldots, s_n \in S_n\}$. Two ordered n-tuples $(s_1, s_2, \ldots, s_n)$ and $(s_1', s_2', \ldots, s_n')$ are considered equal if and only if $s_1 = s_1', s_2, = s_2', \ldots, s_n = s_n'$.

The Cartesian product may also be applied to a single set S to create ordered n-tuples of S, i.e. $S^n = S \times S \times \cdots \times S$ (n times).

Power Set

The power set of a set A (denoted $\mathbb{P}A$) denotes the set of subsets of A. For example, the power set of the set A = $\{1, 2, 3\}$ has eight elements and is given by

$$\mathbb{P}A = \{\varnothing, \{1\}, \{2\}, \{3\}, \{1,2\}, \{1,3\}, \{2,3\}, \{1,2,3\}\}.$$

There are $2^3 = 8$ elements in the power set of $A = \{1, 2, 3\}$ where the cardinality of A is 3. In general, there are $2^{|A|}$ elements in the power set of A.

Theorem 3.1 (Cardinality of Power Set of A) *There are $2^{|A|}$ elements in the power set of A.*

Proof Let $|A| = n$, then the cardinalities of the subsets of A are subsets of size 0, 1, ..., n. There are $\binom{n}{k}$ subsets of A of size k.[6] Therefore, the total number of subsets of A is the total number of subsets of size 0, 1, 2, ... up to n. That is,

$$|\mathbb{P}A| = \sum_{k=0}^{n} \binom{n}{k}$$

The Binomial Theorem states that

---

[5]Cartesian product is named after René Descartes who was a famous seventeenth-century French mathematician and philosopher. He invented the Cartesian coordinates system that links geometry and algebra and allows geometric shapes to be defined by algebraic equations.
[6]Permutations and combinations are discussed in Chap. 8.

$$(1+x)^n = \sum_{k=0}^{n} \binom{n}{k} x^k$$

Therefore, putting $x = 1$ we get that

$$2^n = (1+1)^n = \sum_{k=0}^{n} \binom{n}{k} 1^k = |\mathbb{P}A|$$

Union and Intersection Operations

The union of two sets A and B is denoted by $A \cup B$. It results in a set that contains all of the members of A and of B and is defined by

$$A \cup B = \{r | r \in A \ or \ r \in B\}.$$

For example, suppose $A = \{1, 2, 3\}$ and $B = \{2, 3, 4\}$ then $A \cup B = \{1, 2, 3, 4\}$. Set union is a commutative operation, i.e. $A \cup B = B \cup A$. Venn Diagrams are used to illustrate these operations pictorially.

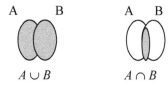

The intersection of two sets A and B is denoted by $A \cap B$. It results in a set containing the elements that A and B have in common and is defined by

$$A \cap B = \{r | r \in A \ and \ r \in B\}.$$

Suppose $A = \{1, 2, 3\}$ and $B = \{2, 3, 4\}$ then $A \cap B = \{2, 3\}$. Set intersection is a commutative operation, i.e. $A \cap B = B \cap A$.

Union and intersection may be extended to more generalized union and intersection operations. For example,

$\bigcup_{i=1}^{n} A_i$ denotes the union of n sets.

$\bigcap_{i=1}^{n} A_i$ denotes the intersection of n sets.

Set Difference Operations

The set difference operation $A\backslash B$ yields the elements in A that are not in B. It is defined by

$$A \backslash B = \{a | a \in A \ and \ a \notin B\}$$

For A and B defined as $A = \{1, 2\}$ and $B = \{2, 3\}$, we have $A\backslash B = \{1\}$ and $B\backslash A = \{3\}$. Clearly, set difference is not commutative, i.e. $A\backslash B \neq B\backslash A$. Clearly, $A\backslash A = \varnothing$ and $A\backslash\varnothing = A$.

The symmetric difference of two sets A and B is denoted by $A \triangle B$ and is given by

$$A \triangle B = A\backslash B \cup B\backslash A$$

The symmetric difference operation is commutative, i.e. $A \triangle B = B \triangle A$. Venn diagrams are used to illustrate these operations pictorially.

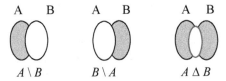

The complement of a set A (with respect to the universal set U) is the elements in the universal set that are not in A. It is denoted by A^c (or A') and is defined as

$$A^c = \{u | u \in U \text{ and } u \notin A\} = U\backslash A$$

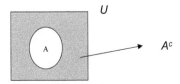

The complement of the set A is illustrated by the shaded area.

3.2.2 Properties of Set-Theoretical Operations

The set union and set intersection properties are commutative and associative. Their properties are listed in Table 3.1.

These properties may be seen to be true with Venn diagrams, and we give a proof of the distributive property (this proof uses logic which is discussed in Chaps. 15–17).

Proof of Properties (*Distributive Property*) To show $A \cap (B \cup C) = (A \cap B) \cup (A \cap C)$

Table 3.1 Properties of set operations

Property	Description
Commutative	Union and intersection operations are commutative, i.e. $S \cup T = T \cup S$ $S \cap T = T \cap S$
Associative	Union and intersection operations are associative, i.e. $R \cup (S \cup T) = (R \cup S) \cup T$ $R \cap (S \cap T) = (R \cap S) \cap T$
Identity	The identity under set union is the empty set $\varnothing$, and the identity under intersection is the universal set U $S \cup \varnothing = \varnothing \cup S = S$ $S \cap U = U \cap S = S$
Distributive	The union operator distributes over the intersection operator and vice versa $R \cap (S \cup T) = (R \cap S) \cup (R \cap T)$ $R \cup (S \cap T) = (R \cup S) \cap (R \cup T)$
De Morgan's (De Morgan's law is named after Augustus De Morgan, a nineteenth-century English mathematician who was a contemporary of George Boole) law	The complement of $S \cup T$ is given by $(S \cup T)^c = S \cap T$ The complement of $S \cap T$ is given by $(S \cap T)^c = S^c \cup T$

Suppose

$$x \in A \cap (B \cup C) \text{ then}$$
$$x \in A \wedge x \in (B \cup C)$$
$$\Rightarrow x \in A \wedge (x \in B \vee x \in C)$$
$$\Rightarrow (x \in A \wedge x \in B) \vee (x \in A \wedge x \in C)$$
$$\Rightarrow x \in (A \cap B) \vee x \in (A \cap C)$$
$$\Rightarrow x \in (A \cap B) \cup (A \cap C)$$

Therefore, $A \cap (B \cup C) \subseteq (A \cap B) \cup (A \cap C)$
Similarly $(A \cap B) \cup (A \cap C) \subseteq A \cap (B \cup C)$
Therefore, $A \cap (B \cup C) = (A \cap B) \cup (A \cap C)$

3.2.3 Russell's Paradox

Bertrand Russell (Fig. 3.1) was a famous British logician, mathematician and philosopher. He was the co-author with Alfred Whitehead of *Principia Mathematica*, which aimed to derive all the truths of mathematics from logic. Russell's Paradox was discovered by Bertrand Russell in 1901 and showed that the system of

Fig. 3.1 Bertrand Russell

logicism being proposed by Frege (discussed in Chap. 15) contained a contradiction.

Question (*Posed by Russell to Frege*) Is the set of all sets that do not contain themselves as members a set?

Russell's Paradox

Let A = {S a set and S ∉ S}. Is A ∈ A? Then A ∈ A ⇒ A ∉ A and vice versa. Therefore, a contradiction arises in either case and there is no such set A.

Two ways of avoiding the paradox were developed in 1908, and these were Russell's theory of types and Zermelo set theory. Russell's theory of types was a response to the paradox by arguing that the set of all sets is ill formed. Russell developed a hierarchy with individual elements the lowest level, sets of elements at the next level, sets of elements at the next level and so on. It is then prohibited for a set to contain members of different types.

A set of elements has a different type from its elements, and one cannot speak of the set of all sets that do not contain themselves as members as these are of different types. The other way of avoiding the paradox was Zermelo's axiomatization of set theory.

Remark Russell's paradox may also be illustrated by the story of a town that has exactly one barber who is male. *The barber shaves all and only those men in town who do not shave themselves.* The question is who shaves the barber.

If the barber does not shave himself, then according to the rule he is shaved by the barber (i.e. himself). If he shaves himself then according to the rule he is not shaved by the barber (i.e. himself).

The paradox occurs due to self-reference in the statement, and a logical examination shows that the statement is a contradiction.

3.2.4 Computer Representation of Sets

Sets are fundamental building blocks in mathematics, and so the question arises as to how is a set is stored and manipulated in a computer. The representation of a set M on a computer requires a change from the normal view that the order of the elements of the set is irrelevant, and we will need to assume a definite order in the underlying universal set $\mathcal{m}$ from which the set M is defined.

That is, a set is defined in a computer program with respect to an underlying universal set, and the elements in the universal set are listed in a definite order. Any set M arising in the program that is defined with respect to this universal set $\mathcal{m}$ is a subset of $\mathcal{m}$. Next, we show how the set M is stored internally on the computer.

The set M is represented in a computer as a string of binary digits $b_1 b_2 \ldots b_n$ where n is the cardinality of the universal set $\mathcal{m}$. The bits b_i (where i ranges over the values 1, 2, ... n) are determined according to the rule:

$$b_i = 1 \text{ if } i^{\text{th}} \text{ element of } \mathcal{m} \text{ is in } M$$
$$b_i = 0 \text{ if } i^{\text{th}} \text{ element of } \mathcal{m} \text{ is not in } M$$

For example, if $\mathcal{m} = \{1, 2, \ldots 10\}$ then the representation of $M = \{1, 2, 5, 8\}$ is given by the bit string 1100100100 where this is given by looking at each element of $\mathcal{m}$ in turn and writing down 1 if it is in M and 0 otherwise.

Similarly, the bit string 0100101100 represents the set $M = \{2, 5, 7, 8\}$, and this is determined by writing down the corresponding element in $\mathcal{m}$ that corresponds to a 1 in the bit string.

Clearly, there is a one-to-one correspondence between the subsets of $\mathcal{m}$ and all possible n-bit strings. Further, the set-theoretical operations of set union, intersection and complement can be carried out directly with the bit strings (provided that the sets involved are defined with respect to the same universal set). This involves a bitwise 'or' operation for set union, a bitwise 'and' operation for set intersection and a bitwise 'not' operation for the set complement operation.

3.3 Relations

A binary relation $R(A, B)$ where A and B are sets is a subset of $A \times B$, i.e. $R \subseteq A \times B$. The domain of the relation is A, and the co-domain of the relation is B. The notation aRb signifies that $(a, b) \in R$.

A binary relation $R(A, A)$ is a relation between A and A (or a relation on A). This type of relation may always be composed with itself, and its inverse is also a binary relation on A. The identity relation on A is defined by $a\ i_A a$ for all $a \in A$.

Example 3.4 There are many examples of relations:

(i)   The relation on a set of students in a class where $(a, b) \in R$ if the height of a is greater than the height of b.

(ii) The relation between A and B where A = {0, 1, 2} and B = {3, 4, 5} with R given by

$$R = \{(0,3),(0,4),(1,4)\}$$

(iii) The relation less than ($<$) between and $\mathbb{R}$ and $\mathbb{R}$ is given by

$$\{(x,y) \in \mathbb{R}^2 : x<y\}$$

(iv) A bank may represent the relationship between the set of accounts and the set of customers by a relation. The implementation of a bank account may be a positive integer with at most eight decimal digits.

      The relationship between accounts and customers may be done with a relation $R \subseteq A \times B$, with the set A chosen to be the set of natural numbers, and the set B chosen to be the set of all human beings alive or dead. The set

$$A \text{ could also be chosen to be } A = \{n \in \mathbb{N} : n < 10^8\}$$

A relation $R(A, B)$ may be represented pictorially. This is referred to as the graph of the relation, and it is illustrated in the diagram below. An arrow from x to y is drawn if (x, y) is in the relation. Thus, for the height relation R given by {(a, p), (a, r), (b, q)} an arrow is drawn from a to p, from a to r and from b to q to indicate that (a, p), (a, r) and (b, q) are in the relation R.

The pictorial representation of the relation makes it easy to see that the height of a is greater than the height of p and r; and that the height of b is greater than the height of q.

An n-ary relation $R\,(A_1, A_2, \ldots A_n)$ is a subset of $(A_1 \times A_2 \times \cdots \times A_n)$. However, an n-ary relation may also be regarded as a binary relation $R(A, B)$ with $A = A_1 \times A_2 \times \cdots \times A_{n-1}$ and $B = A_n$.

3.3.1 Reflexive, Symmetric and Transitive Relations

(i)   A binary relation on A may have additional properties such as being reflexive, symmetric or transitive. These properties are defined as a relation on a set A is *reflexive* if $(a,\ a) \in R$ for all $a \in A$.

(ii) A relation R is *symmetric* if whenever $(a,\ b) \in R$ then $(b,\ a) \in R$.

(iii) A relation is *transitive* if whenever $(a,\ b) \in R$ and $(b,\ c) \in R$ then $(a,\ c) \in R$.

A relation that is reflexive, symmetric and transitive is termed an *equivalence relation*.

Example 3.5 (*Reflexive Relation*) A relation is reflexive if each element possesses an edge looping around on itself. The relation in Fig. 3.2 is reflexive.

Example 3.6 (*Symmetric Relation*) The graph of a symmetric relation will show for every arrow from a to b an opposite arrow from b to a. The relation in Fig. 3.3 is symmetric, i.e. whenever $(a,\ b) \in R$ then $(b,\ a) \in R$.

Example 3.7 (*Transitive relation*) The graph of a transitive relation will show that whenever there is an arrow from a to b and an arrow from b to c that there is an arrow from a to c. The relation in Fig. 3.4 is transitive, i.e. whenever $(a,\ b) \in R$ and $(b,\ c) \in R$ then $(a,\ c) \in R$.

Example 3.8 (*Equivalence relation*) The relation on the set of integers $\mathbb{Z}$ defined by $(a,\ b) \in R$ if $a - b = 2\,k$ for some $k \in \mathbb{Z}$ is an equivalence relation, and it partitions the set of integers into two equivalence classes, i.e. the even and odd integers.

Domain and Range of Relation

The domain of a relation $R\,(A, B)$ is given by $\{a \in A \mid \exists b \in B \text{ and } (a, b) \in R\}$. It is denoted by **dom** R. The domain of the relation R $= \{(a, p), (a, r), (b, q)\}$ is $\{a, b\}$.

The range of a relation $R\,(A, B)$ is given by $\{b \in B \mid \exists a \in A \text{ and } (a, b) \in R\}$. It is denoted by **rng** R. The range of the relation R $= \{(a, p), (a, r), (b, q)\}$ is $\{p, q, r\}$.

Fig. 3.2 Reflexive relation

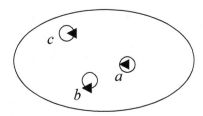

Fig. 3.3 Symmetric relation

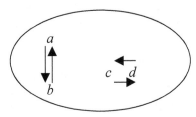

Fig. 3.4 Transitive relation

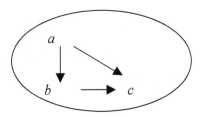

Inverse of a Relation

Suppose $R \subseteq A \times B$ is a relation between A and B then the inverse relation $R^{-1} \subseteq B \times A$ is defined as the relation between B and A and is given by

$$b R^{-1} a \text{ if and only if } a R b$$

That is,

$$R^{-1} = \{(b, a) \in B \times A : (a, b) \in R\}$$

Example 3.9 Let R be the relation between $\mathbb{Z}$ and $\mathbb{Z}^+$ defined by mRn if and only if $m^2 = n$. Then $R = \{(m, n) \in \mathbb{Z} \times \mathbb{Z}^+: m^2 = n\}$ and $R^{-1} = \{(n, m) \in \mathbb{Z}^+ \times \mathbb{Z} : m^2 = n\}$.

For example, $-3 \ R \ 9$, $-4 \ R \ 16$, $0 \ R \ 0$, $16 \ R^{-1} \ -4$, $9 \ R^{-1} \ -3$, etc.

Partitions and Equivalence Relations

An equivalence relation on A leads to a partition of A, and vice versa for every partition of A there is a corresponding equivalence relation.

Let A be a finite set and let $A_1, A_2, \ldots, A_n$ be subsets of $A_i \neq \varnothing$ for all $i, A_i \cap A_j = \varnothing$ if $i \neq j$ and $A = \cup_i^n A_i = A_1 \cup A_2 \cup \ldots \cup A_n$.

The sets A_i partition the set A, and these sets are called the classes of the partition (Fig. 3.5).

Theorem 3.2 (Equivalence Relation and Partitions) *An equivalence relation on A gives rise to a partition of A where the equivalence classes are given by Class (a) = {x | x ∈ A and (a, x) ∈ R}. Similarly, a partition gives rise to an equivalence relation R, where (a, b) ∈ R if and only if a and b are in the same partition.*

Proof Clearly, $a \in$ Class(a) since R is reflexive and clearly the union of the equivalence classes is A. Next, we show that two equivalence classes are either equal or disjoint.

Suppose Class$(a) \cap$ Class$(b) \neq \varnothing$. Let $x \in$ Class$(a) \cap$ Class(b) and so (a, x) and $(b, x) \in$ R. By the symmetric property $(x, b) \in$ R and since R is transitive from (a, x) and (x, b) in R we deduce that $(a, b) \in$ R. Therefore $b \in$ Class(a). Suppose y is an arbitrary member of Class (b) then $(b, y) \in$ R; therefore, from (a, b) and (b, y) in R we deduce that (a, y) is in R. Therefore, since y was an arbitrary member of Class (a) we deduce that Class$(b) \subseteq$ Class(a). Similarly, Class$(a) \subseteq$ Class(b) and so Class $(a) =$ Class(b).

This proves the first part of the theorem and for the second part we define a relation R such that $(a, b) \in$ R if a and b are in the same partition. It is clear that this is an equivalence relation.

3.3.2 Composition of Relations

The composition of two relations $R_1(A, B)$ and $R_2(B, C)$ is given by $R_2 \circ R_1$ where $(a, c) \in R_2 \circ R_1$ if and only there exists $b \in$ B such that $(a, b) \in R_1$ and $(b, c) \in R_2$. The composition of relations is associative, i.e.

$$(R_3 \circ R_2) \circ R_1 = R_3 \circ (R_2 \circ R_1)$$

Fig. 3.5 Partitions of A

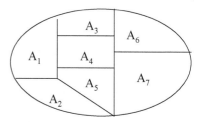

Example 3.10 (*Composition of Relations*) Consider a library that maintains two files. The first file maintains the serial number s of each book as well as the details of the author a of the book. This may be represented by the relation $R_1 = sR_1a$. The second file maintains the library card number c of its borrowers and the serial number s of any books that they have borrowed. This may be represented by the relation $R_2 = c\, R_2s$.

The library wishes to issue a reminder to its borrowers of the authors of all books currently on loan to them. This may be determined by the composition of R_1 o R_2, i.e. $c\, R_1$o R_2 a if there is book with serial number s such that $c\, R_2\, s$ and $s\, R_1\, a$.

Example 3.11 (*Composition of Relations*) Consider sets $A = \{a, b, c\}$, $B = \{d, e, f\}$, $C = \{g, h, i\}$ and relations $R(A, B) = \{(a, d), (a, f), (b, d), (c, e)\}$ and $S(B, C) = \{(d, h), (d, i), (e, g), (e, h)\}$. Then, we graph these relations and show how to determine the composition pictorially.

S o R is determined by choosing $x \in A$ and $y \in C$ and checking if there is a route from x to y in the graph. If so, we join x to y in S o R. For example, if we consider a and h we see that there is a path from a to d and from d to h and therefore (a, h) is in the composition of S and R (Fig. 3.6).

The union of two relations $R_1(A, B)$ and $R_2(A, B)$ is meaningful (as these are both subsets of $A \times B$). The union $R_1 \cup R_2$ is defined as $(a, b) \in R_1 \cup R_2$ if and only if $(a, b) \in R_1$ or $(a, b) \in R_2$.

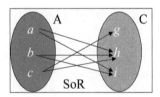

Similarly, the intersection of R_1 and R_2 ($R_1 \cap R_2$) is meaningful and is defined as $(a, b) \in R_1 \cap R_2$ if and only if $(a, b) \in R_1$ and $(a, b) \in R_2$. The relation R_1 is a subset of R_2 ($R_1 \subseteq R_2$) if whenever $(a, b) \in R_1$ then $(a, b) \in R_2$.

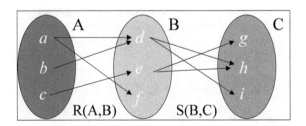

Fig. 3.6 Composition of relations S o R

The inverse of the relation R was discussed earlier and is given by the relation R^{-1} where $R^{-1} = \{(b, a) \mid (a, b) \in R\}$.

The composition of R and R^{-1} yields $R^{-1} \circ R = \{(a, a) \mid a \in \text{dom } R\} = i_A$ and $R \circ R^{-1} = \{(b, b) \mid b \in \text{dom } R^{-1}\} = i_B$.

3.3.3   Binary Relations

A binary relation R on A is a relation between A and A, and a binary relation can always be composed with itself. Its inverse is a binary relation on the same set. The following are all relations on A:

$R^2 = R \circ R$
$R^3 = (R \circ R) \circ R$
$R^0 = i_A$ (identity relation)
$R^{-2} = R^{-1} \circ R^{-1}$

Example 3.12 Let R be the binary relation on the set of all people P such that $(a, b) \in R$ if a is a parent of b. Then the relation R^n is interpreted as

R is the parent relationship.
R^2 is the grandparent relationship.
R^3 is the great-grandparent relationship.
R^{-1} is the child relationship.
R^{-2} is the grandchild relationship.
R^{-3} is the great-grandchild relationship.

This can be generalized to a relation R^n on A where $R^n = R \circ R \circ \cdots \circ R$ (n times). The transitive closure of the relation R on A is given by

$$R^* = \cup_{i=0}^{\infty} R^i = R^0 \cup R^1 \cup R^2 \cup \cdots R^n \cup \cdots$$

where R^0 is the reflexive relation containing only each element in the domain of R, i.e. $R^0 = i_A = \{(a, a) \mid a \in \text{dom } R\}$.

The positive transitive closure is similar to the transitive closure except that it does not contain R^0. It is given by

$$R^+ = \cup_{i=1}^{\infty} R^i = R^1 \cup R^2 \cup \cdots \cup R^n \cup \cdots$$

$a\,R^+\,b$ if and only if $a\,R^n\,b$ for some $n > 0$, i.e. there exists $c_1, c_2 \ldots c_n \in A$ such that

$$aRc_1, c_1Rc_2, \ldots, c_nRb$$

Parnas[7] introduced the concept of the limited domain relation (LD relation), and a LD relation L consists of an ordered pair (R_L, C_L) where R_L is a relation and C_L is a subset of Dom R_L. The relation R_L is on a set U and C_L is termed the competence set of the LD relation L.

The importance of LD relations is that they may be used to describe program execution. The relation component of the LD relation L describes a set of states such that if execution starts in state x it may terminate in state y. The set U is the set of states. The competence set of L is such that if execution starts in a state that is in the competence set then it is guaranteed to terminate. For a more detailed description of LD relations and their properties, see Chap. 2 of Hoffman and Weiss (2001).

3.3.4 Applications of Relations to Databases

A Relational Database Management System (RDBMS) is a system that manages data using the relational model, and a relation is defined as a set of tuples that is usually represented by a table. A table is data organized in rows and columns, with the data in each column of the table of the same data type. Constraints may be employed to provide restrictions on the kinds of data that may be stored in the relations, and these Boolean expressions are a way of implementing business rules in the database.

Relations have one or more keys associated with them, and the *key uniquely identifies the row of the table*. An index is a way of providing fast access to the data in a relational database, as it allows the tuple in a relation to be looked up directly (using the index) rather than checking all tuples in the relation.

The concept of a relational database was first described in a paper '*A Relational Model of Data for Large Shared Data Banks*' by Codd (1970). A relational database is a database that conforms to the relational model, and it may be defined as a set of relations (or tables).

Codd (Fig. 3.7) developed the *relational database model* in the late 1960s, and today, this is the standard way that information is organized and retrieved from computers. Relational databases are at the heart of systems from hospitals' patient records to airline flight and schedule information.

An n-ary relation R $(A_1, A_2, \dots A_n)$ is a subset of the Cartesian product of the n sets, i.e. a subset of $(A_1 \times A_2 \times \dots \times A_n)$. However, an n-ary relation may also be regarded as a binary relation R(A, B) with $A = A_1 \times A_2 \times \dots \times A_{n-1}$ and $B = A_n$.

The data in the relational model is defined as a set of n-tuples and is usually represented by a table. A table is a visual representation of the relation, and the data is organized in rows and columns.

---

[7]Parnas made important contributions to software engineering in the 1970s. He invented information hiding which is used in object-oriented design.

Fig. 3.7 Edgar Codd

The basic relational building block is the domain or data type (often called just type). Each row of the table represents one n-tuple (one tuple) of the relation, and the number of tuples in the relation is the cardinality of the relation. Consider the PART relation taken from Date (1981), where this relation consists of a heading and the body. There are five data types representing part numbers, part names, part colours, part weights and locations where the parts are stored. The body consists of a set of n-tuples, and the PART relation in Fig. 3.8 is of cardinality six.

There is more detailed information on the relational model and databases in O'Regan (2018).

P#	PName	Colour	Weight	City
P1	Nut	Red	12	London
P2	Bolt	Green	17	Paris
P3	Screw	Blue	17	Rome
P4	Screw	Red	14	London
P5	Cam	Blue	12	Paris
P6	Cog	Red	19	London

Fig. 3.8 PART relation

3.4 Functions

A function $f : A \rightarrow B$ is a special relation such that for each element $a \in A$ there is exactly (or at most)[8] one element $b \in B$. This is written as $f(a) = b$.

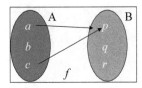

A function is a relation but not every relation is a function. For example, the relation in the diagram below is not a function since there are two arrows from the element $a \in A$.

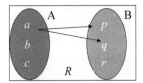

The domain of the function (denoted by **dom** f) is the set of values in A for which the function is defined. The domain of the function is A if f is a total function. The co-domain of the function is B. The range of the function (denoted **rng** f) is a subset of the co-domain and consists of

$$\mathbf{rng}f = \{r|r \in B \text{ such that} f(a) = r \text{ for some } a \in A\}.$$

Functions may be partial or total. A *partial function* (or partial mapping) may be undefined for some values of A, and partial functions arise regularly in the computing field (Fig. 3.9). *Total functions* are defined for every value in A, and many functions encountered in mathematics are total.

Example 3.13 (*Functions*) Functions are an essential part of mathematics and computer science, and there are many well-known functions such as the trigonometric functions $\sin(x)$, $\cos(x)$ and $\tan(x)$; the logarithmic function $\ln(x)$; the exponential functions e^x; and polynomial functions.

(i) Consider the partial function $f : \mathbb{R} \rightarrow \mathbb{R}$ $f(x) = {}^1/x$ (where $x \neq 0$).

Then, this partial function is defined everywhere except for $x = 0$.

(ii) Consider the function $f : \mathbb{R} \rightarrow \mathbb{R}$ where $f(x) = x^2$

Then this function is defined for all $x \in \mathbb{R}$

---

[8]We distinguish between total and partial functions. A total function is defined for all elements in the domain, whereas a partial function may be undefined for one or more elements in the domain.

Fig. 3.9 Domain and range
of a partial function

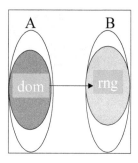

Partial functions often arise in computing as a program may be undefined or fail
to terminate for several values of its arguments (e.g. infinite loops). Care is required
to ensure that the partial function is defined for the argument to which it is to be
applied.

Consider a program P that has one natural number as its input and which fails to
terminate for some input values. It prints a single real result and halts if it terminates. Then P can be regarded as a partial mapping from $\mathbb{N}$ to $\mathbb{R}$.

$$P : \mathbb{N} \to \mathbb{R}$$

Example 3.14 How many total functions $f : A \to B$ are there from A to B (where
A and B are finite sets)?

Each element of A maps to any element of B, i.e. there are $|B|$ choices for each
element $a \in A$. Since there are $|A|$ elements in A the number of total functions is
given by

$$|B|\,|B|\ldots|B| \quad (|A|\text{times})$$
$$= |B|^{|A|} \qquad \text{total functions between A and B.}$$

Example 3.15 How many partial functions $f : A \to B$ are there from A to B (where
A and B are finite sets)?

Each element of A may map to any element of B or to no element of B (as it may
be undefined for that element of A). In other words, there are $|B| + 1$ choices for
each element of A. As there are $|A|$ elements in A, the number of distinct partial
functions between A and B is given by

$$(|B| + 1)(|B| + 1)\ldots(|B| + 1) \quad (|A|\text{times})$$
$$= (|B| + 1)^{|A|}$$

Two partial functions f and g are equal if

1. dom $f =$ dom g
2. $f(a) = g(a)$ for all $a \in$ dom f.

A function f is less defined than a function g ($f \subseteq g$) if the domain of f is a subset of the domain of g, and the functions agree for every value on the domain of f.

1. dom $f \subseteq$ dom g
2. $f(a) = g(a)$ for all $a \in$ dom f.

The composition of functions is similar to the composition of relations. Suppose $f: A \rightarrow B$ and $g : B \rightarrow C$ then $g \circ f: A \rightarrow C$ is a function, and it is written as $g \circ f$ (x) or $g(f(x))$ for $x \in A$.

The composition of functions is not commutative and this can be seen by an example. Consider the function $f : \mathbb{R} \rightarrow \mathbb{R}$ such that $f(x) = x^2$ and the function $g :$ $\mathbb{R} \rightarrow \mathbb{R}$ such that $g(x) = x + 2$. Then

$$g \circ f(x) = g(x^2) = x^2 + 2.$$
$$f \circ g(x) = f(x+2) = (x+2)^2 = x^2 + 4x + 4.$$

Clearly, $g \circ f(x) \neq f \circ g(x)$ and so composition of functions is not commutative. The composition of functions is associative, as the composition of relations is associative and every function is a relation. For $f: A \rightarrow B$, $g : B \rightarrow C$, and $h : C \rightarrow D$ we have

$$h \circ (g \circ f) = (h \circ g) \circ f$$

A function $f : A \rightarrow B$ is *injective* (*one to one*) if

$$f(a_1) = f(a_2) \Rightarrow a_1 = a_2.$$

For example, consider the function $f : \mathbb{R} \rightarrow \mathbb{R}$ with $f(x) = x^2$. Then $f(3) = f(-3) = 9$ and so this function is not one to one.

A function $f: A \rightarrow B$ is *surjective* (*onto*) if given any $b \in B$ there exists an $a \in A$ such that $f(a) = b$. Consider the function $f: \mathbb{R} \rightarrow \mathbb{R}$ with $f(x) = x + 1$. Clearly, given any $r \in \mathbb{R}$ then $f(r - 1) = r$ and so f is onto (Fig. 3.10).

A function is *bijective* if it is one to one and onto (Fig. 3.11). That is, there is a one-to-one correspondence between the elements in A and B, and for each $b \in B$ there is a unique $a \in A$ such that $f(a) = b$.

The inverse of a relation was discussed earlier, and the relational inverse of a function $f: A \rightarrow B$ clearly exists. The relational inverse of the function may or may not be a function.

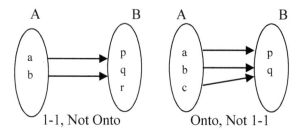

Fig. 3.10 Injective and surjective functions

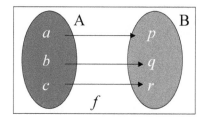

Fig. 3.11 Bijective function (one to one and onto)

However, if the relational inverse is a function it is denoted by $f^{-1} : B \rightarrow A$. A total function has an inverse if and only if it is bijective whereas a partial function has an inverse if and only if it is injective.

The identity function $1_A : A \rightarrow A$ is a function such that $1_A(a) = a$ for all $a \in A$. Clearly, when the inverse of the function exists then we have that $f^{-1} \, o \, f = 1_A$ and $f \, o \, f^{-1} = 1_B$.

Theorem 3.3 (Inverse of Function) *A total function has an inverse if and only if it is bijective.*

Proof Suppose $f : A \rightarrow B$ has an inverse f^{-1}. Then we show that f is bijective.

We first show that f is one to one. Suppose $f(x_1) = f(x_2)$ then

$$f^{-1}(f(x_1)) = f^{-1}(f(x_2))$$
$$\Rightarrow f^{-1} \, o \, f \, (x_1) = f^{-1} \, o \, f \, (x_2)$$
$$\Rightarrow 1_A(x_1) = 1_A(x_2)$$
$$\Rightarrow x_1 = x_2$$

Next, we first show that f is onto. Let $b \in B$ and let $a = f^{-1}(b)$ then

$$f(a) = f(f^{-1}(b)) = 1_B(b) = b \text{ and so } f \text{ is surjective}$$

The second part of the proof is concerned with showing that if $f : A \rightarrow B$ is bijective then it has an inverse f^{-1}. Clearly, since f is bijective we have that for each $a \in A$ there exists a unique $b \in B$ such that $f(a) = b$.

Define $g : B \rightarrow A$ by letting $g(b)$ be the unique a in A such that $f(a) = b$. Then we have that

$$g \circ f(a) = g(b) = a \text{ and } f \circ g(b) = f(a) = b.$$

Therefore, g is the inverse of f.

3.5 Application of Functions to Functional Programming

Functional programming involves the evaluation of mathematical functions, whereas imperative programming involves the execution of sequential (or iterative) commands that change the state. For example, the assignment statement alters the value of a variable, and the value of a given variable x may change during program execution.

There are no changes of state for functional programs, and the fact that the value of x will always be the same makes it easier to reason about functional programs than imperative programs. Functional programming languages provide *referential transparency,* i.e. equals may be substituted for equals, and if two expressions have equal values, then one can be substituted for the other in any larger expression without affecting the result of the computation.

Functional programming languages use higher order functions,[9] recursion, lazy and eager evaluation, monads[10] and Hindley–Milner-type inference systems.[11] These languages are mainly used in academia, but there has been some industrial use, including the use of Erlang for concurrent applications in industry. Alonzo Church developed Lambda calculus in the 1930s, and it provides an abstract framework for describing mathematical functions and their evaluation. It provides the foundation for functional programming languages. Church employed lambda calculus to prove that there is no solution to the decision problem for first-order arithmetic (O'Regan 2016).

---

[9]Higher order functions are functions that take functions as arguments or return a function as a result. They are known as operators (or functionals) in mathematics, and one example is the derivative function dy/dx that takes a function as an argument and returns a function as a result.

[10]Monads are used in functional programming to express input and output operations without introducing side effects. The Haskell functional programming language makes use of uses this feature.

[11]This is the most common algorithm used to perform type inference. Type inference is concerned with determining the type of the value derived from the eventual evaluation of an expression.

The original calculus developed by Church was untyped, but typed lambda calculi have since been developed. Any computable function can be expressed and evaluated using lambda calculus, but there is no general algorithm to determine whether two arbitrary lambda calculus expressions are equivalent. Lambda calculus influenced functional programming languages such as Lisp, ML and Haskell.

Functional programming uses higher order functions which take functions as arguments, and return functions as results. The derivative function $^d/_{dx} f(x) = f'(x)$ is a higher order function, which takes a function as an argument and returns a function as a result. Higher order functions may employ currying (a technique developed by Schönfinkel) which allows a function with several arguments to be applied to each of its arguments one at a time, with each application returning a new (higher order) function that accepts the next argument. This allows a function of n-arguments to be treated as n applications of a function with 1-argument.

John McCarthy developed LISP at MIT in the late 1950s, and this language includes many of the features found in modern functional programming languages.[12] Scheme built upon the ideas in LISP, and Kenneth Iverson developed APL[13] in the early 1960s. APL influenced Backus's FP programming language, and Robin Milner designed the ML programming language in the early 1970s. David Turner developed Miranda in the mid-1980s, and it influenced the Haskell programming language developed by Philip Wadler and others in the late 1980s/early 1990s.

3.5.1 Miranda

Miranda was developed by David Turner at the University of Kent in the mid-1980s (Turner 1985). It is a non-strict functional programming language, i.e. the arguments to a function are not evaluated until they are required within the function being called. This is also known as *lazy evaluation*, and one of its key advantages is that it allows a potentially infinite data structure to be passed as an argument to a function. Miranda is a pure functional language in that there are no side-effect features in the language. The language has been used for

– Rapid prototyping,
– Specification language and
– Teaching language.

A Miranda program is a collection of equations that define various functions and data structures. It is a strongly typed language with a terse notation.

---

[12]Lisp is a multi-paradigm language rather than a functional programming language.
[13]Iverson received the Turing Award in 1979 for his contributions to programming language and mathematical notation. The title of his Turing award paper was 'Notation as a tool of thought'.

$$z = \text{sqr}\,p/\text{sqr}\,q$$
$$\text{sqr}\,k = k * k$$
$$p = a + b$$
$$q = a - b$$
$$a = 10$$
$$b = 5$$

The scope of a formal parameter (e.g. the parameter k above in the function sqr) is limited to the definition of the function in which it occurs.

One of the most common data structures used in Miranda is the list. The empty list is denoted by [ ], and an example of a list of integers is given by [1, 3, 4, 8]. Lists may be appended to by using the '++' operator. For example,

$$[1, 3, 5] + + [2, 4] \text{ is } [1, 3, 5, 2, 4].$$

The length of a list is given by the '#' operator:

$$\#[1, 3] = 2$$

The infix operator ':' is employed to prefix an element to the front of a list. For example,

$$5 : [2, 4, 6] \text{ is equal to } [5, 2, 4, 6]$$

The subscript operator '!' is employed for subscripting: For example,

$$\text{Nums} = [5, 2, 4, 6] \quad \text{then Nums!0 is 5.}$$

The elements of a list are required to be of the same type. A sequence of elements that contains mixed types is called a tuple. A tuple is written as follows:

$$\text{Employee} = (\text{``Holmes''}, \text{``221B Baker St.London''}, 50, \text{``Detective''})$$

A tuple is similar to a record in Pascal, whereas lists are similar to arrays. Tuples cannot be subscripted but their elements may be extracted by pattern matching. Pattern matching is illustrated by the well-known example of the factorial function:

$$\text{fac}\,0 = 1$$
$$\text{fac}(n + 1) = (n + 1) * \text{fac}\,n$$

The definition of the factorial function uses two equations, distinguished by using different patterns in the formal parameters. Another example of pattern matching is the reverse function on lists:

$$\text{reverse} [] = []$$
$$\text{reverse} (a : x) = \text{reverse} x + + [a]$$

Miranda is a higher order language, and it allows functions to be passed as parameters and returned as results. Currying is allowed and this allows a function of n-arguments to be treated as n applications of a function with 1-argument. Function application is left associative, i.e. f x y means (f x) y. That is, the result of applying the function f to x is a function, and this function is then applied to y. Every function with two or more arguments in Miranda is a higher order function.

3.6 Number Theory

Number theory is the branch of mathematics that is concerned with the mathematical properties of the natural numbers and integers. These include properties such as the parity of a number, divisibility, additive and multiplicative properties, whether a number is prime or composite, the prime factors of a number, the greatest common divisor and least common multiple of two numbers, and so on.

Number theory has many applications in computing including cryptography and coding theory. For example, the RSA public-key cryptographic system relies on its security due to the infeasibility of the integer factorization problem for large numbers.

There are several unsolved problems in number theory and especially in prime number theory. For example, Goldbach's[14] Conjecture states that every even integer greater than two is the sum of two primes, and this result has not been proved to date. Fermat's[15] Last Theorem (Fig. 4.12) states that there is no integer solution to $x^n + y^n = z^n$ for $n > 2$, and this result remained unproved for over 300 years until Andrew Wiles finally proved it in the mid-1990s.

The natural numbers $\mathbb{N}$ consist of the numbers $\{1, 2, 3, \ldots\}$. The integer numbers $\mathbb{Z}$ consist of $\{\ldots, -2, -1, 0, 1, 2, \ldots\}$. The rational numbers $\mathbb{Q}$ consist of all numbers of the form $\{{}^p/_q$ where p and q are integers and $q \neq 0\}$. The real numbers $\mathbb{R}$ are defined to be the set of converging sequences of rational numbers and they are a superset of the rational numbers. They contain the rational and irrational numbers. The complex numbers $\mathbb{C}$ consist of all numbers of the form $\{a + bi$ where $a, b \in \mathbb{R}$ and $i = \sqrt{-1}\}$.

---

[14]Goldbach was an eighteenth-century German mathematician and Goldbach's conjecture has been verified to be true for all integers $n < 12 * 10^{17}$.

[15]Pierre de Fermat was a seventeenth-century French civil servant and amateur mathematician. He occasionally wrote to contemporary mathematicians announcing his latest theorem without providing the accompanying proof and inviting them to find the proof. The fact that he never revealed his proofs caused a lot of frustration among his contemporaries, and in his announcement of his famous last theorem he stated that he had a wonderful proof that was too large to include in the margin. He corresponded with Pascal, and they did some early work on the mathematical rules of games of chance and early probability theory.

Pythagorean triples are combinations of three whole numbers that satisfy Pythagoras's equation $x^2 + y^2 = z^2$. There are an infinite number of such triples, and an example of such a triple is 3, 4, 5 since $3^2 + 4^2 = 5^2$.

The Pythagoreans discovered the mathematical relationship between the harmony of music and numbers, and their philosophy was that numbers are hidden in everything from music to science and nature. This led to their philosophy that 'everything is number'. We will discuss number theory in more detail in Chap. 5 and show how it is applied to the field of cryptography in Chap. 10.

3.7 Automata Theory

Automata Theory is the branch of computer science that is concerned with the study of abstract machines and automata. These include finite-state machines, pushdown automata, and Turing machines. Finite-state machines are abstract machines that may be in one of a finite number of states. These machines are in only one state at a time (current state), and the input symbol causes a transition from the current state to the next state. Finite-state machines have limited computational power due to memory and state constraints, but they have been applied to several fields including communication protocols, neurological systems and linguistics.

Warren McCulloch and Walter Pitts published early work on finite-state automata in 1943. They were interested in modelling the thought process for humans and machines. Moore and Mealy developed this work further, and their finite-state machines are referred to as the '*Mealy machine*' and the '*Moore machine*'. The Mealy machine determines its outputs through the current state and the input, whereas the output of Moore's machine is based upon the current state alone.

Finite-state automata can compute only very primitive functions, and so they are not adequate as a model for computing. There are more powerful automata such as the Turing machine that is essentially a finite automaton with a potentially infinite storage (memory). Anything that is computable is computable by a Turing machine.

A finite-state machine can model a system that has a finite number of states, and a finite number of inputs/events that can trigger transitions between states. The behaviour of the system at a point in time is determined from the current state and input, with behaviour defined for the possible input to that state. The system starts in an initial state.

Pushdown automata have greater computational power, and they contain extra memory in the form of a stack from which symbols may be pushed or popped. The state transition is determined from the current state of the machine, the input symbol and the element on the top of the stack. The action may be to change the state and/or push/pop an element from the stack.

The Turing machine is the most powerful model for computation, and this theoretical machine is equivalent to an actual computer in the sense that it can compute the same set of functions. The memory of the Turing machine is a tape that consists of a potentially infinite number of one-dimensional cells. The Turing machine provides a mathematical abstraction of computer execution and storage, as well as providing a mathematical definition of an algorithm. However, Turing machines are not suitable for programming, and therefore they do not provide a good basis for studying programming and programming languages. We discuss finite-state automata in more detail in Chap. 23.

3.8   Graph Theory

Graph theory is a practical branch of mathematics that deals with the arrangements of certain objects known as vertices (or nodes) and the relationships between them. It has been applied to practical problems such as the modelling of computer networks, determining the shortest driving route between two cities, the link structure of a website, the travelling salesman problem and the four-colour problem.[16]

A *graph* is a collection of objects that are interconnected in some way. The objects are typically represented by vertices (or nodes), and the interconnections between them are represented by edges (or lines). We distinguish between directed and adirected graphs, where a *directed graph* is mathematically equivalent to a binary relation, and an *adirected (undirected) graph* is equivalent to a symmetric binary relation.

Graph theory may determine whether or not there is a route from one vertex to another, as well as finding the shortest or most efficient route to the destination vertex. A graph is said to be *connected* if for any two given vertices v_1, v_2 in V there is a path from v_1 to v_2. A *Hamiltonian path* in a graph G = (V, E) is a path that visits every vertex once and once only. We will discuss graph theory in more detail in Chap. 9.

3.9   Computability and Decidability

It is impossible for a human or machine to write out all members of an infinite countable set, such as the set of natural numbers $\mathbb{N}$. However, humans can do something quite useful in the case of certain enumerable infinite sets: they can give explicit instructions (that may be followed by a machine or another human) to produce the n-th member of the set for an arbitrary finite n. The problem remains that for all but a finite number of values of n it will be physically impossible for any human or machine to carry out the computation, due to the limitations on the time available for computation, the speed at which the individual steps in the computation may be carried out, and due to finite materials.

The intuitive meaning of computability is in terms of an algorithm (or effective procedure) that specifies a set of instructions to be followed to complete the task. That is, a function f is *computable* if there exists an algorithm that produces the

---

[16]The four-colour theorem states that given any map it is possible to colour the regions of the map with no more than four colours such that no two adjacent regions have the same colour. This result was finally proved in the mid-1970s.

value of f correctly for each possible argument of f. The computation of f for an argument x just involves following the instructions in the algorithm, and it produces the result $f(x)$ in a finite number of steps if x is in the domain of f. If x is not in the domain of f then the algorithm may produce an answer saying so or it might run forever never halting. A computer program implements an algorithm and we discuss algorithms in more detail in Chap. 4.

The concept of computability may be made precise in several equivalent ways such as Church's *lambda calculus, recursive function theory* or by the theoretical *Turing machines.*[17] These are all equivalent and perhaps the most well known is the Turing machine (O'Regan 2016), where the set of functions that are computable are those that are computable by a Turing machine.

Decidability is an important topic in contemporary mathematics. Church and Turing independently showed in 1936 that mathematics is not decidable. In other words, there is no mechanical procedure (i.e. algorithm) to determine whether an arbitrary mathematical proposition is true or false, and so the only way is to determine the truth or falsity of a statement is to try to solve the problem. We provide a more detailed account of computability and decidability in Chap. 13.

3.10 Review Questions

1. What is a set? A relation? A function?
2. Explain the difference between a partial and a total function.
3. Explain the difference between a relation and a function.
4. Determine $A \times B$ where $A = \{a, b, c, d\}$ and $B = \{1, 2, 3\}$.
5. Determine $A \, \Delta \, B$ where $A = \{a, b, c, d\}$ and $B = \{c, d, e\}$.
6. What is the graph of the relation $\leq$ on the set $A = \{2, 3, 4\}$.
7. What is the domain and range of $R = \{(a, p), (a, r), (b, q)\}$.
8. Determine the inverse relation R^{-1} where $R = \{(a, 2), (a, 5), (b, 3), (b, 4), (c, 1)\}$.
9. Determine the inverse of the function $f : \mathbb{R} \times \mathbb{R} \to \mathbb{R}$ defined by

$$f(x) = \frac{x-2}{x-3} (x \neq 3) \text{ and } f(3) = 1$$

10. Give examples of injective, surjective and bijective functions.
11. Explain the differences between imperative programming languages and functional programming languages.

---

[17]The Church–Turing thesis states that anything that is computable is computable by a Turing Machine.

3.11 Summary

This chapter introduced essential mathematics for computing including set theory, relations and functions. Sets are collections of well-defined objects; a relation between A and B indicates relationships between members of the sets A and B; and functions are a special type of relation where there is at most one relationship for each element $a \in A$ with an element in B.

A binary relation R (A, B) is a subset of the Cartesian product $(A \times B)$ of A and B where A and B are sets. The domain of the relation is A, and the co-domain of the relation is B. An n-ary relation R $(A_1, A_2, \dots A_n)$ is a subset of $(A_1 \times A_2 \times \cdots \times A_n)$.

A total function $f : A \rightarrow B$ is a special relation such that for each element $a \in A$ there is exactly one element $b \in B$. This is written as $f(a) = b$. A function is a relation but not every relation is a function.

Functional programming is quite distinct from imperative programming in that there is no change of state, and the value of the variable x remains the same during program execution. This makes functional programs easier to reason about than imperative programs.

Automata Theory is the branch of computer science that is concerned with the study of abstract machines and automata. These include finite-state machines, pushdown automata and Turing machines. Graph theory is a practical branch of mathematics that deals with the arrangements of certain objects known as vertices (or nodes) and the relationships between them.

References

Codd EF (1970) A relational model of data for large shared data banks. Commun ACM 13(6): 377–387
Date CJ (1981) An introduction to database systems. In: The systems programming series, 3rd edn
Hoffman D, Weiss D (eds) (2001) Software fundamentals: collected papers by David L. Parnas. Addison Wesley
O'Regan G (2016) Guide to discrete mathematics. Springer
O'Regan G (2018) World of computing. Springer
Turner D. Miranda. In: Proceedings IFIP conference, Nancy, France, Springer LNCS (201), Sept 1985

Introduction to Algorithms

4

Key Topics

Euclid's algorithm
Sieve of Eratosthenes algorithm
Early ciphers
Sorting algorithms
Insertion sort and merge sort
Analysis of algorithms
Complexity of algorithms
NP-complete

4.1 Introduction

An *algorithm* is a well-defined procedure for solving a problem, and it consists of a sequence of steps that takes a set of values as input and produces a set of values as output. It is an exact specification of how to solve the problem, and it explicitly defines the procedure so that a computer program may implement the solution. The origin of the word '*algorithm*' is from the name of the ninth-century Persian mathematician, Mohammed Al Khwarizmi.

It is essential that the algorithm is correct, and that it terminates in a reasonable amount of time. This may require mathematical analysis of the algorithm to demonstrate its correctness and efficiency and to show that termination is within an acceptable timeframe. There may be several algorithms to solve a problem, and so the choice of the best algorithm (e.g. fastest/most efficient) needs to be considered. For example, there are several well-known sorting algorithms (e.g. *merge sort* and

© Springer Nature Switzerland AG 2020
G. O'Regan, *Mathematics in Computing*, Undergraduate Topics
in Computer Science, https://doi.org/10.1007/978-3-030-34209-8_4

insertion sort), and the merge sort algorithm is more efficient [o(n lg n)] than the insertion sort algorithm [o(n^2)].

An algorithm may be implemented by a computer program written in some programming language (e.g. C++ or Java). The speed of the program depends on the algorithm employed, the input value(s), how the algorithm has been implemented in the programming language, the compiler, the operating system and the computer hardware.

An algorithm may be described in natural language (care is needed to avoid ambiguity), but it is more common to use a more precise formalism for its description. These include *pseudocode* (an informal high-level language description), flowcharts, a programming language such as C or Java, or a formal specification language such as VDM or Z. We shall mainly use natural language and pseudocode to describe an algorithm. One of the earliest algorithms developed was Euclid's algorithm for determining the greatest common divisor of two natural numbers, and it is described in the next section.

4.2 Early Algorithms

Euclid lived in Alexandria during the early Hellenistic period,[1] and he is considered the father of geometry and the deductive method in mathematics. His systematic treatment of geometry and number theory is published in the thirteen books of the Elements (Heath 1956). It starts from 5 axioms, 5 postulates and 23 definitions to logically derive a comprehensive set of theorems in geometry.

His method of proof was generally constructive, in that as well as demonstrating the truth of the theorem, a construction of the required entity was provided. He employed some indirect proofs and one example was his proof that there are an infinite number of prime numbers. The procedure is to assume the opposite of what one wishes to prove and to show that a contradiction results. This means that the original assumption must be false, and the theorem is established.

1. Suppose there are a finite number of primes (say n primes).
2. Multiply all n primes together and add 1 to form N.

$$(N = p_1 * p_2 * \cdots * p_n + 1)$$

3. N is not divisible by $p_1, p_2, \ldots, p_n$ as dividing by any of these gives a remainder of one.
4. Therefore, N must either be prime or divisible by some other prime that was not included in the original list.
5. Therefore, there must be at least $n + 1$ primes.
6. This is a contradiction (it was assumed that there are exactly n primes).

---

[1]This refers to the period following the conquests of Alexander the Great, which led to the spread of Greek culture throughout the Middle East and Egypt.

7. Therefore, the assumption that there is a finite number of primes is false.
8. Therefore, there are an infinite number of primes.

His proof that there are an infinite number of primes is indirect, and he does not present an algorithm to as such to construct the set of prime numbers. We present the well-known Sieve of Eratosthenes algorithm for determining the prime numbers up to a given number n later in the chapter.

The material in the Euclid's Elements is a systematic development of geometry starting from the small set of axioms, postulates and definitions. It leads to many well-known mathematical results such as Pythagoras's Theorem, Thales Theorem, Sum of Angles in a Triangle, Prime Numbers, Greatest Common Divisor and Least Common Multiple, Euclidean Algorithm, Areas and Volumes, Tangents to a point and Algebra.

4.2.1 Greatest Common Divisors (GCD)

Let a and b be integers not both zero. The *greatest common divisor d* of a and b is a divisor of a and b (i.e. $d \mid a$ and $d \mid b$), and it is the largest such divisor (i.e. if $k \mid a$ and $k \mid b$ then $k \mid d$). It is denoted by gcd (a, b).

Properties of Greatest Common Divisors

(i) Let a and b be integers not both zero, then exist integers x and y such that

$$d = \gcd(a, b) = ax + by.$$

(ii) Let a and b be integers not both zero then the set $S = \{ax + by$ where $x, y \in \mathbb{Z}\}$ is the set of all multiples of $d = $ gcd (a, b).

4.2.2 Euclid's Greatest Common Divisor Algorithm

Euclid's algorithm is one of the oldest known algorithms, and it provides the procedure for finding the greatest common divisor of two numbers a and b. It appears in Book VII of Euclid's Elements, but the algorithm was known prior to Euclid (Fig. 4.1).

The inputs for the *gcd* algorithm consist of two natural numbers a and b, and the output of the algorithm is d (the greatest common divisor of a and b). It is computed as follows:

$$gcd(a, b) = \begin{cases} \text{Check if } b \text{ is zero. If so, then } a \text{ is the } gcd. \\ \text{Otherwise, the } gcd(a, b) \text{ is given by} (b, a \bmod b) \end{cases}$$

Fig. 4.1 Euclid of
Alexandria

It is also possible to determine integers p and q such that $ap + bq = gcd(a, b)$.

The (informal) proof of the Euclidean algorithm is as follows. Suppose a and b are two positive numbers whose greatest common divisor is to be determined, and let r be the remainder when a is divided by b.

1. Clearly, $a = qb + r$ where q is the quotient of the division.
2. Any common divisor of a and b is also a divider or r (since $r = a - qb$).
3. Similarly, any common divisor of b and r will also divide a.
4. Therefore, the greatest common divisor of a and b is the same as the greatest common divisor of b and r.
5. The number r is smaller than b and we will reach $r = 0$ in finitely many steps.
6. The process continues until $r = 0$.

Comment 4.1 *Algorithms are fundamental in computing as they define the procedure by which a problem is solved. A computer program implements the algorithm in some programming language.*

Next, we deal with the Euclidean algorithm more formally, and we start with a basic lemma.

Lemma *Let a, b, q and r be integers with $b > 0$ and $0 \leq r < b$ such that $a = bq + r$. Then $\gcd(a, b) = \gcd(b, r)$.*

Proof Let $K = \gcd(a, b)$ and let $L = \gcd(b, r)$, then we need to show that $K = L$. Suppose m is a divisor of a and b, then as $a = bq + r$ we have m is a divisor of r and so any common divisor of a and b is a divisor of r. Therefore, the greatest common divisor K of a and b is a divisor of r. Similarly, any common divisor n of b and r is a divisor of a. Therefore, the greatest common divisor L of b and r is a divisor of a. That is, K divides L and L divides K and so L = K, and so the greatest common divisor of a and b is equal to the greatest common divisor of b and r.

Euclid's Algorithm (More Formal Proof)

Euclid's algorithm for finding the greatest common divisor of two positive integers a and b involves a repeated application of the division algorithm as follows:

$$
\begin{aligned}
a &= bq_0 + r_1 & 0 < r_1 < b \\
b &= r_1 q_1 + r_2 & 0 < r_2 < r_1 \\
r_1 &= r_2 q_2 + r_3 & 0 < r_3 < r_2 \\
&\cdots \\
&\cdots \\
r_{n-2} &= r_{n-1} q_{n-1} + r_n & 0 < r_n < r_{n-1} \\
r_{n-1} &= r_n q_n
\end{aligned}
$$

Then r_n (i.e. the last non-zero remainder) is the greatest common divisor of a and b, i.e. $\gcd(a, b) = r_n$.

Proof It is clear from the construction that r_n is a divisor of $r_{n-1}, r_{n-2}, \ldots, r_3, r_2, r_1$ and of a and b. Clearly, any common divisor of a and b will also divide r_n. Using the results from the lemma above, we have

$$
\begin{aligned}
gcd(a, b) & \\
&= gcd(b, r_1) \\
&= gcd(r_1 r_2) \\
&= \cdots \\
&= gcd(r_{n-2} r_{n-1}) \\
&= gcd(r_{n-1}, r_n) \\
&= r_n
\end{aligned}
$$

4.2.3 Sieve of Eratosthenes Algorithm

Eratosthenes was a Hellenistic mathematician and scientist who worked in the famous library in Alexandria. He devised a system of latitude and longitude, and he was the first person to estimate of the size of the circumference of the earth. He developed a famous algorithm (the well-known *Sieve of Eratosthenes algorithm*) for determining the prime numbers up to a given number n.

1	2	3	4	5	6	7	8	9	10
	2	3	4	5	6	7	8	9	10
11	12	13	14	15	16	17	18	19	20
21	22	23	24	25	26	27	28	29	30
31	32	33	34	35	36	37	38	39	40
41	42	43	44	45	46	47	48	49	50

Fig. 4.2 Primes between 1 and 50

The algorithm involves listing all numbers from 2 up to n. The first step is to remove all multiples of 2 up to n; the second step is to remove all multiples of 3 up to n; and so on (Fig. 4.2).

The k-th step involves removing multiples of the k-th prime p_k up to n and the steps in the algorithm continue while $p \leq \sqrt{n}$. The numbers remaining in the list are the prime numbers from 2 to n.

1. List the integers from 2 to n.
2. For each prime p_k up to $\sqrt{n}$ remove all multiples of p_k.
3. The numbers remaining are the prime numbers between 2 and n.

The list of primes between 1 and 50 is then given by 2, 3, 5, 7, 11, 13, 17, 19, 23, 29, 31, 37, 41, 43 and 47.

The steps in the algorithm may also be described as follows (in terms of two lists):

1. Write a list of the numbers from 2 to the largest number to be tested. This first list is called A.
2. A second list B is created to list the primes. It is initially empty.
3. The number 2 is the first prime number, and it is added to List B.
4. Strike off (or remove) all multiples of 2 from List A.
5. The first remaining number in List A is a prime number and this prime number is added to List B.
6. Strike off (or remove) this number and all multiples of it from List A.
7. Repeat steps 5 through 7 until no more numbers are left in List A.

4.2.4 Early Cipher Algorithms

Julius Caesar employed a *substitution cipher* on his military campaigns to ensure that important messages were communicated safely (Fig. 4.3). The Caesar cipher is a very simple encryption algorithm, and it involves the substitution of each letter in the *plaintext* (i.e. the original message) by a letter a fixed number of positions down in the alphabet. The Caesar encryption algorithm involves a shift of three positions

Alphabet Symbol	abcde fghij klmno pqrst uvwxyz
Cipher Symbol	dfegh ijklm nopqr stuvw xyzabc

Fig. 4.3 Caesar Cipher

and causes the letter B to be replaced by E, the letter C by F and so on. The Caesar cipher is easily broken, as the frequency distribution of letters may be employed to determine the mapping. The Caesar Cipher is defined as

The process of enciphering a message (i.e. the plaintext) involves mapping each letter in the plaintext to the corresponding cipher letter. For example, the encryption of 'summer solstice' involves

> Plaintext: summer solstice
> Cipher Text   vxpphu vrovwleh

The decryption involves the reverse operation, i.e. for each cipher letter the corresponding plaintext letter is determined from the table.

> Cipher Text   vxpphu vrovwleh
> Plaintext: summer solstice

The Caesar encryption algorithm may be expressed formally using modular arithmetic. The numbers 0–25 represent the alphabet letters, and the algorithm is expressed using addition (modulo 26) to yield the encrypted cipher. The encoding of the plaintext letter x is given by

$$c = x + 3 \pmod{26}$$

Similarly, the decoding of a cipher letter represented by the number c is given by

$$x = c - 3 \pmod{26}$$

The emperor Augustus[2] employed a similar substitution cipher (with a shift key of 1). The Caesar cipher remained in use up to the early twentieth century. However, by then frequency analysis techniques were available to break the cipher. The *Vigenère cipher* uses a Caesar cipher with a different shift at each position in the text. The value of the shift to be employed with each plaintext letter is defined using a repeating keyword.

---

[2]Augustus was the first Roman emperor and his reign ushered in a period of peace and stability following the bitter civil war that occurred after the assassination of Julius Caesar. He was the adopted son of Julius Caesar (he was called Octavian before he became emperor). The civil war broke out between Mark Anthony and Octavian, and Anthony and Cleopatra were defeated by Octavian and Agrippa at the battle of Actium in 31B.C.

4.3   Sorting Algorithms

One of the most common tasks to be performed in a computer program is that of sorting (e.g. consider the problem of sorting a list of names or numbers). This has led to the development of many sorting algorithms (e.g. selection sort, bubble sort, insertion sort, merge sort and quicksort) as sorting is a fundamental task to be performed.

For example, consider the problem of specifying the algorithm for sorting a sequence of n numbers. Then, the input to the algorithm is $\langle x_1, x_2, \dots x_n \rangle$, and the output is $\langle x'_1, x'_2, \dots x'_n \rangle$, where $x'_1 \leq x'_2 \leq \dots \leq x'_n$. Further, $\langle x'_1, x'_2, \dots x'_n \rangle$ is a permutation of $\langle x_1, x_2, \dots x_n \rangle$, i.e. the same numbers are in both sequences except that the sorted sequence is in ascending order, whereas no order is imposed on the original sequence.

Insertion sort is an efficient algorithm for sorting a small list of elements. It iterates over the input sequence, examines the next input element during the iteration and builds up the sorted output sequence. During the iteration, Insertion sort removes the next element from the input data, and it then finds and inserts it into the location where it belongs in the sorted list. This continues until there are no more input elements to process.

We first give an example of Insertion Sort and then give a more formal definition of the algorithm (Fig. 4.4). The example considered is that of the algorithm applied to the sequence A = $\langle 5,3 \ 1,4 \rangle$. The current input element for each iteration is highlighted and the arrow point to the location where it is inserted in the sorted sequence. For each iteration, the elements to the left of the current element are already in increasing order, and the operation of insertion sort is to move the current element to the appropriate position in the ordered sequence.

We shall assume that we have an unsorted array A with n elements that we wish to sort. The operation of Insertion Sort is to rearrange the elements of A within the array, and the output is that the array A contains the sorted output sequence.

```
Insertion Sort
for i from 2 to n do
    C ←A[i]
    j ← i-1
    while j > 0 and A[j] > C do
        A[j+1]← A[j]
        j ← j-1
    A[j+1]← C
```

The analysis of an algorithm involves determining its efficiency and establishing the resources that it requires (e.g. memory and bandwidth), as well as determining the computational time required. The time taken by the Insertion Sort algorithm depends on the size of the input sequence (clearly, a large sequence will take longer to sort than a short sequence), and on the extent to which the sequences are already sorted. The worst-case running time for the Insertion Sort algorithm is of order n^2—

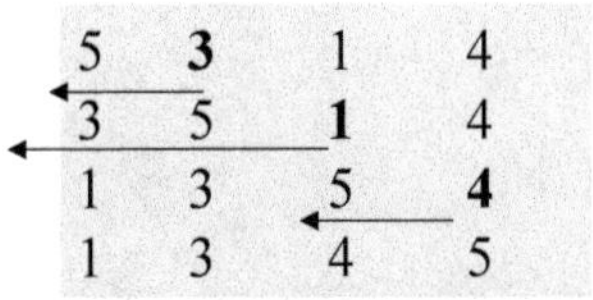

5	**3**	1	4
3	5	**1**	4
1	3	5	4
1	3	4	5

Fig. 4.4 Insertion sort example

i.e. o(n^2), where n is the size of the sequence to be sorted (the average case is also of order n^2 with the best case linear).

There are a number of ways to design sorting algorithms, and the Insertion Sort algorithm uses an incremental approach, with the sub-array A[1 ... $i - 1$] already sorted and the element A[i] is then inserted into its correct place to yield the sorted array A[1 ... i].

Another approach is to employ divide-and-conquer techniques, and this technique is used in the Merge Sort algorithm. This is a more efficient algorithm, and it involves breaking a problem down into several sub-problems and then solving each problem separately. The problem-solving may involve recursion or directly solving the sub-problem (if it is small enough), and then combining the solutions to the sub-problems into the solution for the original problem. The Merge Sort algorithm involves three steps (Divide, Conquer and Combine):

1. *Divide* the list A (with n elements) to be sorted into two sub-sequences (each with $^n/_2$ elements).
2. Sort each of the sub-sequences by calling Merge Sort recursively (*Conquer*).
3. Merge the two sorted sub-sequences to produce a single sorted list (*Combine*).

The recursive part of the Merge Sort algorithm bottoms out when the sequence to be sorted is of length 1, as for this case the sequence is of length 1 which is already (trivially) sorted. The key operation then (where all the work is done) is the combine step that merges two sorted sequences to produce a single sorted sequence. The Merge Sort algorithm may also be described as follows:

1. Divide the sequence (with n elements) to be sorted into n sub-sequences each with one element (a sequence with one element is sorted).
2. Repeatedly merge sub-sequences to form new sub-sequences (each new sub-sequence is sorted), until there is only one remaining sub-sequence (the sorted sequence).

First, we consider an example (Fig. 4.5) to illustrate how the Merge Sort algorithm operates, and we then give a formal definition.

It may be seen from the example that the list is repeatedly divided into equal halves with each iteration, until we get to the atomic values that can no longer be divided. The lists are then combined in the order in which they were broken down, and this involves comparing the elements of both lists and combining them to form

Fig. 4.5 Merge sort example

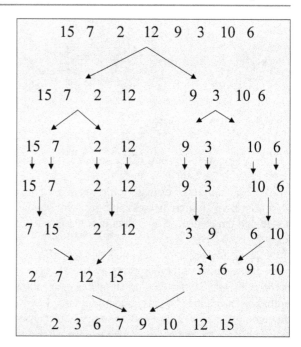

a sorted list. The merging continues in this way until there are no more lists to merge, and the list remaining is the sorted list. The formal definition of Merge Sort is as follows:

Merge Sort (A, *m*, *n*)
If *m* < *n* then
    r ← (*m* + *n*) div 2
    Merge Sort (A, *m*, *r*)
    Merge Sort (A, *r*+1, *n*)
    Merge (A, *m*, *r*, *n*)

The worst-case and average-case running time for the Merge Sort algorithm is of order $n \lg n$, i.e. o($n \lg n$), where n is the size of the sequence to be sorted (the average case and best case is also of order o($n \lg n$)).

The Merge procedure merges two sorted lists to produce a single sorted list. Merge (A, *p*, *q*, *r*) merges A[*p* … *q*] with A[*q* + 1 … *r*] to yield the sorted list A [*p* … *r*]. We use a temporary working array B[*p* … *r*] with the same index range as A. The indices *i* and *j* point to the current element of each sub-array, and we move the smaller element into the next position in B (indicated by index *k*) and then increment either *i* or *j*. When we run out of entries in one array, then we copy the rest of the other array into B. Finally, we copy the entire contents of B back to A.

Merge (A, *p*, *q*, *r*)
    Array B[*p* .. *r*]
    $i \leftarrow p$
    $k \leftarrow p$
    $j \leftarrow q+1$
    while ($i \leq q \wedge j \leq r$)   ... i.e., while both subarrays are non-empty
        if A[*i*] ≤ A[*j*]
            B[*k*] ← A[*i*]
            $i \leftarrow i+1$
        else
            B[*k*] ← A[*j*]
            $j \leftarrow j+1$
        $k \leftarrow k+1$
    while ($i \leq q$)   ... copy any leftover to B
      B[*k*] ← A[*i*]
      $i \leftarrow i+1$
      $k \leftarrow k+1$
    while ($j \leq r$)   ... copy any leftover to B
      B[*k*] ← A[*j*]
      $j \leftarrow j+1$
      $k \leftarrow k+1$
    for *i* = *p* to *r* do   ... copy B back to A
      A[*i*] = B[*i*]

4.4 Binary Trees and Graph Theory

A *binary tree* (Fig. 4.6) is a tree in which each node has at most two child nodes (termed left and right child node). A node with children is termed a *parent node*, and the top node of the tree is termed the root node. Any node in the tree can be reached by starting from the root node, and by repeatedly taking either the left branch (left child) or right branch (right child) until the node is reached. Binary trees are often used in computing to implement efficient searching algorithms.

Fig. 4.6 Sorted binary tree

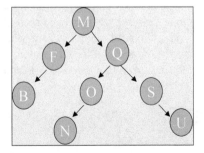

The *depth* of a node is the length of the path (i.e. the number of edges) from the root to the node. The depth of a tree is the length of the path from the root to the deepest node in the tree. A *balanced* binary tree is a binary tree in which the depth of the two subtrees of any node never differs by more than one.

Tree traversal is a systematic way of visiting each node in the tree exactly once, and we distinguish between *breadth-first search* algorithms in which every node at a particular level is visited before going to a lower level, and *depth-first search* algorithms where one starts at the root and explores as far as possible along each branch before backtracking. The traversal in depth-first search may be in *preorder*, *inorder* or *postorder*.

Graph algorithms are employed to solve various problems in graph theory including network cost minimization problems, construction of spanning trees, shortest path algorithms, longest path algorithms and timetable construction problems.

The reader may consult the many texts on graph theory (e.g. Piff 1991) to explore many well-known graph algorithms such as Dijkstra's shortest path algorithm and longest path algorithm, Kruskal's minimal spanning tree algorithm and Prim's minimal spanning tree algorithms.

4.5 Modern Cryptographic Algorithms

A cryptographic system is concerned with the secure transmission of messages. The message is encrypted prior to its transmission, and any unauthorized interception and viewing of the message is meaningless to anyone other than the intended recipient. The recipient uses a key to decrypt the encrypted text to retrieve the original message.

- M represents the message (plaintext).
- C represents the encrypted message (ciphertext).
- e_k represents the encryption key.
- d_k represents the decryption key.
- E represents the encryption.
- D represents the decryption.

There are essentially two different types of cryptographic systems, namely, the public-key cryptosystems and secret-key cryptosystems. A *public-key cryptosystem* is an asymmetric cryptosystem where two different keys are employed: one for encryption and one for decryption. The fact that a person can encrypt a message does not mean that the person is able to decrypt a message.

The same key is used for both encryption and decryption in a *secret-key cryptosystem*, and anyone who has knowledge on how to encrypt messages has sufficient knowledge to decrypt messages. The encryption and decryption algorithms satisfy the following equation:

$$Dd_k(C) = Dd_k(Ee_k(M)) = M$$

There are two different keys employed in a public-key cryptosystem. These are the encryption key e_k and the decryption key d_k with $e_k \neq d_k$. It is called asymmetric as the encryption key differs from the decryption key.

A symmetric-key cryptosystem (Fig. 10.6) uses the same secret key for encryption and decryption, and so the sender and the receiver first need to agree on a shared key prior to communication. This needs to be done over a secure channel to ensure that the shared key remains secret. Once this has been done, they can begin to encrypt and decrypt messages using the secret key.

The encryption of a message is in effect a transformation from the space of messages m to the space of cryptosystems $\mathbb{C}$. That is, the encryption of a message with key k is an invertible transformation f such that

$$f: m \xrightarrow{k} \mathbb{C}$$

The ciphertext is given by $C = E_k(M)$ where $M \in m$ and $C \in \mathbb{C}$. The legitimate receiver of the message knows the secret key k (as it will have transmitted previously over a secure channel), and so the ciphertext C can be decrypted by the inverse transformation f^{-1} defined by

$$f^{-1}: \mathbb{C} \xrightarrow{k} m$$

Therefore, we have that $D_k(C) = D_k(E_k(M)) = M$ the original plaintext message.

A public-key cryptosystem (Fig. 10.7) is an asymmetric-key system where there is a separate key e_k for encryption and d_k decryption with $e_k \neq d_k$. The fact that a person is able to encrypt a message does not mean that the person has sufficient information to decrypt messages. There is a more detailed account of cryptography in Chap. 10.

4.6   Computational Complexity

An algorithm is of little practical use if it takes millions of years to compute the solution to a problem. That is, the fact that there is an algorithm to solve a problem is not sufficient, as there is also the need to consider the efficiency of the algorithm. The security of the RSA encryption algorithm relies on the fact that there is no known efficient algorithm to determine the prime factors of a large number.

There are often slow and fast algorithms for the same problem, and a measure of the complexity of an algorithm is the number of steps in its computation. An algorithm is of *time complexity* $f(n)$ if for all n and all inputs of length n the execution of the algorithm takes at most $f(n)$ steps.

An algorithm is said to be *polynomially bounded* if there is a polynomial $p(n)$ such that for all n and all inputs of length n the execution of the algorithm takes at most $p(n)$ steps. The notation P is used for all problems that can be solved in polynomial time.

A problem is said to be *computationally intractable* if it may not be solved in polynomial time: that is, there is no known algorithm to solve the problem in polynomial time.

A problem L is said to be in the set *NP* (non-deterministic polynomial-time problems) if any given solution to L can be verified quickly in polynomial time. A non-deterministic Turing machine may have several possibilities for its behaviour, and an input may give rise to several computations.

A problem is *NP-complete* if it is in the set NP of non-deterministic polynomial-time problems and it is also in the class of *NP-hard* problems. A key characteristic to NP-complete problems is that there is no known fast solution to them, and the time required to solve the problem using known algorithms increases quickly as the size of the problem grows. Often, the time required to solve the problem is in billions of years. That is, although any given solution may be verified quickly, there is no known efficient way to find a solution.

4.7 Review Questions

1. What is an algorithm?
2. Explain why the efficiency of an algorithm is important.
3. Investigate the principles underlying modern cryptography, and show how they are related to modern computer algorithms.
4. What factors should be considered in the choice of algorithm where several algorithms exist for solving the particular problem?
5. Investigate some of the early algorithms developed by the Babylonians (e.g. finding square roots and factorization).
6. Explain the difference between the Insertion sort algorithm and merge sort.
7. Investigate famous computer algorithms such as Dijkstra's shortest path, Prim's algorithm and Kruskal's algorithm.

4.8 Summary

This chapter gave a short introduction to computer algorithms, where an algorithm is a well-defined procedure for solving a problem. It consists of a sequence of steps that takes a set of input values and produces a set of output values. It is an exact specification of how to solve the problem, and a computer program implements the algorithm in some programming language.

It is essential that the algorithm is correct, and that it terminates in a reasonable period of time. There may be several algorithms for a problem, and so the choice of the best algorithm (e.g. fastest/most efficient) needs to be considered.

This may require mathematical analysis of the algorithm to demonstrate its correctness and efficiency, and to show that it terminates in a finite period of time. An algorithm may be implemented by a computer program, and the speed of the program depends on the algorithm employed, the input value(s), how the algorithm has been implemented in the programming language, the compiler, the operating system and the computer hardware.

An algorithm may be described in natural language, *pseudocode*, flowchart, a programming language or a formal specification language.

References

Heath T (trans) (1956) The thirteen books of the elements, vol 1. Dover Publications. (First published in 1925)

Piff M (1991) Discrete mathematics. An introduction for software engineers. Cambridge University Press

Number Theory

5

Key Topics

Square, rectangular and triangular numbers
Prime numbers
Pythagorean triples
Mersenne primes
Division algorithm
Perfect and Amicable numbers
Greatest common divisor
Least common multiples
Euclid's algorithm
Modular arithmetic
Binary numbers
Computer representation of numbers

5.1 Introduction

Number theory is the branch of mathematics that is concerned with the mathematical properties of the natural numbers and integers. These include properties such as the parity of a number, divisibility, additive and multiplicative properties, whether a number is prime or composite, the prime factors of a number, the greatest common divisor and least common multiple of two numbers, and so on.

Number theory has many applications in computing including cryptography and coding theory. For example, the RSA public-key cryptographic system relies on its security due to the infeasibility of the integer factorization problem for large numbers.

© Springer Nature Switzerland AG 2020
G. O'Regan, *Mathematics in Computing*, Undergraduate Topics
in Computer Science, https://doi.org/10.1007/978-3-030-34209-8_5

Fig. 5.1 Pierre de Fermat

There are several unsolved problems in number theory and especially in prime number theory. For example, Goldbach's[1] Conjecture states that every even integer greater than two is the sum of two primes, and this result has not been proved to date. Fermat's[2] Last Theorem states that there is no integer solution to $x^n + y^n = z^n$ for $n > 2$, and this result remained unproved for over 300 years until Andrew Wiles finally proved it in the mid-1990s (Fig. 5.1).

The natural numbers $\mathbb{N}$ consist of the numbers $\{1, 2, 3, \ldots\}$. The integer numbers $\mathbb{Z}$ consist of $\{\ldots -2, -1, 0, 1, 2, \ldots\}$. The rational numbers $\mathbb{Q}$ consist of all numbers of the form $\{^p/_q$ where p and q are integers and $q \neq 0\}$. The real numbers $\mathbb{R}$ are defined to be the set of converging sequences of rational numbers and they are a superset of the set of rational numbers. They contain the rational and irrational numbers. The complex numbers $\mathbb{C}$ consist of all numbers of the form $\{a + bi$ where $a, b \in \mathbb{R}$ and $i = \sqrt{-1}\}$.

---

[1]Goldbach was an eighteenth-century German mathematician and Goldbach's conjecture has been verified to be true for all integers $n < 12 * 10^{17}$.

[2]Pierre de Fermat was a seventeenth-century French civil servant and amateur mathematician. He occasionally wrote to contemporary mathematicians announcing his latest theorem without providing the accompanying proof and inviting them to find the proof. The fact that he never revealed his proofs caused a lot of frustration among his contemporaries, and in his announcement of his famous last theorem he stated that he had a wonderful proof that was too large to include in the margin. He corresponded with Pascal, and they did some early work on the mathematical rules of games of chance and early probability theory.

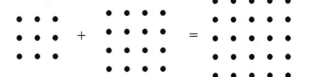

Fig. 5.2 Pythagorean triples

Pythagorean triples (Fig. 5.2) are combinations of three whole numbers that satisfy Pythagoras's equation $x^2 + y^2 = z^2$. There are an infinite number of such triples, and an example of such a triple is 3, 4, 5 since $3^2 + 4^2 = 5^2$.

The Pythagoreans discovered the mathematical relationship between the harmony of music and numbers, and their philosophy was that numbers are hidden in everything from music to science and nature. This led to their philosophy that 'everything is number'.

5.2 Elementary Number Theory

A square number is an integer that is the square of another integer. For example, the number 4 is a square number since $4 = 2^2$. Similarly, the number 9 and the number 16 are square numbers. A number n is a square number if and only if one can arrange the n points in a square. For example, the square numbers 4, 9 and 16 are represented in squares as follows (Fig. 5.3).

The square of an odd number is odd, whereas the square of an even number is even. This is clear since an even number is of the form $n = 2k$ for some k, and so $n^2 = 4k^2$ which is even. Similarly, an odd number is of the form $n = 2k + 1$ and so $n^2 = 4k^2 + 4k + 1$ which is odd.

A rectangular number n may be represented by a vertical and horizontal rectangle of n points. For example, the number 6 may be represented by a rectangle with length 3 and breadth 2, or a rectangle with length 2 and breadth 3. Similarly, the number 12 can be represented by a 4×3 or a 3×4 rectangle (Fig. 5.4).

A triangular number n may be represented by an equilateral triangle of n points. It is the sum of k natural numbers from 1 to k (Fig. 5.5). That is,

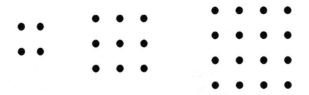

Fig. 5.3 Square numbers

Fig. 5.4 Rectangular numbers

Fig. 5.5 Triangular numbers $n = 1 + 2 + \ldots + k$

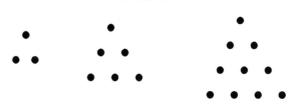

$$n = 1 + 2 + \cdots + k$$

Parity of Integers

The parity of an integer refers to whether the integer is odd or even. An integer n is odd if there is a remainder of one when it is divided by two, and it is of the form $n = 2k + 1$. Otherwise, the number is even and of the form $n = 2k$.

The sum of two numbers is even if both are even or both are odd. The product of two numbers is even if at least one of the numbers is even. These properties are expressed as

$$\text{even} \pm \text{even} = \text{even}$$
$$\text{even} \pm \text{odd} = \text{odd}$$
$$\text{odd} \pm \text{odd} = \text{even}$$
$$\text{even} \times \text{even} = \text{even}$$
$$\text{even} \times \text{odd} = \text{even}$$
$$\text{odd} \times \text{odd} = \text{odd}$$

Divisors

Let a and b be integers with $a \neq 0$, then a is said to be a divisor of b (denoted by $a \mid b$) if there exists an integer k such that $b = ka$.

A divisor of n is called a *trivial divisor* if it is either 1 or n itself; otherwise, it is called a *non-trivial divisor*. A *proper divisor* of n is a divisor of n other than n itself.

Fig. 5.6 Marin Mersenne

Definition (*Prime Number*) A *prime number* is a number whose only divisors are trivial. There are an infinite number of prime numbers.

The *fundamental theorem of arithmetic* states that every integer number can be factored as the product of prime numbers.

Mersenne Primes

Mersenne primes are prime numbers of the form $2^p - 1$ where p is a prime. They are named after Marin Mersenne who was a seventeenth-century French monk, philosopher and mathematician (Fig. 5.6). Mersenne did some early work in identifying primes of this format, and there are 51 known Mersenne primes. It remains an open question as to whether there are an infinite number of Mersenne primes.

Properties of Divisors

(i) $a|b$ and $a|c$ then $a|b + c$
(ii) $a|b$ then $a|bc$
(iii) $a|b$ and $b|c$ then $a|c$.

Proof (of i) Suppose $a \mid b$ and $a \mid c$ then $b = k_1 a$ and $c = k_2 a$.

Then $b + c = k_1 a + k_2 a = (k_1 + k_2)a$ and so $a|b + c$.

Proof (of iii) Suppose $a \mid b$ and $b \mid c$ then $b = k_1 a$ and $c = k_2 b$.

Then $c = k_2 b = (k_2 k_1)\, a$ and thus $a \mid c$.

Perfect and Amicable Numbers

Perfect and amicable numbers have been studied for millennia. A positive integer m is said to be *perfect* if it is the sum of its proper divisors. Two positive integers m and n are said to be an *amicable pair* if m is equal to the sum of the proper divisors of n and vice versa.

A *perfect number* is a number whose divisors add up to the number itself. For example, the number 6 is perfect since it has divisors 1, 2, 3 and $1 + 2 + 3 = 6$.

Perfect numbers are quite rare and Euclid showed that $2^{p-1}(2^p - 1)$ is an even perfect number whenever $(2^p - 1)$ is prime. Euler later showed that all even perfect numbers are of this form. It is an open question as to whether there are any odd perfect numbers, and if such an odd perfect number N was to exist then $N > 10^{1500}$.

A prime number of the form $(2^p - 1)$ where p is prime is called a *Mersenne prime*. Mersenne primes are quite rare, and each Mersenne prime generates an even perfect number and vice versa. That is, there is a one-to-one correspondence between the number of Mersenne primes and the number of even perfect numbers.

It remains an open question as to whether there are an infinite number of Mersenne primes and perfect numbers.

An *amicable pair* of numbers is a pair of numbers such that each number is the sum of divisors of the other number. For example, the numbers 220 and 284 are an amicable pair since the divisors of 220 are 1, 2, 4, 5, 10, 11, 20, 22, 44, 55 and 110, which have sum 284, and the divisors of 284 are 1, 2, 4, 71 and 142, which have sum 220.

Theorem 5.1 (Division Algorithm) *For any integer a and any positive integer b there exists unique integers q and r such that*

$$a = bq + r \quad 0 \leq r < b$$

Proof The first part of the proof is to show the existence of integers q and r such that the equality holds, and the second part of the proof is to prove uniqueness of q and r.

Consider … $-3b, -2b, -b, 0, b, 2b, 3b, \ldots$ then there must be an integer q such that

$$qb \leq a < (q+1)b$$

Then $a - qb = r$ with $0 \leq r < b$ and so $a = bq + r$ and the existence of q and r is proved.

The second part of the proof is to show the uniqueness of q and r. Suppose q_1 and r_1 also satisfy $a = bq_1 + r_1$ with $0 \leq r_1 < b$ and suppose $r < r_1$. Then $bq +$

$r = bq_1 + r_1$ and so $b(q - q_1) = r_1 - r$ and clearly $0 < (r_1 - r) < b$. Therefore, $b \mid (r_1 - r)$ which is impossible unless $r_1 - r = 0$. Hence, $r = r_1$ and $q = q_1$.

Theorem 5.2 (Irrationality of Square Root of Two) *The square root of two is an irrational number (i.e. it cannot be expressed as the quotient of two integer numbers).*

Proof The Pythagoreans[3] discovered this result and it led to a crisis in their community as number was considered to be the essence of everything in their world. The proof is indirect, i.e. the opposite of the desired result is assumed to be correct and it is showed that this assumption leads to a contradiction. Therefore, the assumption must be incorrect and so the result is proved.

Suppose $\sqrt{2}$ is rational then it can be put in the form p/q where p and q are integers and $q \neq 0$. Therefore, we can choose p, q to be co-prime (i.e. without any common factors) and so

$$(p/q)^2 = 2$$
$$\Rightarrow p^2/q^2 = 2$$
$$\Rightarrow p^2 = 2q^2$$
$$\Rightarrow 2 \mid p^2$$
$$\Rightarrow 2 \mid p$$
$$\Rightarrow p = 2k$$
$$\Rightarrow p^2 = 4k^2$$
$$\Rightarrow 4k^2 = 2q^2$$
$$\Rightarrow 2k^2 = q^2$$
$$\Rightarrow 2 \mid q^2$$
$$\Rightarrow 2 \mid q$$

---

[3]Pythagoras of Samos (a Greek island in the Aegean Sea) was an influential ancient mathematician and philosopher of the sixth century B.C. He gained his mathematical knowledge from his travels throughout the ancient world (especially in Egypt and Babylon). He became convinced that everything is number, and he and his followers discovered the relationship between mathematics and the physical world as well as relationships between numbers and music. On his return to Samos, he founded a school and he later moved to Croton in southern Italy to set up a school. This school and the Pythagorean brotherhood became a secret society with religious beliefs such as reincarnation and they were focused on the study of mathematics. They maintained secrecy of the mathematical results that they discovered. Pythagoras is remembered today for Pythagoras's Theorem, which states that for a right-angled triangle that the square of the hypotenuse is equal to the sum of the square of the other two sides. The Pythagoreans discovered the irrationality of the square root of two and as this result conflicted in a fundamental way with their philosophy that number is everything, and they suppressed the truth of this mathematical result.

This is a contradiction as we have chosen p and q to be co-prime, and our assumption that there is a rational number that is the square root of two results in a contradiction. Therefore, this assumption must be false and we conclude that there is no rational number whose square is two.

5.3 Prime Number Theory

A positive integer $n > 1$ is called prime if its only divisors are n and 1. A number that is not a prime is called composite.

Properties of Prime Numbers

(i)   There are an infinite number of primes.
(ii)   There is a prime number p between n and $n! + 1$ such that $n < p \leq n! + 1$.
(iii) If n is composite, then n has a prime divisor p such that $p \leq \sqrt{n}$.
(iv)   There are arbitrary large gaps in the series of primes (given any $k > 0$ there exists k consecutive composite integers).

Proof (i) Suppose there are a finite number of primes and they are listed as p_1, p_2, p_3, ..., p_k. Then consider the number N obtained by multiplying all known primes and adding one. That is,

$$N = p_1 p_2 p_3 \ldots p_k + 1$$

Clearly, N is not divisible by any of p_1, p_2, p_3, ..., p_k since they all leave a remainder of 1. Therefore, N is either a new prime or divisible by a prime q (that is not in the list of p_1, p_2, p_3, ..., p_k.).

This is a contradiction since this was the list of all the prime numbers, and so the assumption that there are a finite number of primes is false, and we deduce that there are an infinite number of primes.

Proof (ii) Consider the integer $N = n! + 1$. If N is prime then we take $p = N$. Otherwise, N is composite and has a prime factor p. We will show that $p > n$.

Suppose, $p \leq n$ then $p|n!$ and since $p|N$ we have $p|n! + 1$ and therefore $p \mid 1$, which is impossible. Therefore, $p > n$ and the result is proved.

Proof (iii) Let p be the smallest prime divisor of n. Since n is composite $n = uv$ and clearly $p \leq u$ and $p \leq v$. Then $p^2 \leq uv = n$ and so $p \leq \sqrt{n}$.

Proof *(iv)* Consider the k consecutive integers $(k+1)!+2,\ (k+1)!+3,\ \ldots,$ $(k + 1)! + k,\ (k + 1)! + k + 1$. Then each of these is composite since $j \mid$ $(k + 1)! + j$ where $2 \leq j \leq k + 1$.

Algorithm for Determining Primes

We discussed the *Sieve of Eratosthenes algorithm* for determining the prime numbers up to a given number n in Chap. 4. It was developed by the Hellenistic mathematician, Eratosthenes.

The algorithm involves first listing all of the numbers from 2 to n. The first step is to remove all multiples of two up to n; the second step is to remove all multiples of three up to n; and so on.

The k-th step involves removing multiples of the k-th prime p_k up to n and the steps in the algorithm continue while $p \leq \sqrt{n}$. The numbers remaining in the list are the prime numbers from 2 to n.

1. List the integers from 2 to n.
2. For each prime p_k up to $\sqrt{n}$ remove all multiples of p_k.
3. The numbers remaining are the prime numbers between 2 and n.

The list of primes between 1 and 50 is given in Fig. 4.2. They are 2, 3, 5, 7, 11, 13, 17, 19, 23, 29, 31, 37, 41, 43 and 47.

Theorem 5.3 (Fundamental Theorem of Arithmetic) *Every natural number $n > 1$ may be written uniquely as the product of primes:*

$$n = p_1^{\alpha_1} p_2^{\alpha_2} p_3^{\alpha_3} \cdots p_k^{\alpha}$$

Proof There are two parts to the proof. The first part shows that there is a factorization, and the second part shows that the factorization is unique.

Part(a)
If n is prime, then it is a product with a single prime factor. Otherwise, n can be factored into the product of two numbers ab where $a > 1$ and $b > 1$. The argument can then be applied to each of a and b each of which is either prime or can be factored as the product of two numbers both of which are greater than one. Continue in this way with the numbers involved decreasing with every step in the process until eventually all of the numbers must be prime (This argument can be made more rigorous using induction).

Part(b)
Suppose the factorization is not unique and let $n > 1$ be the smallest number that has more than one factorization of primes. Then n may be expressed as follows:

$$n = p_1 p_2 p_3 \ldots p_k = q_1 q_2 q_3 \ldots q_r$$

Clearly, $k > 1$ and $r > 1$ and $p_i \neq q_j$ for $(i = 1, \ldots k)$ and $(j = 1, \ldots, r)$ as otherwise we could construct a number smaller than n (e.g. n/p_i where $p_i = q_j$) that has two distinct factorizations. Next, without loss of generality take $p_1 < q_1$ and define the number N by

$$N = (q_1 - p_1)q_2 q_3 \ldots q_r$$
$$= p_1 p_2 p_3 \ldots p_k - p_1 q_2 q_3 \ldots q_r$$
$$= p_1(p_2 p_3 \ldots p_k - q_2 q_3 \ldots q_r)$$

Clearly $1 < N < n$ and so N is uniquely factorizable into primes. However, clearly p_1 is not a divisor of $(q_1 - p_1)$, and so N has two distinct factorizations, which is a contradiction of the choice of n.

5.3.1   Greatest Common Divisors (GCD)

Let a and b be integers not both zero. The *greatest common divisor d* of a and b is a divisor of a and b (i.e. $d \mid a$ and $d \mid b$), and it is the largest such divisor (i.e. if $k \mid a$ and $k \mid b$ then $k \mid d$). It is denoted by gcd (a, b).

Properties of Greatest Common Divisors

(i)   Let a and b be integers not both zero then exists integers x and y such that

$$d = \gcd(a, b) = ax + by$$

(ii) Let a and b be integers not both zero then the set S = $\{ax + by$ where $x, y \in \mathbb{Z}\}$ is the set of all multiples of d = gcd (a, b).

Proof *(of* i) Consider the set of all linear combinations of a and b forming the set $\{ka + nb: k, n \in \mathbb{Z}\}$. Clearly, this set includes positive and negative numbers. Choose x and y such that $m = ax + by$ is the smallest positive integer in the set. Then we shall show that m is the greatest common divisor.

We know from the division algorithm (Theorem 5.1) that $a = mq + r$ where $0 \leq r < m$. Thus

$$r = a - mq = a - (ax + by)q = (1 - qx)a + (-yq)b$$

r is a linear combination of a and b and so r must be 0 from the definition of m. Therefore, $m \mid a$ and similarly $m \mid b$ and so m is a common divisor of a and b. Since the greatest common divisor d is such that $d \mid a$ and $d \mid b$ and $d \leq m$, we must have $d = m$.

Proof *(of* ii*)* This follows since $d \mid a$ and $d \mid b \Rightarrow d \mid ax + by$ for all integers x and y and so every element in the set $S = \{ax + by$ where $x, y \in \mathbb{Z}\}$ is a multiple of d.

Relatively Prime

Two integers a, b are relatively prime if $\gcd(a, b) = 1$.

Properties

If p is a prime and $p \mid ab$ then $p \mid a$ or $p \mid b$.

Proof Suppose $p \mid a$ then from the results on the greatest common divisor we have $\gcd (a, p) = 1$. That is,

$$ra + sp = 1$$
$$\Rightarrow rab + spb = b$$
$$\Rightarrow p|b(\text{ since } p|rab \text{ and } p|spb \text{ and so } p|rab + spb)$$

5.3.2 Least Common Multiple (LCM)

If m is a multiple of a and m is a multiple of b, then it is said to be a *common multiple* of a and b. The least common multiple is the smallest of the common multiples of a and b and it is denoted by lcm (a, b).

Properties

If x is a common multiple of a and b then $m \mid x$. That is, every common multiple of a and b is a multiple of the least common multiple m.

Proof We assume that both a and b are non-zero as otherwise the result is trivial (since all common multiples are 0). Clearly, by the division algorithm, we have

$$x = mq + r \text{ where } 0 \leq r < m$$

Since x is a common multiple of a and b we have $a \mid x$ and $b \mid x$ and also that $a \mid m$ and $b \mid m$. Therefore, $a \mid r$ and $b \mid r$ and so r is a common multiple of a and b and since m is the least common multiple we have r is 0. Therefore, x is a multiple of the least common multiple m as required,

5.3.3 Euclid's Algorithm

Euclid's[4] algorithm was discussed in Chap. 4 and it is one of the oldest known algorithms. It provides a procedure for finding the greatest common divisor of two numbers, and it is described in Book VII of Euclid's Elements.

Lemma *Let a, b, q and r be integers with $b > 0$ and $0 \leq r < b$ such that $a = bq + r$. Then $\gcd(a, b) = \gcd(b, r)$.*

Proof Let $K = \gcd(a, b)$ and let $L = \gcd(b, r)$ and we therefore need to show that $K = L$. Suppose m is a divisor of a and b then as $a = bq + r$ we have m is a divisor of r and so any common divisor of a and b is a divisor of r.

Similarly, any common divisor n of b and r is a divisor of a. Therefore, the greatest common divisor of a and b is equal to the greatest common divisor of b and r.

Theorem 5.4 (Euclid's Algorithm) *Euclid's algorithm for finding the greatest common divisor of two positive integers a and b involves applying the division algorithm repeatedly as follows:*

$$
\begin{aligned}
a &= bq_0 + r_1 & 0 < r_1 < b \\
b &= r_1 q_1 + r_2 & 0 < r_2 < r_1 \\
r_1 &= r_2 q_2 + r_3 & 0 < r_3 < r_2 \\
& \cdots \\
& \cdots \\
r_{n-2} &= r_{n-1} q_{n-1} + r_n & 0 < r_n < r_{n-1} \\
r_{n-1} &= r_n q_n
\end{aligned}
$$

Then r_n is the greatest common divisor of a and b, i.e. $\gcd(a, b) = r_n$.

Proof This was proved in Chap. 4.

Lemma *Let n be a positive integer greater than one then the positive divisors of n are precisely those integers of the form:*

$$d = p_1^{\beta_1} p_2^{\beta_2} p_3^{\beta_3} \ldots p_k^{\beta_k} \quad (where\ 0 \leq \beta_i \leq \alpha_i)$$

where the unique factorization of n is given by

$$n = p_1^{\alpha_1} p_2^{\alpha_2} p_3^{\alpha_3} \ldots p_k^{\alpha_k}$$

---

[4]Euclid was a third-century B.C. Hellenistic mathematician and is considered the father of geometry.

Proof Suppose d is a divisor of n then $n = dq$. By the unique factorization theorem, the prime factorization of n is unique, and so the prime numbers in the factorization of d must appear in the prime factors $p_1, p_2, p_3, \ldots, p_k$ of n.

Clearly, the power β_i of p_i must be less than or equal to α_i, i.e. $\beta_i \leq \alpha_i$. Conversely, whenever $\beta_i \leq \alpha_i$ then clearly d divides n.

5.3.4 Distribution of Primes

We already have shown that there are an infinite number of primes. However, most integer numbers are composite and a reasonable question to ask is how many primes are there less than a certain number. The number of primes less than or equal to x is known as the prime distribution function (denoted by $\pi(x)$) and it is defined by

$$\pi(x) = \sum_{p \leq x} 1 \quad (\text{where } p \text{ is prime})$$

The prime distribution function satisfies the following properties:

(i) $\quad \lim_{x \to \infty} \dfrac{\pi(x)}{x} = 0$

(ii) $\quad \lim_{x \to \infty} \pi(x) = \infty$

The first property expresses the fact that most integer numbers are composite, and the second property expresses the fact that there are an infinite number of prime numbers.

There is an approximation of the prime distribution function in terms of the logarithmic function ($^x/_{\ln x}$) as follows:

$$\lim_{x \to \infty} \frac{\pi(x)}{x/\ln x} = 1 \quad (\text{Prime Number Theorem})$$

The approximation $x/\ln x$ to $\pi(x)$ gives an easy way to determine the approximate value of $\pi(x)$ for a given value of x. This result is known as the *Prime Number Theorem*, and Gauss originally conjectured this theorem.

Palindromic Primes

A palindromic prime is a prime number that is also a palindrome (i.e. it reads the same left to right as right to left). For example, 11, 101 and 353 are all palindromic primes.

All palindromic primes (apart from 11) have an odd number of digits. It is an open question as to whether there are an infinite number of palindromic primes.

Let $\sigma(m)$ denote the sum of all the positive divisors of m (including m):

$$\sigma(m) = \sum_{d\mid m} d$$

Let $s(m)$ denote the sum of all the positive divisors of m (excluding m):

$$s(m) = \sigma(m) - m.$$

Clearly, $s(m) = m$ and $\sigma(m) = 2m$ when m is a perfect number.

Theorem 5.5 (Euclid–Euler Theorem) *The positive integer n is an even perfect number if and only if $n = 2^{p-1}(2^p - 1)$ where $2^p - 1$ is a Mersenne prime.*

Proof Suppose $n = 2^{p-1}(2^p - 1)$ where $2^p - 1$ is a Mersenne prime, then

$$\begin{aligned}
\sigma(n) &= \sigma\big(2^{p-1}(2^p - 1)\big) \\
&= \sigma\big(2^{p-1}\big)\sigma(2^p - 1) \\
&= \sigma\big(2^{p-1}\big)2^p \quad (2^p - 1 \text{ is prime with 2 divisors:} 1 \text{ and itself}) \\
&= (2^p - 1)2^p \quad (\text{Sum of arithmetic series} \\
&= (2^p - 1)2.2^{p-1} \\
&= 2.2^{p-1}(2^p - 1) \\
&= 2n
\end{aligned}$$

Therefore, n is a perfect number since $\sigma(n) = 2n$.

The next part of the proof is to show that any even perfect number must be of the form above. Let n be an arbitrary even perfect number ($n = 2^{p-1}q$) with q odd and so the gcd $(2^{p-1}, q) = 1$ and so

$$\begin{aligned}
\sigma(n) &= \sigma\big(2^{p-1}q\big) \\
&= \sigma\big(2^{p-1}\big)\sigma(q) \\
&= (2^p - 1)\sigma(q) \\
\sigma(n) &= 2n \quad (\text{since } n \text{ is perfect}) \\
&= 2.2^{p-1}q \\
&= 2^p q
\end{aligned}$$

Fig. 5.7 Leonard Euler

Therefore,

$$2^p q = (2^p - 1)\sigma(q)$$
$$= (2^p - 1)(s(q) + q)$$
$$= (2^p - 1)s(q) + (2^p - 1)q$$
$$= (2^p - 1)s(q) + 2^p q - q$$

Therefore, $(2^p - 1)\, s(q) = q$.

Therefore, $d = s(q)$ is a proper divisor of q. However, $s(q)$ is the sum of all the proper divisors of q including d, and so d is the only proper divisor of q and $d = 1$. Therefore, $q = (2^p - 1)$ is a Mersenne prime.

Euler φ Function

The Euler[5] φ function (also known as the *totient function*) is defined for a given positive integer n to be the number of positive integers k less than n that are relatively prime to n (Fig. 5.7). Two integers a, b are relatively prime if $\gcd(a, b) = 1$.

$$\phi(n) = \sum_{1 \leq k < n} 1 \quad \text{where } \gcd(k, n) = 1$$

---

[5]Euler was an eighteenth-century Swiss mathematician who made important contributions to mathematics and physics. His contributions include the well-known formula V − E + F = 2 in graph theory, and he also made contributions to the calculus, infinite series, the exponential function for complex numbers and the totient function.

5.4 Theory of Congruences[6]

Let a be an integer and n a positive integer greater than 1, then $(a \bmod n)$ is defined to be the remainder r when a is divided by n. That is,

$$a = kn + r \quad \text{where } 0 \leq r < n.$$

Definition Suppose a, b are integers and n a positive integer, then a is said to be congruent to b modulo n denoted by $a \equiv b \pmod{n}$ if they both have the same remainder when divided by n.

This is equivalent to n being a divisor of $(a - b)$ or $n \mid (a - b)$ since we have $a = k_1 n + r$ and $b = k_2 n + r$ and so $(a - b) = (k_1 - k_2) n$ and so $n \mid (a - b)$.

Theorem 5.6 *Congruence modulo n is an equivalence relation on the set of integers, i.e. it is a reflexive, symmetric and transitive relation.*

Proof

(i) *Reflexive*
     For any integer a, it is clear that $a \equiv a \pmod{n}$ since $a - a = 0.n$.
(ii) *Symmetric*
     Suppose $a \equiv b \pmod{n}$ then $a - b = kn$. Clearly, $b - a = -kn$ and so $b \equiv a \pmod{n}$.
(iii) *Transitive.*

$$\text{Suppose } a \equiv b (\bmod\, n) \quad \text{and} \quad b \equiv c (\bmod\, n)$$
$$\Rightarrow a - b = k_1 n \quad \text{and} \quad b - c = k_2 n$$
$$\Rightarrow a - c = (a - b) + (b - c)$$
$$= k_1 n + k_2 n$$
$$= (k_1 + k_2) n$$
$$\Rightarrow a \equiv c (\bmod\, n).$$

Therefore, congruence modulo n is an equivalence relation, and an equivalence relation partitions a set S into equivalence classes (Theorem 3.2). The integers are partitioned into n equivalence classes for the congruence modulo n equivalence relation, and these are called *congruence classes* or *residue classes*.

The residue class of a modulo n is denoted by $[a]_n$ or just $[a]$ when n is clear. It is the set of all those integers that are congruent to a modulo n.

$$[a]_n = \{x : x \in \mathbb{Z} \text{ and } x \equiv a(\bmod\, n)\} = \{a + kn : k \in \mathbb{Z}\}$$

---

[6]The theory of congruences was introduced by the German mathematician, Carl Friedrich Gauss.

Any two equivalence classes $[a]$ and $[b]$ are either equal or disjoint, i.e. we have $[a] = [b]$ or $[a] \cap [b] = \emptyset$. The set of all residue classes modulo n is denoted by

$$\mathbb{Z}/n\mathbb{Z} = \mathbb{Z}_n = \{[a]_n : 0 \leq a \leq n - 1\} = \{[0]_n, [1]_n, \ldots, [n-1]_n\}$$

For example, consider $\mathbb{Z}_4$ the residue classes mod 4 then

$$[0]_4 = \{\ldots, -8, -4, 0, 4, 8, \ldots\}$$
$$[1]_4 = \{\ldots, -7, -3, 1, 5, 9, \ldots\}$$
$$[2]_4 = \{\ldots, -6, -2, 2, 6, 10, \ldots\}$$
$$[3]_4 = \{\ldots, -5, -1, 3, 7, 11, \ldots\}$$

The *reduced residue class* is a set of integers r_i such that $(r_i, n) = 1$ and r_i is not congruent to r_j (mod n) for $i \neq j$, and such that every x relatively prime to n is congruent modulo n to for some element r_i of the set. There are $\varphi(n)$ elements $\{r_1, r_2, \ldots, r_{\varphi(n)}\}$ in the reduced residue class set S.

Modular Arithmetic

Addition, subtraction and multiplication may be defined in $\mathbb{Z}/n\mathbb{Z}$ and are similar to these operations in $\mathbb{Z}$. Given a positive integer n and integers a, b, c, d such that $a \equiv b$ (mod n) and $c \equiv d$ (mod n), then the following are properties of modular arithmetic:

(i) $a + c \equiv b + d \pmod{n}$ and $a - c \equiv b - d \pmod{n}$
(ii) $ac \equiv bd \pmod{n}$
(iii) $a^m \equiv b^m \pmod{n} \forall m \in \mathbb{N}$

Proof *(of ii)* Let $a = kn + b$ and $c = ln + d$ for some $k, l \in \mathbb{Z}$ then

$$ac = (kn + b)(ln + d)$$
$$= (kn)(ln) + (kn)d + b(ln) + bd$$
$$= (knl + kd + bl)n + bd$$
$$= sn + bd \quad (\text{where } s = knl + kd + bl)$$

and so $ac \equiv bd$ (mod n)

The three properties above may be expressed in the following equivalent formulation:

(i) $[a + c]_n = [b + d]_n$ and $[a - c]_n = [b - d]_n$
(ii) $[ac]_n = [bd]_n$
(iii) $[a^m]_n = [b^m]_n \quad \forall m \in \mathbb{N}$

Two integers x, y are said to be multiplicative inverses of each other modulo n if

$$xy \equiv 1 \pmod{n}$$

However, x does not always have an inverse modulo n, and this is clear since, for example, $[3]_6$ is a zero divisor modulo 6, i.e. $[3]_6 \cdot [2]_6 = [0]_6$ and so it does not have a multiplicative inverse. However, if n and x are relatively prime then it is easy to see that x has an inverse (mod n) since we know that there are integers k, l such that $kx + ln = 1$.

Given $n > 0$, there are $\varphi(n)$ numbers b that are relatively prime to n and so there are $\varphi(n)$ numbers that have an inverse modulo n. Therefore, for p prime there are $p - 1$ elements that have an inverse (mod p).

Theorem 5.7 (Euler's Theorem) *Let a and n be positive integers with* $\gcd(a, n) = 1$. *Then*

$$a^{\phi(n)} \equiv 1 \pmod{n}$$

Proof Let $\{r_1, r_2, \ldots, r_{\phi(n)}\}$ be the reduced residue system (mod n). Then $\{ar_1, ar_2, \ldots, ar_{\varphi(n)}\}$ is also a reduced residue system (mod n) since $ar_i \equiv ar_j$ (mod n) and $(a, n) = 1$ implies that $r_i \equiv r_j$ (mod n).

For each r_i, there is exactly one r_j such that $ar_i \equiv r_j$ (mod n), and different r_i will have different corresponding ar_j Therefore, $\{ar_1, ar_2, \ldots, ar_{\varphi(n)}\}$ are just the residues module n of $\{r_1, r_2, \ldots, r_{\phi(n)}\}$ but not necessarily in the same order. Multiplying, we get

$$\prod_{j=1}^{\phi(n)} (ar_j) \equiv \prod_{i=1}^{\phi(n)} r_i \pmod{n}$$

$$a^{\phi(n)} \prod_{j=1}^{\phi(n)} (r_j) \equiv \prod_{i=1}^{\phi(n)} r_i \pmod{n}$$

Since $(r_j, n) = 1$ we can deduce that $a^{\phi(n)} \equiv 1$ (mod n) from the result that $ax \equiv ay$ (mod n) and $(a, n) = 1$ then $x \equiv y$ (mod n).

Theorem 5.8 (Fermat's Little Theorem) *Let a be a positive integer and p a prime. If* $\gcd(a, p) = 1$ *then*

$$a^{p-1} \equiv 1 \pmod{p}$$

Proof This result is an immediate corollary to Euler's Theorem as $\varphi(p) = p - 1$.

Theorem 5.9 (Wilson's Theorem) *If p is a prime then $(p - 1)! \equiv -1$ (mod p).*

Proof Each element $a \in 1, 2, \ldots p - 1$ has an inverse a^{-1} such that $aa^{-1} \equiv 1$ (mod p). Exactly, two of these elements 1 and $p - 1$ are their own inverse (i.e. $x^2 \equiv 1$ (mod p) has two solutions 1 and $p - 1$). Therefore, the product 1.2. ... $p - 1$ (mod p) $= p - 1$ (mod p) $\equiv -1$ (mod p).

Diophantine equations

The word '*Diophantine*' is derived from the name of the third-century mathematician, Diophantus, who lived in the city of Alexandria in Egypt. Diophantus studied various polynomial equations of the form $f(x, y, z, \ldots) = 0$ with integer coefficients to determine which of them had integer solutions.

A Diophantine equation may have no solution, a finite number of solutions or an infinite number of solutions. The integral solutions of a Diophantine equation $f(x, y) = 0$ may be interpreted geometrically as the points on the curve with integral coordinates.

Example A linear Diophantine equation $ax + by = c$ is an algebraic equation with two variables x and y, and the problem is to find integer solutions for x and y.

5.5 Binary System and Computer Representation of Numbers

The binary number system was discussed in Chap. 2 and Leibniz[7] was one of the earliest people to recognize its potential. This system uses just two digits, namely, '0' and '1', and Leibniz showed how binary numbers may be added, subtracted, multiplied and divided.

The number two is represented by 10, the number four by 100 and so on. A table of values for the first 15 binary numbers is given in Table 2.2. The binary number system (base 2) is a positional number system, which uses two binary digits 0 and 1, and an example binary number is 1001.01_2 which represents $1 \times 2^3 + 1 + 1 \times 2^{-2} = 8 + 1 + 0.25 = 9.25$.

The binary system is ideally suited to the digital world of computers, as Claude Shannon showed in his Master's thesis (Shannon 1937) that a binary digit may be implemented by an *on–off switch*. This allows binary arithmetic and more complex mathematical operations to be performed by relay circuits and provided the foundation of digital computing. The digital world consists of devices that store information or data on permanent storage media such as discs and CDs, or

---

[7]Wilhelm Gottfried Leibniz was a German philosopher, mathematician and inventor in the field of mechanical calculators. He developed the binary number system used in digital computers and invented the Calculus independently of Sir Isaac Newton. He was embroiled in a bitter dispute towards the end of his life with Newton, as to who developed the calculus first.

temporary storage media such as random access memory (RAM) that consist of a large number of memory elements that may be in one of two states (i.e. on or off).

There are algorithms to convert numbers from decimal to binary and vice versa (see Chap. 2). The base 2 is written on the left, and the number to be converted to binary is placed in the first column. At each stage in the conversion, the number in the first column is divided by 2 to form the quotient and remainder, which are then placed on the next row. The process continues until the quotient is 0, and the binary representation result is then obtained by reading the second column from the bottom up.

Similarly, there are algorithms to convert decimal fractions to binary representation (to a defined number of binary digits as the representation may not terminate), and the conversion of a number that contains an integer part and a fractional part involves converting each part separately and then combining them.

Numbers are represented in a digital computer as sequences of bits of fixed length (e.g. 16 bits, 32 bits). There is a difference in the way in which integers and real numbers are represented, with the representation of real numbers being more complicated.

An integer number is represented by a sequence (usually 2 or 4) bytes where each byte is 8 bits. For example, a 2-byte integer has 16 bits with the first bit used as the sign bit (the sign is 1 for negative numbers and 0 for positive integers), and the remaining 15 bits represent the number. This means that 2 bytes may be used to represent all integer numbers between -32768 and 32767. A positive number is represented by the normal binary representation discussed earlier, whereas a negative number is represented using 2's complement of the original number (i.e. 0 changes to 1 and 1 changes to 0 and the sign bit is 1). All of the standard arithmetic operations may then be carried out (using modulo 2 arithmetic).

The representation of floating-point real numbers is with a representation to a fixed number of significant digits (the significand) and scaled using an exponent in some base (usually 2). That is, the number is represented (approximated) as

$$\text{significand} \times \text{base}^{\text{exponent}}$$

The significand (also called mantissa) and exponent have a sign bit. For example, in simple floating-point representation (4 bytes), the mantissa is generally 24 bits and the exponent 8 bits, whereas for double precision (8 bytes) the mantissa is generally 53 bits and the exponent 11 bits.

5.6   Review Questions

1. Show that

   (i)   if $a|b$ then $a|bc$
   (ii) If $a|b$ and $c|d$ then $ac|bd$

2. Show that 1184 and 1210 are an amicable pair.
3. Use the Euclidean Algorithm to find $g = \gcd(b, c)$ where $b = 42823$ and $c = 6409$, and find integers x and y such that $bx + cy = g$.
4. List all integers x in the range $1 \le x \le 100$ such that $x \equiv 7 \pmod{17}$.
5. Evaluate $\phi(m)$ for $m = 1, 2, 3, \ldots 12$.
6. Determine a complete and reduced residue system modulo 12.
7. Convert 767 to binary, octal and hexadecimal.
8. Convert (you may need to investigate) 0.32_{10} to binary (to 5 places).
9. Explain the difference between binary, octal and hexadecimal.
10. Find the 16-bit integer representation of -4961.

5.7 Summary

Number theory is concerned with the mathematical properties of the natural numbers and integers. These include properties such as, whether a number is prime or composite, the prime factors of a number, the greatest common divisor and least common multiple of two numbers, and so on.

The natural numbers $\mathbb{N}$ consist of the numbers $\{1, 2, 3, \ldots\}$. The integer numbers $\mathbb{Z}$ consist of $\{\ldots -2, -1, 0, 1, 2, \ldots\}$. The rational numbers $\mathbb{Q}$ consist of all numbers of the form $\{^p/_q$ where p and q are integers and $q \ne 0\}$. Number theory has been applied to cryptography in the computing field.

Prime numbers have no factors apart from themselves and one, and there are an infinite number of primes. The Sieve of Eratosthenes algorithm may be employed to determine prime numbers, and the approximation to the distribution of prime numbers less than a number n is given by the prime distribution function $\pi(n) = n/\ln n$. Prime numbers are the key building blocks in number theory, and the fundamental theorem of arithmetic states that every number may be written uniquely as the product of factors of prime numbers.

Mersenne primes and perfect numbers were considered and it was shown that there is a one-to-one correspondence between the Mersenne primes and the even perfect numbers.

Modulo arithmetic including addition, subtraction and multiplication were defined, and the residue classes and reduced residue classes were discussed. There are unsolved problems in number theory such as Goldbach's conjecture that states that every even integer is the sum of two primes. Other open questions include whether there are an infinite number of Mersenne primes and palindromic primes.

We discussed the binary number system, which is ideally suited for digital computers. We discussed the conversion between binary and decimal systems, as well as the octal and hexadecimal systems. Finally, we discussed the representation of integers and real numbers on a computer. For more detailed information on number theory, see Yan (1998).

References

Shannon C (1937) A symbolic analysis of relay and switching circuits. Masters thesis. Massachusetts Institute of Technology
Yan SY (1998) Number theory for computing, 2nd ed. Springer

Algebra

6

Key Topics

Simultaneous equations
Quadratic equations
Polynomials
Indices
Logs
Abstract algebra
Groups
Rings
Fields
Vector spaces

6.1 Introduction

Algebra is the branch of mathematics that uses letters in the place of numbers, where the letters stand for variables or constants that are used in mathematical expressions. Algebra is the study of such mathematical symbols and the rules for manipulating them, and it is a powerful tool for problem-solving in science and engineering.

The origins of algebra are in work done by Islamic mathematicians during the Golden age in Islamic civilization, and the word '*algebra*' comes from the Arabic '*al-jabr*', which appears as part of the title of a book by the Islamic mathematician, Al Khwarizmi, in the ninth century A.D. The third-century A.D. Hellenistic mathematician, Diophantus, also did early work on algebra.

© Springer Nature Switzerland AG 2020
G. O'Regan, *Mathematics in Computing*, Undergraduate Topics
in Computer Science, https://doi.org/10.1007/978-3-030-34209-8_6

Algebra covers many areas such as elementary algebra, linear algebra and abstract algebra. Elementary algebra includes the study of symbols and rules for manipulating them to form valid mathematical expressions, simultaneous equations, quadratic equations, polynomials, indices and logarithms. Linear algebra is concerned with the solution of a set of linear equations, the study of matrices (see Chap. 14) and vectors. Abstract algebra is concerned with the study of abstract algebraic structures such as monoids, groups, rings, integral domains, fields and vector spaces, and we show in Chap. 11 how these abstract structures are applicable to coding theory.

6.2   Simple and Simultaneous Equations

A simple equation is an equation with one unknown, and the unknown may be on both the left-hand side and right-hand side of the equation. The method of solving such equations is to bring the unknowns to one side of the equation, and the values to the other side.

Simultaneous equations are equations with two (or more) unknowns. There are a number of methods to finding a solution to two simultaneous equations such as elimination, substitution and graphical techniques. The solution of n linear equations with n unknowns may be done using Gaussian elimination and matrix theory (see Chap. 14).

Example (*Simple Equation*) Solve the simple equation $4 - 3x = 2x - 11$

Solution (*Simple Equation*)

$$4 - 3x = 2x - 11$$
$$4 - (-11) = 2x - (3x)$$
$$4 + 11 = 2x + 3x$$
$$15 = 5x$$
$$3 = x$$

Example (*Simultaneous Equation—Substitution Method*) Solve the following simultaneous equations by the method of substitution:

$$x + 2y = -1$$
$$4x - 3y = 18$$

Solution (*Simultaneous Equation—Substitution Method*) The method of substitution involves expressing x in terms of y and substituting it in the other equation (or vice versa expressing y in terms of x and substituting it in the other equation). For this example, we use the first equation to *express x in terms of y*.

$$x + 2y = -1$$
$$x = -1 - 2y$$

We then substitute for x, i.e. $(-1 - 2y)$ in the second equation, and we get a simple equation involving just the unknown y.

$$4(-1 - 2y) - 3y = 18$$
$$\Rightarrow -4 - 8y - 3y = 18$$
$$\Rightarrow -11y = 18 + 4$$
$$\Rightarrow -11y = 22$$
$$\Rightarrow y = -2$$

We then obtain the value of x from the substitution:

$$x = -1 - 2y$$
$$\Rightarrow x = -1 - 2(-2)$$
$$\Rightarrow x = -1 + 4$$
$$\Rightarrow x = 3$$

We can then verify that our solution is correct by checking our answer for both equations.

$$3 + 2(-2) = -1 \quad \checkmark$$
$$4(3) - 3(-2) = 18 \quad \checkmark$$

Example (*Simultaneous Equation—Method of Elimination*) Solve the following simultaneous equations by the method of elimination:

$$3x + 4y = 5$$
$$2x - 5y = -12$$

Solution (*Simultaneous Equation—Method of Elimination*) The approach is to manipulate both equations so that we may eliminate either x or y, and so reduce to a simple equation with just x or y. For this example, we are going to eliminate x, and so we multiply equation (1) by 2 and equation (2) by -3 and this yields two equations with the opposite coefficient of x.

$$6x + 8y = 10$$
$$-6x + 5y = 36$$
$$\text{-------------------}$$
$$0x + 23y = 46$$
$$y = 2$$

We then add both equations together and conclude that $y = 2$. We then determine the value of x by replacing y with 2 in equation (1).

$$3x + 4(2) = 5$$
$$3x + 8 = 5$$
$$3x = 5 - 8$$
$$3x = -3$$
$$x = -1$$

We can then verify that our solution is correct as before by checking our answer for both equations.

Example (*Simultaneous Equation—Graphical Techniques*) Find the solution to the following simultaneous equations using graphical techniques:

$$x + 2y = -1$$
$$4x - 3y = 18$$

Solution (*Simultaneous Equation—Graphical Techniques*) Each simultaneous equation represents a straight line, and so the solution to the two simultaneous equations is the point of intersection of both lines (if there is such a point). Therefore, the solution involves drawing each line and finding the point of intersection of both lines (Fig. 6.1).

First we find two points on line 1, e.g. (0, −0.5) and (−1, 0) are on line 1, since when $x = 0$ we have $2y = -1$ and so $y = -0.5$. Similarly, when $y = 0$ we have $x = -1$. We find two points on line 2 in a similar way, e.g. when x is 0 y is −6 and when y is 0 we have $x = 4.5$ and so the points (0, −6) and (4.5, 0) are on line 2.

We then draw the X-axis and the Y-axis, draw the scales on the axes, label the axes, plot the points and draw both lines. Finally, we find the point of intersection of both lines (if there is such a point), and this is our solution to the simultaneous equations.

Fig. 6.1 Graphical solution to simultaneous equations

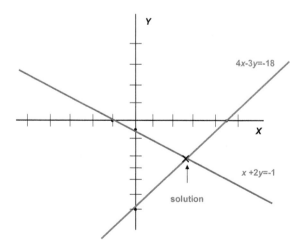

For this example, there is a point of intersection for the lines, and so we determine the x- and y-coordinates and the solution is then given by $x = 3$ and $y = -2$. The solution using graphical techniques requires care (as inaccuracies may be introduced from poor drawing) and graph paper is required to give more accurate results.

6.3   Quadratic Equations

A quadratic equation is an equation of the form $ax^2 + bx + c = 0$, and solving the quadratic equation is concerned with finding the unknown value x (roots of the quadratic equation). There may be no solution to the quadratic equation, one solution or two solutions. There are several techniques to solve quadratic equations such as factorization, completing the square, the quadratic formula and graphical techniques.

Example (*Quadratic Equations—Factorization*) Solve the quadratic equation $3x^2 - 11x - 4 = 0$ by factorization.

Solution (*Quadratic Equations—Factorization*) The approach taken is to find the factors of the quadratic equation. Sometimes, this is easy, but often other techniques will need to be employed. For the above quadratic equation, we note immediately that its factors are $(3x + 1)(x - 4)$ since

$$
\begin{aligned}
&(3x + 1)(x - 4) \\
&= 3x^2 - 12x + x - 4 \\
&= 3x^2 - 11x - 4
\end{aligned}
$$

Next, we note the property that if the product of two numbers A and B is 0 then either A is 0 or B is 0. In other words, $AB = 0 \Rightarrow A = 0$ or $B = 0$. We conclude from this property that as

$$
\begin{aligned}
&3x^2 - 11x - 4 = 0 \\
&\Rightarrow (3x + 1)(x - 4) = 0 \\
&\Rightarrow (3x + 1) = 0 \text{ or } (x - 4) = 0 \\
&\Rightarrow 3x = -1 \text{ or } x = 4 \\
&\Rightarrow x = -0.33 \text{ or } x = 4
\end{aligned}
$$

Therefore, the solutions (or roots) of the quadratic equation $3x^2 - 11x - 4 = 0$ are $x = -0.33$ or $x = 4$.

Example (*Quadratic Equations—Completing the Square*) Solve the quadratic equation $2x^2 + 5x - 3 = 0$ by completing the square.

Solution (*Quadratic Equations—Completing the Square*) First, we convert the quadratic equation to an equivalent quadratic with a unary coefficient of x^2. This involves division by 2. Next, we examine the coefficient of x (in this case $^5/_2$) and we add the square of half the coefficient of x to both sides. This allows us to complete the square, and we then to take the square root of both sides. Finally, we solve for x.

$$2x^2 + 5x - 3 = 0$$
$$\Rightarrow \quad x^2 + {}^5/_2 x - {}^3/_2 = 0$$
$$\Rightarrow \quad x^2 + {}^5/_2 x = {}^3/_2$$
$$\Rightarrow \quad x^2 + {}^5/_2 x + ({}^5/_4)^2 = {}^3/_2 + ({}^5/_4)^2$$
$$\Rightarrow \quad (x + {}^5/_4)^2 = {}^3/_2 + ({}^{25}/_{16})$$
$$\Rightarrow \quad (x + {}^5/_4)^2 = {}^{24}/_{16} + ({}^{25}/_{16})$$
$$\Rightarrow \quad (x + {}^5/_4)^2 = {}^{49}/_{16}$$
$$\Rightarrow \quad (x + {}^5/_4) = \pm {}^7/_4$$
$$\Rightarrow \quad x = -{}^5/_4 \pm {}^7/_4$$
$$\Rightarrow \quad x = -{}^5/_4 - {}^7/_4 \text{ or } x = -{}^5/_4 + {}^7/_4$$
$$\Rightarrow \quad x = -{}^{12}/_4 \text{ or } x = {}^2/_4$$
$$\Rightarrow \quad x = -3 \text{ or } x = 0.5$$

Example 1 (*Quadratic Equations—Quadratic Formula*) Establish the quadratic formula for solving quadratic equations.

Solution (*Quadratic Equations—Quadratic Formula*) We complete the square and the result will follow.

$$ax^2 + bx + c = 0$$
$$\Rightarrow \quad x^2 + {}^b/_a x + {}^c/_a = 0$$
$$\Rightarrow \quad x^2 + {}^b/_a x = -{}^c/_a$$
$$\Rightarrow \quad x^2 + {}^b/_a x + ({}^b/_{2a})^2 = -{}^c/_a + ({}^b/_{2a})^2$$
$$\Rightarrow \quad (x + {}^b/_{2a})^2 = -{}^c/_a + ({}^b/_{2a})^2$$
$$\Rightarrow$$
$$\Rightarrow \quad (x + {}^b/_{2a})^2 = \frac{-4ac}{4a^2} + \frac{b^2}{4a^2}$$
$$\Rightarrow \quad (x + {}^b/_{2a})^2 = \frac{b^2 - 4ac}{4a^2}$$
$$\Rightarrow \quad (x + {}^b/_{2a}) = \pm\frac{\sqrt{b^2 - 4ac}}{2a}$$
$$\Rightarrow \quad x = \frac{-b \pm \sqrt{b^2 - 4ac}}{2a}$$

Example 2 (*Quadratic Equations—Quadratic Formula*) Solve the quadratic equation $2x^2 + 5x - 3 = 0$ using the quadratic formula.

Solution (*Quadratic Equations—Quadratic Formula*) For this example, $a = 2$, $b = 5$ and $c = -3$, and we put these values into the quadratic formula.

$$x = \frac{-5 \pm \sqrt{5^2 - 4 \cdot 2 \cdot (-3)}}{2 \cdot 2} = \frac{-5 \pm \sqrt{25 + 24}}{4}$$
$$x = \frac{-5 \pm \sqrt{49}}{4} = \frac{-5 \pm 7}{4}$$
$$x = 0.5 \text{ or } x = -3.$$

Example (*Quadratic Equations—Graphical Techniques*) Solve the quadratic equation $2x^2 - x - 6 = 0$ using graphical techniques given that the roots of the quadratic equation lie between $x = -3$ and $x = 3$.

Solution (*Quadratic Equations—Graphical Techniques*) The approach is first to create a table of values for the curve $y = 2x^2 - x - 6$, and to draw the X- and Y-axes and scales, and then to plot the points from the table of values (Table 6.1), and to join the points together to form the curve (Fig. 6.2).

The graphical solution to the quadratic equation is then given by the points where the curve intersects the X-axis (i.e. $y = 0$ on the X-axis). There may be no solution (i.e. the curve does not intersect the X-axis), one solution (a double root) or two solutions.

Table 6.1 Table of values for quadratic equation

x	-3	-2	-1	0	1	2	3
$y = 2x^2 - x - 6$	15	4	-3	-6	-5	0	9

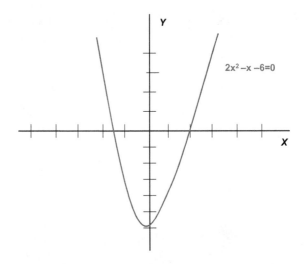

Fig. 6.2 Graphical solution to quadratic equation

$2x^2 - x - 6 = 0$

The graph for the curve $y = 2x^2 - x - 6$ is given in Fig. 6.2, and so the points where the curve intersects the X-axis are determined. We note from the graph that the curve intersects the X-axis at two distinct points, and we see from the graph that the roots of the quadratic equation are given by $x = -1.5$ and $x = 2$.

The solution to quadratic equations using graphical techniques requires care (as with the solution to simultaneous equations using graphical techniques), and graph paper is required for accuracy.

6.4 Indices and Logarithms

The product $a.a.a.a \ldots a$ (n times) is denoted by a^n, and the number n is the index of a. The following are properties of indices:

$$a^0 = 1$$
$$a^{m+n} = a^m \cdot a^n$$
$$a^{mn} = (a^m)^n$$
$$a^{-n} = \frac{1}{a^n}$$
$$a^{\frac{1}{n}} = \sqrt[n]{a}$$

Logarithms are closely related to indices, and if the number b can be written in the form $b = a^x$, then we say that log to the base a of b is x, i.e. $\log_a b = x \Leftrightarrow a^x = b$. Clearly, $\log_{10} 100 = 2$ since $10^2 = 100$. The following are properties of logarithms:

$$\log_a AB = \log_a A + \log_a B$$
$$\log_a A^n = n \, \log_a A$$
$$\log \frac{A}{B} = \log A - \log B$$

We will prove the first property of logarithms. Suppose $\log_a A = x$ and $\log_a B = y$. Then $A = a^x$ and $B = a^y$ and so $AB = a^x a^y = a^{x+y}$ and so $\log_a AB = x + y = \log_a A + \log_a B$.

The law of logarithms may be used to solve certain equations involving powers (called indicial equations). We illustrate this by an example.

Example (*Indicial Equations*) Solve the equation $2^x = 3$, correct to four significant places.

Solution (*Indicial Equations*)

$$2^x = 3$$
$$\Rightarrow \log_{10} 2^x = \log_{10} 3$$
$$\Rightarrow x \log_{10} 2 = \log_{10} 3$$
$$\Rightarrow x = \frac{\log_{10} 3}{\log_{10} 2}$$
$$= \frac{0.4771}{0.3010}$$
$$\Rightarrow x = 1.585$$

6.5 Horner's Method for Polynomials

Horner's Method is a computationally efficient way to evaluate a polynomial function. It is named after William Horner who was a nineteenth-century British mathematician and schoolmaster. Chinese mathematicians were familiar with the method in the third century A.D.

The normal method for the evaluation of a polynomial involves computing exponentials, and this is computationally expensive. Horner's method has the advantage that fewer calculations are required, and it eliminates all exponentials by using nested multiplication and addition. It also provides a computationally efficient way to determine the derivative of the polynomial.

Horner's Method and Algorithm

Consider a polynomial $P(x)$ of degree n defined by

$$P(x) = a_n x^n + a_{n-1} x^{n-1} + a_{n-2} x^{n-2} + \cdots + a_1 x + a_0$$

The Horner method to evaluate $P(x_0)$ essentially involves writing $P(x)$ as

$$P(x) = ((((a_n x + a_{n-1})x + a_{n-2})x + \cdots + a_1)x + a_0$$

The computation of $P(x_0)$ involves defining a set of coefficients b_k such that

$$b_n = a_n$$
$$b_{n-1} = a_{n-1} + b_n x_0$$
$$\ldots$$
$$b_k = a_k + b_{k+1} x_0$$
$$\ldots$$
$$b_1 = a_1 + b_2 x_0$$
$$b_0 = a_0 + b_1 x_0$$

Then, the computation of $P(x_0)$ is given by

$$P(x_0) = b_0$$

Further, if $Q(x) = b_n x^{n-1} + b_{n-1} x^{n-2} + b_{n-2} x^{n-3} + \cdots + b_1$ then it is easy to verify that

$$P(x) = (x - x_0)Q(x) + b_0$$

This also allows the derivative of $P(x)$ to be easily computed for x_0 since

$$P'(x) = Q(x) + (x - x_0)Q'(x)$$
$$P'(x_0) = Q(x_0)$$

Algorithm (To evaluate polynomial and its derivative)

(i)   Initialize y to a_n and z to a_n (Compute b_n for P and b_{n-1} for Q).
(ii) For each j from $n-1$, $n-2$ to 1 compute b_j for P and b_{j-1} for Q by
      Set y to $x_0 y + a_j$ (i.e. b_j for P) and z to $x_0 z + y$ (i.e. b_{j-1} for Q).
(iii) Compute b_0 by setting y to $x_0 y + a_0$.

Then $P(x_0) = y$ and $P'(x_0) = z$.

6.6 Abstract Algebra

One of the important features of modern mathematics is the power of abstraction. This has opened up whole new areas of mathematics, and it has led to a large body of new results and problems. The term '*abstract*' is subjective, as what is abstract to one person may be quite concrete to another. We shall introduce some important algebraic structures in this section including monoids, groups, rings, fields and vector spaces.

6.6.1 Monoids and Groups

A non-empty set M together with a binary operation '$*$' is called a *monoid* if for all elements $a, b, c \in M$ the following properties hold:

(1) $a * b \in M$ (Closure property)
(2) $a * (b * c) = (a * b) * c$ (Associative property)
(3) $\exists u \in M$ such that: $a * u = u * a = a (\forall a \in M)$ (Identity Element)

A monoid is commutative if $a * b = b * a$ for all $a, b \in M$. A *semi-group* $(M, *)$ is a set with a binary operation '*' such that the closure and associativity properties hold (but it may not have an identity element).

Example 6.1 (*Monoids*)

(i)   The set of sequences $\Sigma*$ under concatenation where the empty sequence Λ is the identity element.

(ii) The set of integers under addition forms an infinite monoid in which 0 is the identity element.

A non-empty set G together with a binary operation '*' is called a *group* if for all elements $a, b, c \in G$ the following properties hold:

$(1)\, a * b \in G$ (Closure property)
$(2)\, a * (b * c) = (a * b) * c$ (Associative property)
$(3)\, \exists e \in G$ such that $a * e = e * a = a (\forall a \in G)$ (Identity Element)
$(4)\,$ For every $a \in G, \exists a^{-1} \in G$, such that
     $a^* a^{-1} = a^{-1} * a = e$ (Inverse Element)

The identity element is unique, and the inverse a^{-1} of an element a is unique (see Exercise 5). A *commutative group* has the additional property that $a * b = b * a$ for all $a, b \in G$. The order of a group G is the number of elements in G and is denoted by $o(G)$. If the order of G is finite, then G is said to be a finite group.

Example 6.1 (*Groups*)

(i)   The set of integers under addition $(\mathbb{Z}, +)$ forms an infinite group in which 0 is the identity element.

(ii)   The set of integer 2×2 matrices under addition is formed, where the identity element is $\begin{pmatrix} 0 & 0 \\ 0 & 0 \end{pmatrix}$.

(iii) The set of integers under multiplication $(\mathbb{Z}, \times)$ forms an infinite monoid with 1 as the identity element.

A *cyclic group* is a group where all elements $g \in G$ are obtained from the powers a^i of one element $a \in G$, with $a^0 = e$. The element 'a' is termed the generator of the cyclic group G. A finite cyclic group with n elements is of the form $\{a^0, a^1, a^2, ..., a^{n-1}\}$.

A non-empty subset H of a group G is said to be a *subgroup* of G if for all $a, b \in H$ then $a * b \in H$, and for any $a \in H$ then $a^{-1} \in H$. A subgroup N is termed a *normal subgroup* of G if $gng^{-1} \in N$ for all $g \in G$ and all $n \in N$. Further, if G is a group and N is a normal subgroup of G, then the *quotient group* G/N may be formed.

Lagrange's theorem states the relationship between the order of a subgroup H of G, and the order of G. The theorem states that if G is a finite group, and H is a subgroup of G, then $o(H)$ is a divisor of $o(G)$.

We may also define mapping between similar algebraic structures termed *homomorphism*, and these mapping preserve structure. If the homomorphism is one to one and onto, it is termed an *isomorphism*, which means that the two structures are identical in some sense (apart from a relabelling of elements).

6.6.2 Rings

A *ring* is a non-empty set R together with two binary operations '+' and '×' where $(R, +)$ is a commutative group; $(R, ×)$ is a semi-group; and the left and right distributive laws hold. Specifically, for all elements $a, b, c \in R$ the following properties hold:

$(1)\, a + b \in R$ (Closure property)
$(2)\, a + (b + c) = (a + b) + c$ (Associative property)
$(3)\, \exists 0 \in R$ such that $\forall a \in R : a + 0 = 0 + a = a$ (Identity Element)
$(4)\, \forall a \in R : \exists(-a) \in R : a + (-a) = (-a) + a = 0$ (Inverse Element)
$(5)\, a + b = b + a$ (Commutativity)
$(6)\, a × b \in R$ (Closure property)
$(7)\, a × (b × c) = (a × b) × c$ (Associative property)
$(8)\, a × (b + c) = a × b + a × c$ (Distributive Law)
$(9)\, (b + c) × a = b × a + c × a$ (Distributive Law)

The element 0 is the identity element under addition, and the additive inverse of an element a is given by $-a$. If a ring $(R, ×, +)$ has a multiplicative identity 1 where $a × 1 = 1 × a = a$ for all $a \in R$, then R is termed a ring with a unit element. If $a × b = b × a$ for all $a, b \in R$, then R is termed a *commutative ring*.

An element $a \neq 0$ in a ring R is said to be a *zero divisor* if there exists $b \in R$, with $b \neq 0$ such that $ab = 0$. A commutative ring is an *integral domain* if it has no zero divisors. A ring is said to be a *division ring* if its non-zero elements form a group under multiplication.

Example 6.2 (*Rings*)

(i)   The set of integers $(\mathbb{Z}, +, ×)$ forms an infinite commutative ring with multiplicative unit element 1. Further, since it has no zero divisors it is an integral domain.

(ii)   The set of integers mod 4 (i.e. $\mathbb{Z}_4$ where addition and multiplication are performed modulo 4)[1] is a finite commutative ring with unit element $[1]_4$. Its elements are $\{[0]_4, [1]_4, [2]_4, [3]_4\}$. It has zero divisors since $[2]_4[2]_4 = [0]_4$ and so it is not an integral domain.

---

[1]Recall that $\mathbb{Z}/n\mathbb{Z} = \mathbb{Z}_n = \{[a]_n : 0 \leq a \leq n - 1\} = \{[0]_n, [1]_n, ..., [n - 1]_n\}$

(iii)   The Quaternions (discussed in Chap. 26) are an example of a non-commutative ring (they form a division ring).

(iv)   The set of integers mod 5 (i.e. $\mathbb{Z}_5$ where addition and multiplication are performed modulo 5) is a finite commutative division ring[2] and it has no zero divisors.

6.6.3 Fields

A *field* is a non-empty set F together with two binary operation '+' and '×' where $(F, +)$ is a commutative group; $(F \setminus \{0\}, \times)$ is a commutative group; and the distributive properties hold. The properties of a field are

(1) $a + b \in F$	(Closure property)
(2) $a + (b + c) = (a + b) + c$	(Associative property)
(3) $\exists 0 \in F$ such that $\forall a \in F \, a + 0 = 0 + a = a$	(Identity Element)
(4) $\forall a \in F \, \exists (-a) \in F \, a + (-a) = (-a) + a = 0$	(Inverse Element)
(5) $a + b = b + a$	(Commutativity)
(6) $a \times b \in F$	(Closure property)
(7) $a \times (b \times c) = (a \times b) \times c$	(Associative property)
(8) $\exists 1 \in F$ such that $\forall a \in F \, a \times 1 = 1 \times a = a$	(Identity Element)
(10) $\forall a \in F \setminus \{0\} \, \exists a^{-1} \in F \, a \times 1 = a^{-1} \times a = 1$	(Inverse Element)
(11) $a \times b = b \times a$	(Commutativity)
(12) $a \times (b + c) = a \times b + a \times c$	(Distributive Law)
(13) $(b + c) \times a = b \times a + c \times a$	(Distributive Law)

The following are examples of fields:

Example 6.3 (*Fields*)

(i)   The set of rational numbers $(\mathbb{Q}, +, \times)$ forms an infinite commutative field. The additive identity is 0, and the multiplicative identity is 1.

(ii) The set of real numbers $(\mathbb{R}, +, \times)$ forms an infinite commutative field. The additive identity is 0, and the multiplicative identity is 1.

(iii) The set of complex numbers $(\mathbb{C}, +, \times)$ forms an infinite commutative field. The additive identity is 0, and the multiplicative identity is 1.

(iv) The set of integers mod 7 (i.e. $\mathbb{Z}_7$ where addition and multiplication are performed mod 7) is a finite field.

(v)   The set of integers mod p where p is a prime (i.e. $\mathbb{Z}_p$ where addition and multiplication are performed mod p) is a finite field with p elements. The additive identity is [0] and the multiplicative identity is [1].

---

[2]A finite division ring is actually a field (i.e. it is commutative under multiplication), and this classic result was proved by Wedderburn.

A field is a commutative division ring but not every division ring is a field. For example, the quaternions (discovered by Hamilton) are an example of a division ring, which is not a field. If the number of elements in the field F is finite, then F is called a finite field, and F is written as F_q where q is the number of elements in F. In fact, every finite field has $q = p^k$ elements for some prime p, and some $k \in \mathbb{N}$ and $k > 0$.

6.6.4 Vector Spaces

A non-empty set V is said to be a *vector space* over a field F if V is a commutative group under vector addition +, and if for every $\alpha \in F$, $v \in V$ there is an element αv in V such that the following properties hold for $v, w \in$ V and $\alpha, \beta \in$ F:

1. $u + v \in V$	(Closure Property)
2. $u + (v + w) = (u + v) + w$	(Associative)
3. $\exists 0 \in V$ such that $\forall v \in Vv + 0 = 0 + v = v$	(Identity element)
4. $\forall v \in V \exists(-v) \in V$ such that $v + (-v) = (-v) + v = 0$	(Inverse)
5. $v + w = w + v$	(Commutative)
6. $\alpha(v + w) = \alpha v + \alpha w$	
7. $(\alpha + \beta)v = \alpha v + \beta v$	
8. $\alpha(\beta v) = (\alpha \beta)v$	
9. $1v = v$	

The elements in V are referred to as *vectors* and the elements in F are referred to as *scalars*. The element 1 refers to the identity element of the field F under multiplication.

Application of Vector Spaces to Coding Theory

The representation of codewords in coding theory (discussed in Chap. 11) is by n-dimensional vectors over the finite field F_q. A codeword vector v is represented as the n-tuple:

$$v = (a_0, a_1, \ldots a_{n-1})$$

where each $a_i \in F_q$. The set of all n-dimensional vectors is the n-dimensional vector space F_q^n with q^n elements. The addition of two vectors v and w, where $v = (a_0, a_1, \ldots a_{n-1})$ and $w = (b_0, b_1, \ldots b_{n-1})$ is given by

$$v + w = (a_0 + b_0, a_1 + b_1, \ldots a_{n-1} + b_{n-1})$$

The scalar multiplication of a vector $v = (a_0, a_1, \ldots a_{n-1}) \in F_q^n$ by a scalar $\beta \in F_q$ is given by

$$\beta v = (\beta a_0, \beta a_1, \ldots \beta a_{n-1})$$

The set F_q^n is called the vector space over the finite field F_q, if the vector space properties above hold. A finite set of vectors $v_1, v_2, \ldots v_k$ is said to be *linearly independent* if

$$\beta_1 v_1 + \beta_2 v_2 + \cdots + \beta_k v_k = 0 \quad \Rightarrow \quad \beta_1 = \beta_2 = \cdots \beta_k = 0$$

Otherwise, the set of vectors $v_1, v_2, \ldots v_k$ is said to be *linearly dependent*.

A non-empty subset W of a vector space V ($W \subseteq V$) is said to be a *subspace* of V, if W forms a vector space over F under the operations of V. This is equivalent to W being closed under vector addition and scalar multiplication, i.e. $w_1, w_2 \in W$, α, $\beta \in F$ then $\alpha w_1 + \beta w_2 \in W$.

The *dimension* (dim W) of a subspace $W \subseteq V$ is k if there are k linearly independent vectors in W but every $k + 1$ vectors are linearly dependent. A subset of a vector space is a *basis* for V if it consists of linearly independent vectors, and its linear span is V (i.e. the basis generates V). We shall employ the basis of the vector space of codewords (see Chap. 11) to create the generator matrix to simplify the encoding of the information words. The linear span of a set of vectors $v_1, v_2, \ldots, v_k$ is defined as $\beta_1 v_1 + \beta_2 v_2 + \cdots + \beta_k v_k$.

Example 6.4 (*Vector Spaces*)

(i) The Real coordinate space $\mathbb{R}^n$ forms an n-dimensional vector space over $\mathbb{R}$. The elements of $\mathbb{R}^n$ are the set of all n-tuples of elements of $\mathbb{R}$, where an element x in $\mathbb{R}^n$ is written as

$$x = (x_1, x_2, \ldots x_n)$$

where each $x_i \in \mathbb{R}$ and vector addition and scalar multiplication are given by

$$\alpha x = (\alpha x_1, \alpha x_2, \ldots \alpha x_n)$$
$$x + y = (x_1 + y_1, x_2 + y_2 \cdots x_n + y_n)$$

(ii) The set of $m \times n$ matrices over the real numbers forms a vector space, with vector addition given by matrix addition, and the multiplication of a matrix by a scalar given by the multiplication of each entry in the matrix by the scalar.

6.7 Review Questions

1. Solve the simple equation: $4(3x + 1) = 7(x + 4) - 2(x + 5)$.
2. Solve the following simultaneous equations by

$$x + 2y = -1$$
$$4x - 3y = 18$$

   (a) Graphical techniques,
   (b) Method of substitution and
   (c) Method of Elimination.

3. Solve the quadratic Eq. $3x^2 + 5x - 2 = 0$ given that the solution is between $x = -3$ and $x = 3$ by

   (a) Graphical techniques,
   (b) Factorization and
   (c) Quadratic Formula.

4. Solve the following indicial equation using logarithms:

$$2^{x=1} = 3^{2x-1}$$

5. Explain the differences between semigroups, monoids and groups.
6. Show that the following properties are true for groups.

   (i) The identity element is unique in a group.
   (ii) The inverse of an element is unique in a group.

7. Explain the differences between rings, commutative rings, integral domains, division rings and fields.
8. What is a vector space?
9. Explain how vector spaces may be applied to coding theory (see Chap. 11 for more details).

6.8 Summary

This chapter provided a brief introduction to algebra, which is the branch of mathematics that studies mathematical symbols and the rules for manipulating them. Algebra is a powerful tool for problem-solving in science and engineering.

Elementary algebra includes the study of simultaneous equations (i.e. two or more equations with two or more unknowns); the solution of quadratic equations $ax^2 + bx + c = 0$; and the study of polynomials, indices and logarithms. Linear algebra is concerned with the solution of a set of linear equations, and the study of matrices and vector spaces.

Abstract algebra is concerned with the study of abstract algebraic structures such as monoids, groups, rings, integral domains, fields and vector spaces. The abstract approach in modern mathematics has opened up whole new areas of mathematics as well as applications in areas such as coding theory in the computing field.

Sequences, Series and Permutations and Combinations

7

Key Topics

Arithmetic sequence
Arithmetic series
Geometric sequence
Geometric series
Simple and compound interest
Annuities
Present value
Permutations and combinations
Counting principle

7.1 Introduction

The goal of this chapter is to give an introduction to sequences and series, including arithmetic and geometric sequences, and arithmetic and geometric series. We derive formulae for the sum of an arithmetic series and geometric series, and we discuss the convergence of a geometric series when $|r| < 1$, and the limit of its sum as n gets larger and larger.

We discuss the calculation of simple and compound interest, and the concept of the time value of money. We then show that this is applied to determine the present value of a payment to be made in the future. We discuss annuities, which are a series of payments made at regular intervals over a period of time, and then determine the present value of an annuity.

© Springer Nature Switzerland AG 2020
G. O'Regan, *Mathematics in Computing*, Undergraduate Topics
in Computer Science, https://doi.org/10.1007/978-3-030-34209-8_7

We consider the counting principle where one operation has m possible outcomes and a second operation has n possible outcomes. We determine that the total number of outcomes after performing the first operation followed by the second operation to be $m \times n$. A permutation is an arrangement of a given number of objects, by taking some or all of them at a time. The order of the arrangement is important, as the arrangement 'abc' is different from 'cba'. A combination is a selection of a number of objects in any order, where the order of the selection is unimportant. That is, the selection 'abc' is the same as the selection 'cba'.

7.2 Sequences and Series

A sequence $a_1, a_2, ... a_n ...$ is any succession of terms (usually numbers). For example, each term in the Fibonacci sequence (apart from the first two terms) is obtained from the sum of the previous two terms in the sequence (see Sect. 8.3 for a formal definition of the Fibonacci sequence).

$$1, 1, 2, 3, 5, 8, 13, 21, ...$$

A sequence may be finite (with a fixed number of terms) or infinite. The Fibonacci sequence is infinite, whereas the sequence 2, 4, 6, 8, 10 is finite. We distinguish between convergent and divergent sequences, where a *convergent* sequence approaches a certain value as n gets larger and larger (technically we say that $\lim_{n \to \infty} a_n$ exists) (i.e. the limit of a_n exists). Otherwise, the sequence is said to be *divergent*.

Often, there is a mathematical expression for the n-th term in a sequence (e.g. for the sequence of even integers 2, 4, 6, 8, ... the general expression for a_n is given by $a_n = 2n$). Clearly, the sequence of the even integers is divergent, as it does not approach a particular value, as n gets larger and larger. Consider the following sequence:

$$1, -1, 1, -1, 1, -1, ...$$

Then this sequence is divergent since it does not approach a certain value, as n gets larger and larger, since it continues to alternate between 1 and −1. The formula for the n-th term in the sequence may be given by

$$(-1)^{n+1}$$

The sequence $1, \frac{1}{2}, \frac{1}{3}, \frac{1}{4}, ... \frac{1}{n} ...$ is convergent and it converges to 0. The n-th term in the sequence is given by $\frac{1}{n}$, and as n gets larger and larger it gets closer and closer to 0.

A series is the sum of the terms in a sequence, and the sum of the first n terms of the sequence $a_1, a_2, ... a_n ...$ is given by $a_1 + a_2 + \cdots + a_n$ which is denoted by

$$\sum_{k=1}^{n} a_k$$

A series is convergent if its sum approaches a certain value S as n gets larger and larger, and this is written formally as

$$\lim_{n \to \infty} \sum_{k=1}^{n} a_k = S$$

Otherwise, the series is said to be divergent.

7.3 Arithmetic and Geometric Sequences

Consider the sequence 1, 4, 7, 10, ... where each term is obtained from the previous term by adding the constant value 3. This is an example of an arithmetic sequence, and there is a difference of 3 between any term and the previous one. The general form of a term in this sequence is $a_n = 3n - 2$.

The general form of an *arithmetic sequence* is given by

$$a, a+d, a+2d, a+3d, \ldots a+(n-1)d, \ldots$$

The value a is the initial term in the sequence, and the value d is the constant difference between a term and its successor. For the sequence, 1, 4, 7, ..., we have $a = 1$ and $d = 3$, and the sequence is not convergent. In fact, all arithmetic sequences (apart from the constant sequence a, a, ... a which converges to a) are divergent.

Consider, the sequence 1, 3, 9, 27, 81, ..., where each term is achieved from the previous term by multiplying by the constant value 3. This is an example of a geometric sequence, and the general form of a geometric sequence is given by

$$a, ar, ar^2, ar^3, \ldots, ar^{n-1}$$

The first term in the geometric sequence is a and r is the common ratio. Each term is obtained from the previous one by multiplying by the common ratio r. For the sequence 1, 3, 9, 27, the value of a is 1 and r is 3.

A geometric sequence is convergent if $r < 1$, and for this case it converges to 0. It is also convergent if $r = 1$, as for this case it is simply the constant sequence a, a, a, ..., which converges to a. For the case where $r > 1$ the sequence is divergent.

7.4   Arithmetic and Geometric Series

An arithmetic series is the sum of the terms in an arithmetic sequence, and a geometric sequence is the sum of the terms in a geometric sequence. It is possible to derive a simple formula for the sum of the first n terms in an arithmetic and geometric series.

Arithmetic Series

We write the series in two ways: first the normal left to right addition, and then the reverse, and then we add both series together.

$$Sn = a \quad + (a+d) + (a+2d) + (a+3d) + \cdots + (a + (n-1)d)$$
$$Sn = a + (n-1)d + a + (n-2)d + \cdots + \quad + (a+d) + a$$

_ _

$$2Sn = [2a + (n-1)d] + [2a + (n-1)d] + \cdots + [2a + (n-1)d] \quad (n \text{ times})$$
$$2Sn = n \times [2a + (n-1)d]$$

Therefore, we conclude that

$$S_n = \frac{n}{2}[2a + (n-1)d]$$

Example (*Arithmetic Series*) Find the sum of the first n terms in the following arithmetic series 1, 3, 5, 7, 9, ...

Solution Clearly, $a = 1$ and $d = 2$. Therefore, applying the formula, we get

$$S_n = \frac{n}{2}[2.1 + (n-1)2] = \frac{2n^2}{2} = n^2$$

Geometric Series

For a geometric series, we have

$$S_n = a + ar + ar^2 + ar^3 + \cdots + ar^{n-1}$$
$$\Rightarrow rS_n = ar + ar^2 + ar^3 + \cdots + ar^{n-1} + ar^n$$

_ _ _ _ _ _ _ _ _ _ _ _ _ _ _ _ _

$$\Rightarrow rS_n - S_n = ar^n - a$$
$$= a(r^n - 1)$$
$$\Rightarrow (r-1)S_n = a(r^n - 1)$$

Therefore, we conclude that (where $r \neq 1$)

$$S_n = \frac{a(r^n - 1)}{r - 1} = \frac{a(1 - r^n)}{1 - r}$$

The case of when $r = 1$ corresponds to the arithmetic series $a + a + \cdots + a$, and the sum of this series is simply na. The geometric series converges when $|r| < 1$ as $r^n \to 0$ as $n \to \infty$, and so

$$S_n \to \frac{a}{1 - r} \quad \text{as } n \to \infty$$

Example (*Geometric Series*) Find the sum of the first n terms in the following geometric series 1, $^1/_2, ^1/_4, ^1/_8, \ldots$ What is the sum of the series?

Solution Clearly, $a = 1$ and $r = {}^1/_2$. Therefore, applying the formula we get

$$S_n = \frac{1(1 - 1/2^n)}{1 - 1/2} = \frac{(1 - 1/2^n)}{1 - 1/2} = 2(1 - 1/2^n)$$

The sum of the series is the limit of the sum of the first n terms as n approaches infinity. This is given by

$$\lim_{n \to \infty} S_n = \lim_{n \to \infty} 2(1 - 1/2^n) = 2$$

7.5   Simple and Compound Interest

Savers receive interest on placing deposits at the bank for a period of time, whereas lenders pay interests on their loans to the bank. We distinguish between simple and compound interest, where *simple interest* is always calculated on the original principal, whereas for *compound interest*, the interest is added to the principal sum, so that interest is also earned on the added interest for the next compounding period.

For example, if Euro 1000 is placed on deposit at a bank with an interest rate of 10% per annum for 2 years, it would earn a total of Euro 200 in simple interest. The interest amount is calculated by

$$\frac{1000 * 10 * 2}{100} = \text{Euro } 200$$

The general formula for calculating simple interest on principal P, at a rate of interest I, and for time T (in years:) is

$$A = \frac{P \times I \times T}{100}$$

The calculation of compound interest is more complicated as may be seen from the following example.

Example (*Compound Interest*) Calculate the interest earned and what the new principal will be on Euro 1000, which is placed on deposit at a bank, with an interest rate of 10% per annum (compound) for 3 years.

Solution At the end of year 1, Euro 100 of interest is earned, and this is capitalized making the new principal at the start of year 2 Euro 1100. At the end of year 2, Euro 110 is earned in interest, making the new principal at the start of year 3 Euro 1210. Finally, at the end of year 3 a further Euro 121 is earned in interest, and so the new principal is Euro 1331 and the total interest earned for the 3 years is the sum of the interest earned for each year (i.e. Euro 331). This may be seen from Table 7.1.

The new principal each year is given by the geometric sequence with $a = 1000$ and $r = {}^{10}/_{100} = 0.1$.

$$1000, 1000(1.1), 1000(1.1)^2, 1000(1.1)^3, \ldots$$

In general, if a principal amount P is invested for T years at a rate R of interest (r is expressed as a proportion, i.e., $r = {}^{R}/_{100}$) then it will amount to

$$A = P(1+r)^T$$

For our example above, $A = 1000$, $T = 3$ and $r = 0.1$. Therefore,

$$A = 1000(1.1)^3$$
$$= 1331 (\text{as before})$$

Table 7.1 Calculation of compound interest

Year	Principal	Interest earned
1	1000	100
2	1100	110
3	1210	121

There are variants of the compound interest formula to cover situations where there are m-compounding periods per year, and so the reader may consult the available texts.

7.6 Time Value of Money and Annuities

The time value of money is concerned with the concept that the earlier that cash is received the greater value it has to the recipient. Similarly, the later that a cash payment is made, the lower its value to the recipient, and the lower its cost to the payer.

This is clear if we consider the example of a person who receives $1000 now and a person who receives $1000 5 years from now. The person who receives $1000 now is able to invest it and to receive annual interest on the principal, whereas the other person who receives $1000 in 5 years earns no interest during the period. Further, the inflation during the period means that the purchasing power of $1000 is less in 5 years time is less than it is today.

We presented the general formula for what the future value of a principal P invested for n years at a compound rate r of interest as $A = P (1 + r)^n$.

We can determine the present value of an amount A received in n years time at a discount rate r by

$$P = \frac{A}{(1+r)^n}$$

An annuity is a series of equal cash payments made at regular intervals over a period of time, and so there is a need to calculate the present value of the series of payments made over the period. The actual method of calculation is clear from Table 7.2.

Example (*Annuities*)
Calculate the present value of a series of payments of $1000 (made at the end of each year) with the payments made for 5 years at a discount rate of 10%.

Table 7.2 Calculation of present value of annuity

Year	Amount	Present value ($r = 0.1$)
1	1000	909.91
2	1000	826.44
3	1000	751.31
4	1000	683.01
5	1000	620.92

Solution The regular payment A is 1000, the rate r is 0.1 and $n = 5$. The present value of the payment received at the end of year 1 is $1000/1.1 = 909.91$; at the end of year 2 it is $1000/(1.1)^2 = 826.45$; and so on. The total present value of the payments over the 5 years is given by the sum of the individual present values and is $3791 (Table 7.2).

We may derive a formula for the present value of a series of payments A over a period of n years at a discount rate of r as follows: Clearly, the present value is given by

$$\frac{A}{(1+r)} + \frac{A}{(1+r)^2} + \cdots + \frac{A}{(1+r)^n}$$

This is a geometric series where the constant ratio is $\frac{1}{1+r}$ and the present value of the annuity is given by its sum

$$PV = \frac{A}{r}\left[1 - \frac{1}{(1+r)^n}\right]$$
$$PV = \frac{1000}{0.1}\left[1 - \frac{1}{(1.1)^5}\right]$$

For the example above, we apply the formula and get

$$= 10000\,(0.3791)$$
$$= \$3791$$

7.7   Permutations and Combinations

A permutation is an arrangement of a given number of objects, by taking some or all of them at a time. A combination is a selection of a number of objects where the order of the selection is unimportant. Permutations and combinations are defined in terms of the factorial function, which is defined as

$$n! = n(n-1)\cdots 3.2.1.$$

Principles of Counting

(a) Suppose one operation has m possible outcomes and a second operation has n possible outcomes, then the total number of possible outcomes when performing the first operation **followed by** the second operation is $m \times n$ (**Product Rule**).

(b) Suppose one operation has m possible outcomes and a second operation has n possible outcomes, then the possible outcome of the first operation **or** the second operation is given by $m + n$ (***Sum Rule***).

Example (*Counting Principle (a)*) Suppose a dice is thrown and a coin is then tossed. How many different outcomes are there and what are they?

Solution There are six possible outcomes from a throw of the dice: 1, 2, 3, 4, 5 or 6, and two possible outcomes from the toss of a coin: H or T. Therefore, the total number of outcomes is determined from the product rule as $6 \times 2 = 12$. The outcomes are given by

$$(1, H), (2, H), (3, H), (4, H), (5, H), (6, H), (1, T), (2, T), (3, T), (4, T), (5, T), (6, T)$$

Example (*Counting Principle (b)*) Suppose a dice is thrown and if the number is even a coin is tossed and if it is odd then there is a second throw of the dice. How many different outcomes are there?

Solution There are two experiments involved with the first experiment involving an even number and a toss of a coin. There are 3 possible outcomes that result in an even number and 2 outcomes from the toss of a coin. Therefore, there are $3 \times 2 = 6$ outcomes from the first experiment.

The second experiment involves an odd number from the throw of a dice and the further throw of the dice. There are 3 possible outcomes that result in an odd number and 6 outcomes from the throw of a dice. Therefore, there are $3 \times 6 = 18$ outcomes from the second experiment.

Finally, there are 6 outcomes from the first experiment and 18 outcomes from the second experiment, and so from the sum rule there are a total of $6 + 18 = 24$ outcomes.

Pigeonhole Principle

The pigeonhole principle states that if n items are placed into m containers (with $n > m$) then at least one container must contain more than one item.

Examples (*Pigeonhole Principle*)

(a) Suppose there is a group of 367 people, then there must be at least two people with the same birthday.

   This is clear as there are 365 days in a year (with 366 days in a leap year), and so as there are at most 366 possible birthdays in a year. The group size is 367 people, and so there must be at least two people with the same birthday.[1]

(b) Suppose that a class of 102 students is assessed in an examination (the outcome from the exam is a mark between 0 and 100). Then, there are at least two students who receive the same mark.

This is clear as there are 101 possible outcomes from the test (as the mark that a student may achieve is between 0 and 100), and as there are 102 students in the class and 101 possible outcomes from the test, then there must be at least two students who receive the same mark.

Permutations

A permutation is an arrangement of a number of objects in a definite order.

Consider the three letters A, B and C. If these letters are written in a row, then there are six possible arrangements:

$$ABC\ ACB\ BAC\ BCA\ CAB\ CBA$$

There is a choice of 3 letters for the first place, then there is a choice of 2 letters for the second place, and there is only 1 choice for the third place. Therefore, there are $3 \times 2 \times 1 = 6$ arrangements.

If there are n different objects to arrange, then the total number of arrangements (permutations) of n objects is given by $n! = n(n-1)(n-2) \cdots 3.2.1$.

Consider the four letters A, B, C and D. How many arrangements (taking 2 letters at a time with no repetition) of these letters can be made?

There are 4 choices for the first letter and 3 choices for the second letter, and so there are 12 possible arrangements. These are given by

$$AB\ AC\ AD\ BA\ BC\ BD\ CA\ CB\ CD\ DA\ DB\ DC$$

The total number of arrangements of n different objects taking r at a time ($r \leq n$) is given by $^nP_r = n(n-1)(n-2) \cdots (n-r+1)$. It may also be written as

---

[1] The *birthday paradox* is the unintuitive result that in a group as small as 23 people the probability that there is a pair of people with the same birthday is above 0.5 (over 50%).

$$^nP_r = \frac{n!}{(n-r)!}$$

Example (*Permutations*) Suppose A, B, C, D, E and F are six students. How many ways can they be seated in a row if

(i)   There is no restriction on the seating.
(ii)   A and B must sit next to one another.
(iii) A and B must not sit next to one another.

Solution For unrestricted seating, the number of arrangements is given by $6.5.4.3.2.1 = 6! = 720$.

For the case where A and B must be seated next to one another, then consider A and B as one person, and then the five people may be arranged in $5! = 120$ ways. There are $2! = 2$ ways in which AB may be arranged, and so there are $2! \times 5! = 240$ arrangements.

AB	C	D	E	F

For the case where A and B must not be seated next to one another, then this is given by the difference between the total number of arrangements and the number of arrangements with A and B together, i.e. $720 - 240 = 480$.

Combinations

A combination is a selection of a number of objects in any order, and the order of the selection is unimportant, in which both **AB** and **BA** represent the same selection.

The total number of arrangements of n different objects taking r at a time is given by ^nP_r, and the number of ways that r objects can be selected from n different objects may be determined from this, since each selection may be permuted $r!$ times.

That is, the total number of arrangements is $r! \times$ total number of combinations. That is, $^nP_r = r! \times^n C_r$, and we may also write this as

$$\binom{n}{r} = \frac{n!}{r!(n-r)!} = \frac{n(n-1)\cdots(n-r+1)}{r!}$$

It is clear from the definition that

$$\binom{n}{r} = \binom{n}{n-r}$$

Example 7.1 (*Combinations*) How many ways are there to choose a team of 11 players from a panel of 15 players?

Solution Clearly, the number of ways is given by $\binom{15}{11} = \binom{15}{4}$

That is, $15.14.13.12/4.3.2.1 = 1365$.

Example 7.2 (*Combinations*) How many ways can a committee of 4 people be chosen from a panel of 10 people where

(i)   There is no restriction on membership of the panel.
(ii)   A certain person must be a member.
(iii) A certain person must not be a member.

Solution For (*i*) with no restrictions on membership, the number of selections of a committee of 4 people from a panel of 10 people is given by $\binom{10}{4} = 210$.

For (*ii*) where one person must be a member of the committee, then this involves choosing 3 people from a panel of 9 people and is given by $\binom{9}{3} = 84$.

For (*iii*) where one person must not be a member of the committee, then this involves choosing 4 people from a panel of 9 people and is given by $\binom{9}{4} = 126$.

7.8   Review Questions

1. Determine the formula for the general term and the sum of the following arithmetic sequence:

$$1, 4, 7, 10, \ldots$$

2. Write down the formula for the n-th term in the following sequence: $\frac{1}{4}$, $\frac{1}{12}$, $\frac{1}{36}$, $\frac{1}{108}$, ...

3. Find the sum of the following geometric sequence: $\frac{1}{3}, \frac{1}{6}, \frac{1}{12}, \frac{1}{24}, \ldots$
4. How many years will it take a principal of $5000 to exceed $10,000 at a constant annual growth rate of 6% compound interest?
5. What is the present value of $5000 to be received in 5 years time at a discount rate of 7%?
6. Determine the present value of a 20-year annuity of an annual payment of $5000 per year at a discount rate of 5%.
7. How many different five-digit numbers can be formed from the digits 1, 2, 3, 4, 5 where

   (i)   No restrictions on digits and repetitions allowed.
   (ii) The number is odd and no repetitions are allowed.
   (iii) The number is even and repetitions are allowed.

8. (i)   How many ways can a group of five people be selected from nine people?
   (ii) How many ways can a group be selected if two particular people are always included?
   (iii) How many ways can a group be selected if two particular people are always excluded?

7.9 Summary

This chapter provided a brief introduction to sequences and series, including arithmetic and geometric sequences, and arithmetic series and geometric series. We derived formulae for the sum of an arithmetic series and geometric series, and we discussed the convergence of a geometric series when $|r| < 1$.

We discussed the calculation of simple and compound interest, and the concept of the time value of money, and its application to determine the present value of a payment to be made in the future. We discussed annuities, which are a series of payments made at regular intervals over a period of time, and we calculated the present value of an annuity.

We considered counting principles including the product and sum rules. The product rule is concerned with where one operation has m possible outcomes and a second operation has n possible outcomes; then, the total number of possible outcomes when performing the first operation followed by the second operation is $m \times n$.

We discussed the pigeonhole principle, which states that if n items are placed into m containers (with $n > m$) then at least one container must contain more than one item. We discussed permutations and combinations where permutations are an arrangement of a given number of objects, by taking some or all of them at a time. A combination is a selection of a number of objects in any order, and the order of the selection is unimportant.

Mathematical Induction and Recursion

8

Key Topics

Mathematical induction
Strong and weak induction
Base case
Inductive step
Recursion
Recursive definition
Structural induction

8.1 Introduction

Mathematical induction is an important proof technique used in mathematics, and it is often used to establish the truth of a statement for all natural numbers. There are two parts to a proof by induction, and these are the base step and the inductive step. The *base case* involves showing that the statement is true for some natural number (usually the number 1). The second step is termed the *inductive step*, and it involves showing that if the statement is true for some natural number $n = k$, then the statement is true for its successor $n = k + 1$. This is often written as $P(k) \rightarrow P(k + 1)$.

The statement $P(k)$ that is assumed to be true when $n = k$ is termed the *inductive hypothesis*. From the base step and the inductive step, we infer that the statement is true for all natural numbers (that are greater than or equal to the number specified in

© Springer Nature Switzerland AG 2020

G. O'Regan, *Mathematics in Computing*, Undergraduate Topics
in Computer Science, https://doi.org/10.1007/978-3-030-34209-8_8

the base case). Formally, the proof technique used in mathematical induction is of the form[1]:

$$(P(1) \land (P(k) \rightarrow P(k+1))) \rightarrow \forall n P(n)$$

Mathematical induction (weak induction) may be used to prove a wide variety of theorems, and especially theorems of the form $\forall n\ P(n)$. It may be used to provide a proof of theorems about summation formulae, inequalities, set theory and the correctness of algorithms and computer programs. One of the earliest inductive proofs was the sixteenth-century proof that the sum of the first n odd integers is n^2, which was proved by Francesco Maurolico in 1575. Later mathematicians made the method of mathematical induction more precise.

We distinguish between *strong induction* and *weak induction*, where strong induction also has a base case and an inductive step, but the inductive step is a little different. It involves showing that if the statement is true for **all** natural numbers less than or equal to an arbitrary number k, then the statement is true for its successor $k + 1$. Weak induction involves showing that if the statement is true for some natural number $n = k$, then the statement is true for its successor $n = k + 1$. *Structural induction* is another form of induction and this mathematical technique is used to prove properties about recursively defined sets and structures.

Recursion is often used in mathematics to define functions, sequences and sets. However, care is required with a recursive definition to ensure that it actually defines something, and that what is defined makes sense. Recursion defines a concept in terms of itself, and we need to ensure that the definition is not circular (i.e. that it does not lead to a vicious circle).

Recursion and induction are closely related and are often used together. Recursion is extremely useful in developing algorithms for solving complex problems, and induction is a useful technique in verifying the correctness of such algorithms.

Example 8.1 Show that the sum of the first n natural numbers is given by the formula:

$$1 + 2 + 3 + \cdots + n = \frac{n(n+1)}{2}$$

Proof

Base Case

We consider the case where $n = 1$ and clearly $1 = \frac{1(1+1)}{2}$ and so the base case $P(1)$ is true.

---

[1]This definition of mathematical induction covers the base case of $n = 1$, and would need to be adjusted if the number specified in the base case is higher.

Inductive Step

Suppose the result is true for some number k then we have $P(k)$

$$1 + 2 + 3 + \cdots + k = \frac{k(k+1)}{2}$$

Then consider the sum of the first $k + 1$ natural numbers, and we use the inductive hypothesis to show that its sum is given by the formula:

$$
\begin{aligned}
& 1 + 2 + 3 + \cdots + k + (k+1) \\
&= \frac{k(k+1)}{2} + (k+1) \quad \text{(by inductive hypothesis)} \\
&= \frac{k^2 + k}{2} + \frac{(2k+2)}{2} \\
&= \frac{k^2 + 3k + 2}{2} \\
&= \frac{(k+1)(k+2)}{2}
\end{aligned}
$$

Thus, we have shown that if the formula is true for an arbitrary natural number k, then it is true for its successor $k + 1$. That is, $P(k) \rightarrow P(k + 1)$. We have shown that $P(1)$ is true, and so it follows from mathematical induction that $P(2), P(3), \ldots$ are true, and so $P(n)$ is true, for all natural numbers and the theorem is established.

Note 8.1 There are opportunities to make errors in proofs with induction, and the most common mistakes are not to complete the base case or inductive step correctly. These errors can lead to strange results and so care is required. It is important to be precise in the statements of the base case and inductive step.

Example 8.2 (*Binomial Theorem*) Prove the binomial theorem using induction (permutations and combinations were discussed in Chap. 7). That is,

$$(1+x)^n = 1 + \binom{n}{1}x + \binom{n}{2}x^2 + \ldots + \binom{n}{r}x^r + \ldots + \binom{n}{n}x^n$$

Proof

Base Case

We consider the case where $n = 1$ and clearly $(1+x)^1 = (1+x) = 1 + \binom{1}{1}x^1$

and so the base case $P(1)$ is true.

Inductive Step

Suppose the result is true for some number k then we have $P(k)$

$$(1+x)^k = 1 + \binom{k}{1}x + \binom{k}{2}x^2 + \cdots + \binom{k}{r}x^r + \cdots + \binom{k}{k}x^k$$

Then consider $(1+x)^{k+1}$ and we use the inductive hypothesis to show that it is given by the formula:

$$(1+x)^{k+1} = (1+x)^k(1+x)$$

$$= \left(1+\binom{k}{1}x+\binom{k}{2}x^2+\cdots+\binom{k}{r}x^r+\cdots+\binom{k}{k}x^k\right)(1+x)$$

$$= \left(1+\binom{k}{1}x+\binom{k}{2}x^2+\cdots+\binom{k}{r}x^r+\cdots+\binom{k}{k}x^k\right)$$

$$+x+\binom{k}{1}x^2+\cdots+\binom{k}{r}x^{r+1}+\cdots+\binom{k}{k}x^{k+1}$$

$$= \begin{array}{l} 1+\binom{k}{1}x+\binom{k}{2}x^2+\cdots+\binom{k}{r}x^r+\cdots+\binom{k}{k}x^k \\ +\binom{k}{0}x+\binom{k}{1}x^2+\cdots+\binom{k}{r-1}x^r+\cdots+\binom{k}{k-1}x^k+\binom{k}{k}x^{k+1} \end{array}$$

$$= 1+\binom{k+1}{1}x+\cdots+\binom{k+1}{r}x^r+\cdots+\binom{k+1}{k}x^k+\binom{k+1}{k+1}x^{k+1}$$

(which follows from Exercise 7 in Sect 8.5).

Thus, we have shown that if the binomial theorem is true for an arbitrary natural number k, then it is true for its successor $k + 1$. That is, $P(k) \rightarrow P(k + 1)$. We have shown that $P(1)$ is true, and so it follows from mathematical induction that $P(n)$ is true, for all natural numbers, and so the theorem is established.

The standard formula of the binomial theorem $(x + y)^n$ follows immediately from the formula for $(1 + x)^n$, by noting that $(x+y)^n = \{x(1+y/x)\}^n = x^n(1+y/x)^n$.

8.2   Strong Induction

Strong induction is another form of mathematical induction, which is often employed when we cannot prove a result with (weak) mathematical induction. It is similar to weak induction in that there is a base step and an inductive step. The base step is identical to weak mathematical induction, and it involves showing that the statement is true for some natural number (usually the number 1). The inductive step is a little different, and it involves showing that if the statement is true for all natural numbers less than or equal to an arbitrary number k, then the statement is true for its successor $k + 1$. This is often written as $(P(1) \wedge P(2) \wedge \cdots \wedge P(k)) \rightarrow P(k + 1)$.

From the base step and the inductive step, we infer that the statement is true for all natural numbers (that are greater than or equal to the number specified in the base case). Formally, the proof technique used in mathematical induction is of the form[2]:

$$(P(1) \wedge [(P(1) \wedge P(2) \wedge \cdots \wedge P(k)) \rightarrow P(k+1)]) \rightarrow \forall n P(n)$$

Strong and weak mathematical inductions are equivalent in that any proof done by weak mathematical induction may also be considered a proof using strong induction, and a proof conducted with strong induction may also be converted into a proof using weak induction.

Weak mathematical induction is generally employed when it is reasonably clear how to prove $P(k + 1)$ from $P(k)$, with strong mathematical induction typically employed where it is not so obvious. The validity of both forms of mathematical induction follows from the *well-ordering property* of the natural numbers, which states that every non-empty set has a least element.

Well-Ordering Principle

Every non-empty set of natural numbers has a least element. The well-ordering principle is equivalent to the principle of mathematical induction.

Example 8.3 Show that every natural number greater than one is divisible by a prime number.

Proof

Base Case

We consider the case of $n = 2$ which is trivially true, since 2 is a prime number and is divisible by itself.

Inductive Step (strong induction)

Suppose that the result is true for every number less than or equal to k. Then we consider $k + 1$, and there are two cases to consider. If $k + 1$ is prime then it is divisible by itself. Otherwise, it is composite and it may be factored as the product of two numbers each of which is less than or equal to k. Each of these numbers is divisible by a prime number by the strong inductive hypothesis, and so $k + 1$ is divisible by a prime number.

Thus, we have shown that if all natural numbers less than or equal to k are divisible by a prime number, then $k + 1$ is divisible by a prime number. We have shown that the base case $P(2)$ is true, and so it follows from strong mathematical induction that every natural number greater than one is divisible by some prime number.

---

[2]As before, this definition covers the base case of $n = 1$ and would need to be adjusted if the number specified in the base case is higher.

8.3   Recursion

Some functions (or objects) used in mathematics (e.g. the Fibonacci sequence) are difficult to define explicitly, and are best defined by a *recurrence relation* (i.e. an equation that recursively defines a sequence of values, once one or more initial values are defined). Recursion may be employed to define functions, sequences and sets.

There are two parts to a recursive definition, namely, the *base case* and the *recursive step*. The base case usually defines the value of the function at $n = 0$ or $n = 1$, whereas the recursive step specifies how the application of the function to a number may be obtained from its application to one or more smaller numbers.

It is important that care is taken with the recursive definition, to ensure that it is not circular, and does not lead to an infinite regress. The argument of the function on the right-hand side of the definition in the recursive step is usually smaller than the argument on the left-hand side to ensure termination (there are some unusual recursively defined functions such as the *McCarthy* 91 *function* where this is not the case).

It is natural to ask when presented with a recursive definition whether it means anything at all, and in some cases, the answer is negative. Fixed-point theory provides the mathematical foundations for recursion, and ensures that the functions/objects are well defined.

Chapter 12 (see Sect. 12.6) discusses various mathematical structures such as partial orders, complete partial orders and lattices, which may be employed to give a secure foundation for recursion. A precise mathematical meaning is given to recursively defined functions in terms of domains and fixed-point theory, and it is essential that the conditions in which recursion may be used safely be understood. The reader is referred to Meyer (1990) for more detailed information.

A recursive definition will include at least one non-recursive branch with every recursive branch occurring in a context that is different from the original, and brings it closer to the non-recursive case. Recursive definitions are a powerful and elegant way of giving the denotational semantics of language constructs.

Next, we present examples of the recursive definition of the factorial function and Fibonacci numbers.

Example 8.4 (*Recursive Definition of Functions*) The factorial function $n!$ is very common in mathematics and its well-known definition is $n! = n(n - 1)(n - 2) \cdots 3.2.1$ and $0! = 1$. The formal definition in terms of a base case and inductive step is given as follows:

$$
\begin{aligned}
&\textit{Base Step} \qquad \text{fac } (0) = 1 \\
&\textit{Recursive Step} \quad \text{fac } (n) = n^* \text{ fac } (n - 1)
\end{aligned}
$$

This recursive definition defines the procedure by which the factorial of a number is determined from the base case or by the product of the number by the factorial of its predecessor. The definition of the factorial function is built up in a sequence: fac(0), fac(1), fac(2), ...

The Fibonacci sequence[3] is named after the Italian mathematician Fibonacci, who introduced it in the thirteenth century. It had been previously described in Indian mathematics, and the Fibonacci numbers are the numbers in the following integer sequence:

$$1, 1, 2, 3, 5, 8, 13, 21, 34$$

Each Fibonacci number (apart from the first two in the sequence) is obtained by adding the two previous Fibonacci numbers in the sequence together. Formally, the definition is given by

Base Step $F_1 = 1, F_2 = 1$
Recursive Step $F_n = F_{n-1} + F_{n-2}$ (Definition for when $n > 2$)

Example 8.5 (*Recursive Definition of Sets and Structures*) Sets and sequences may also be defined recursively, and there are two parts to the recursive definition (as before). The base case specifies an initial collection of elements in the set, whereas the recursive step provides rules for adding new elements to the set based on those already there. Properties of recursively defined sets may often be proved by a technique called structural induction.

Consider the subset S of the natural numbers defined by

Base Step $5 \in S$
Recursive Step   For $x \in S$ then $x + 5 \in S$

Then the elements in S are given by the set of all multiples of 5, as clearly $5 \in S$; therefore, by the recursive step $5 + 5 = 10 \in S$, $5 + 10 = 15 \in S$ and so on.

The recursive definition of the set of strings Σ^* over an alphabet Σ is given by

Base Step $\Lambda \in \Sigma^*$ (Λ is the empty string)
Recursive Step   For $\sigma \in \Sigma^*$ and $v \in \Sigma$ then $\sigma v \in \Sigma^*$

Clearly, the empty string is created from the base step. The recursive step states that a new string is obtained by adding a letter from the alphabet to the end of an existing string in Σ^*. Each application of the inductive step produces a new string that contains one additional character. For example, if $\Sigma = \{0,1\}$ then the strings in Σ^* are the set of bit strings Λ, 0, 1, 00, 01, 10, 11, 000, 001, 010, etc.

---

[3]We are taking the Fibonacci sequence as starting at 1, whereas others take it as starting at 0.

We can define an operation to determine the length of a string (len: $\Sigma^* \to \mathbb{N}$) recursively.

Base Step $\text{len}(\Lambda) = 0$
Recursive Step $\text{len}(\sigma v) = \text{len}(\sigma) + 1$ (where $\sigma \in \Sigma^*$ and $v \in \Sigma$)

A binary tree[4] is a well-known data structure in computer science, and it consists of a root node together with left and right binary trees. A binary tree is defined as a finite set of nodes (starting with the root node), where each node consists of a data value and a link to a left subtree and a right subtree. Recursion is often used to define the structure of a binary tree.

Base Step   A single node is a binary tree (root)

Recursive Step

(i) Suppose X and Y are binary trees and x is a node then XxY is a binary tree, where X is the left subtree, Y is the right subtree and x is the new root node.
(ii) Suppose X is a binary tree and x is a node then xX and Xx are binary trees, which consist of the root node x and a single child left or right subtree.

That is, a binary tree has a root node and it may have no subtrees; it may consist of a root node with a left subtree only, a root node with a right subtree only or a root node with both left and right subtrees.

8.4 Structural Induction

Structural induction is a mathematical technique that is used to prove properties about recursively defined sets and structures. It may be used to show that all members of a recursively defined set have a certain property, and there are two parts to the proof (as before), namely, the base case and the recursive (inductive) step.

The first part of the proof is to show that the property holds for all elements specified in the base case of the recursive definition. The second part of the proof involves showing that if the property is true for all elements used to construct the new elements in the recursive definition, then the property holds for the new elements. From the base case and the recursive step, we deduce that the property holds for all elements of the set (structure).

Example 4.6 (*Structural Induction*) We gave a recursive definition of the subset S of the natural numbers that consists of all multiples of 5. We did not prove that all

---

[4]We will give an alternate definition of a tree in terms of a connected acyclic graph in Chap. 9 on graph theory.

elements of the set S are divisible by 5, and we use structural induction to prove this.

Base Step $5 \in S$ (and clearly the base case is divisible by 5).

Inductive Step Suppose $q \in S$ then $q = 5 \, k$ for some k. From the inductive
hypothesis $q + 5 \in S$ and $q + 5 = 5 \, k + 5 = 5(k + 1)$ and so $q + 5$ is
divisible by 5.

Therefore, all elements of S are divisible by 5.

8.5 Review Questions

1. Show that $9^n + 7$ is always divisible by 8.
2. Show that the sum of $1^2 + 2^2 + \cdots + n^2 = n(n + 1)(2n + 1)/6$.
3. Explain the difference between strong and weak inductions.
4. What is structural induction?
5. Explain how recursion is used in mathematics.
6. Investigate the recursive definition of the McCarthy 91 function, and explain how it differs from usual recursive definitions.
7. Show that $\dbinom{r}{r} + \dbinom{n}{r-1} = \dbinom{n+1}{r}$.
8. Determine the standard formula for the binomial theorem $(x + y)^n$ from the formula for $(1 + x)^n$.

8.6 Summary

Mathematical induction is an important proof technique that is used to establish the truth of a statement for all natural numbers. There are two parts to a proof by induction, and these are the base case and the inductive step. The base case involves showing that the statement is true for some natural number (usually for the number $n = 1$). The inductive step involves showing that if the statement is true for some natural number $n = k$, then the statement is true for its successor $n = k + 1$.

From the base step and the inductive step, we infer that the statement is true for all natural numbers (that are greater than or equal to the number specified in the base case). Mathematical induction may be used to prove a wide variety of theorems, such as theorems about summation formulae, inequalities, set theory and the correctness of algorithms and computer programs.

Strong induction is often employed when we cannot prove a result with (weak) mathematical induction. It also has a base case and an inductive step, where the inductive step is a little different, and it involves showing that if the statement is true for all natural numbers less than or equal to an arbitrary number k, then the statement is true for its successor $k + 1$.

Recursion may be employed to define functions, sequences and sets in mathematics, and there are two parts to a recursive definition, namely, the base case and the recursive step. The base case usually defines the value of the function at $n = 0$ or $n = 1$, whereas the recursive step specifies how the application of the function to a number may be obtained from its application to one or more smaller numbers. It is important that care is taken with the recursive definition, to ensure that it is not circular, and does not lead to an infinite regress.

Structural induction is a mathematical technique that is used to prove properties about recursively defined sets and structures. It may be used to show that all members of a recursively defined set have a certain property, and there are two parts to the proof, namely, the base case and the recursive (inductive) step.

Reference

Meyer B (1990) Introduction to the theory of programming languages. Prentice Hall

Graph Theory

9

9.1 Introduction

Graph theory is a practical branch of mathematics that deals with the arrangements of certain objects known as vertices (or nodes) and the relationships between them. It has been applied to practical problems such as the modelling of computer networks, determining the shortest driving route between two cities, the link structure of a website, the travelling salesman problem and the four-colour problem.[1]

Consider a map of the London underground, which is issued to users of the underground transport system in London. Then this map does not represent every feature of the city of London, as it includes only material that is relevant to the users

---

[1]The four-colour theorem states that given any map it is possible to colour the regions of the map with no more than four colours such that no two adjacent regions have the same colour. This result was finally proved in the mid-1970s.

© Springer Nature Switzerland AG 2020

141

G. O'Regan, *Mathematics in Computing*, Undergraduate Topics
in Computer Science, https://doi.org/10.1007/978-3-030-34209-8_9

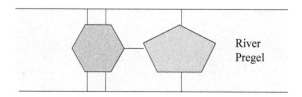

Fig. 9.1 Königsberg seven bridges problem

of the London underground transport system. In this map, the exact geographical location of the stations is unimportant, and the essential information is how the stations are interconnected to one another, as this allows a passenger to plan a route from one station to another. That is, the map of the London underground is essentially a model of the transport system that shows how the stations are interconnected.

The seven bridges of Königsberg[2] (Fig. 9.1) is one of the earliest problems in graph theory. The city was set on both sides of the Pregel River in the early eighteenth century, and it consisted of two large islands that were connected to each other and the mainland by seven bridges. The problem was to find a walk through the city that would cross each bridge once and once only.

Euler showed that the problem had no solution, and his analysis helped to lay the foundations for graph theory as a discipline. The Königsberg problem in graph theory is concerned with the question as to whether it is possible to travel along the edges of a graph starting from a vertex and returning to it and travelling along each edge exactly once. An Euler path in a graph G is a simple path containing every edge of G.

Euler noted that a walk through a graph traversing each edge exactly once depends on the *degree* of the nodes (i.e. the number of edges touching it). He showed that a necessary and sufficient condition for the walk is that the graph is connected and has zero or two nodes of odd degree. For the Königsberg graph, the four nodes (i.e. the land masses) have odd degree (Fig. 9.2).

A *graph* is a collection of objects that are interconnected in some way. The objects are typically represented by vertices (or nodes), and the interconnections between them are represented by edges (or lines). We distinguish between directed and adirected graphs, where a *directed graph* is mathematically equivalent to a binary relation, and an *adirected (undirected) graph* is equivalent to symmetric binary relations.

---

[2]Königsberg was founded in the thirteenth century by Teutonic knights and was one of the cities of the Hanseatic League. It was the historical capital of East Prussia (part of Germany), and it was annexed by Russia at the end of the Second World War. The German population either fled the advancing Red Army or were expelled by the Russians in 1949. The city is now called Kaliningrad. The famous German philosopher, Immanuel Kant, spent all his life in the city, and is buried there.

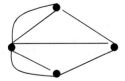

Fig. 9.2 Königsberg graph

9.2 Undirected Graphs

An *undirected graph* (*adirected graph*) (Fig. 9.3) G is a pair of finite sets (V, E) such that E is a binary symmetric relation on V. The set of vertices (or nodes) is denoted by V(G) and the set of edges is denoted by E(G).

A *directed graph* (Fig. 9.4) is a pair of finite sets (V, E) where E is a binary relation (that may not be symmetric) on V. A *directed acyclic graph* (*dag*) is a directed graph that has no cycles. The example in Fig. 9.4 is of a directed graph with three edges and four vertices.

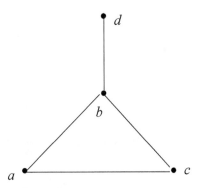

Fig. 9.3 Undirected graph

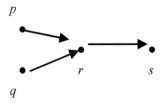

Fig. 9.4 Directed graph

An edge $e \in E$ consists of a pair $<x, y>$ where x, y are adjacent nodes in the graph. The *degree* of x is the number of nodes that is adjacent to x. The set of edges is denoted by $E(G)$ and the set of vertices is denoted by $V(G)$.

A *weighted graph* is a graph $G = (V, E)$ together with a weighting function w: $E \rightarrow \mathbb{N}$, *which* associates a weight with every edge in the graph. A weighting function may be employed in modelling computer networks, for example, the weight of an edge may be applied to model the bandwidth of a telecommunication link between two nodes. Another application of the weighting function is in determining the distance (or shortest path) between two nodes in the graph (where such a path exists).

For an adirected graph, the weight of the edge is the same in both directions, i.e. $w(v_i, v_j) = w(v_j, v_i)$ for all edges $<v_i, v_j>$ in the graph G, whereas the weights may be different for a directed graph.

Two vertices x, y are adjacent if $xy \in E$, and x and y are said to be incident to the edge xy. A matrix may be employed to represent the adjacency relationship.

Example 9.1

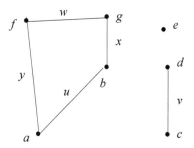

Consider the graph $G = (V, E)$ where $E = \{u = ab, \; v = cd, \; w = fg, \; x = bg, \; y = af\}$.

An adjacency matrix (Fig. 9.5) may be employed to represent the relationship of adjacency in the graph. Its construction involves listing the vertices in the rows and columns, and an entry of 1 is made in the table if the two vertices are adjacent and 0 otherwise.

Similarly, we can construct a table describing the incidence of edges and vertices by constructing an incidence matrix (Fig. 9.6). This matrix lists the vertices and edges in the rows and columns, and an entry of 1 is made if the vertex is one of the nodes of the edge and 0 otherwise.

Two graphs $G = (V, E)$ and $G' = (V', E')$ are said to be isomorphic if there exists a bijection f: $V \rightarrow V'$ such that for any $u, v \in V$, $uv \in E$, $f(u)f(v) \in E'$. The mapping f is called an isomorphism. Two graphs that are isomorphic are essentially equivalent apart from a re-labelling of the nodes and edges.

Fig. 9.5 Adjacency matrix

	a	b	c	d	e	f	g
a	0	1	0	0	0	1	0
b	1	0	0	0	0	0	1
c	0	0	0	1	0	0	0
d	0	0	1	0	0	0	0
e	0	0	0	0	0	0	0
f	1	0	0	0	0	0	1
g	0	1	0	0	0	1	0

Fig. 9.6 Incidence matrix

	u	v	w	x	y
a	1	0	0	0	1
b	1	0	0	1	0
c	0	1	0	0	0
d	0	1	0	0	0
e	0	0	0	0	0
f	0	0	1	0	1
g	0	0	1	1	0

Let $G = (V, E)$ and $G' = (V', E')$ be two graphs then G' is a *subgraph* of G if $V' \subseteq V$ and $E' \subseteq E$. Given $G = (V, E)$ and $V' \subseteq V$ then we can induce a subgraph $G' = (V', E')$ by restricting G to V' (denoted by $G|_{V'}$). The set of edges in E' is defined as

$$E' = \{e \in E : e = uv \text{ and } u, v \in V'\}$$

The *degree* of a vertex v is the number of distinct edges incident to v. It is denoted by deg v, where

$$\deg v = |\{e \in E : e = vx \text{ for some } x \in V\}|$$
$$= |\{x \in V : vx \in E\}|$$

A vertex of degree 0 is called an isolated vertex.

Theorem 9.1 Let $G = (V, E)$ be a graph, then

$$\Sigma_{v \in V}\deg v = 2|E|$$

Proof This result is clear since each edge contributes one to each of the vertex degrees. The formal proof is by induction based on the number of edges in the graph, and the basis case is for a graph with no edges (i.e. where every vertex is isolated), and the result is immediate for this case.

The inductive step (strong induction) is to assume that the result is true for all graphs with k or fewer edges. We then consider a graph $G = (V, E)$ with $k + 1$ edges.

Choose an edge $e = xy \in E$ and consider the graph $G' = (V, E')$ where $E' = E \setminus \{e\}$. Then G' is a graph with k edges, and therefore letting $\text{deg}'\, v$ represent the degree of a vertex in G', we have

$$\Sigma_{v \in V}\, \text{deg}'\, v = 2|E'| = 2(|E| - 1) = 2|E| - 2$$

The degrees of x and y are one less in G' than they are in G. That is,

$$\Sigma_{v \in V}\text{deg}\, v - 2 = \Sigma_{v \in V}\text{deg}'v = 2|E| - 2$$
$$\Rightarrow \Sigma_{v \in V}\text{deg}\, v = 2|E|$$

A graph $G = (V, E)$ is said to be *complete* if all the vertices are adjacent, i.e. $E = V \times V$. A graph $G = (V, E)$ is said to be *simple graph* if each edge connects two different vertices, and no two edges connect the same pair of vertices. Similarly, a graph that may have multiple edges between two vertices is termed as *multigraph*.

A common problem encountered in graph theory is determining whether or not there is a route from one vertex to another. Often, once a route has been identified the problem then becomes that of finding the shortest or most efficient route to the destination vertex. A graph is said to be *connected* if for any two given vertices v_1, v_2 in V there is a path from v_1 to v_2.

Consider a person walking in a forest from A to B where the person does not know the way to B. Often, the route taken will involve the person wandering around aimlessly, and often retracing parts of the route until eventually the destination B is reached. This is an example of a *walk* from v_1 to v_k where in a walk there may be repetition of edges.

If all of the edges of a walk are distinct then it is called a *trail*. A *path* $v_1, v_2, ...,$ v_k from vertex v_1 to v_k is of length $k - 1$ and consists of the sequence of edges $<v_1,$ $v_2>,<v_2, v_3>, ..., <v_{k-1}, v_k>$ where each $<v_i, v_{i+1}>$ is an edge in E. The vertices in the path are all distinct apart from possibly v_1 and v_k. The path is said to be a cycle if $v_1 = v_k$. A graph is said to be *acyclic* if it contains no cycles.

Theorem 9.2 Let $G = (V, E)$ be a graph and $W = v_1, v_2, ..., v_k$ be a walk from v_1 to v_k. Then there is a path from v_1 to v_k using only edges of W.

Proof The walk W may be reduced to a path by successively replacing redundant parts in the walk of the form $v_i, v_{i+1} ..., v_j$ where $v_i = v_j$ with v_i. That is, we successively remove cycles from the walk and this clearly leads to a path (not necessarily the shortest path) from v_1 to v_k.

Theorem 9.3 Let G = (V, E) be a graph and let $u, v \in V$ with $u \neq v$. Suppose that there exists two different paths from u to v in G, then G contains a cycle.

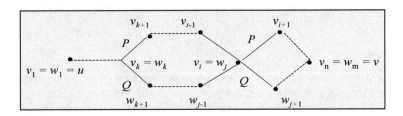

Suppose that P = $v_1, v_2, \ldots, v_n$ and Q = $w_1, w_2, \ldots, w_m$ are two distinct paths from u to v (where $u \neq v$), and $u = v_1 = w_1$ and $v = v_n = w_m$.

Suppose P and Q are identical for the first k vertices (k could be 1), and then differ (i.e. $v_{k+1} \neq w_{k+1}$). Then Q crosses P again at $v_n = w_m$, and possibly several times before then. Suppose the first occurrence is at $v_i = w_j$ with $k < i \leq n$. Then $w_k, w_{k+1}, w_{k+2}, \ldots, w_j v_{i-1}, v_{i-2}, \ldots, v_k$ is a closed path (i.e. a cycle) since the vertices are all distinct.

If there is a path from v_1 to v_2 then it is possible to define the *distance* between v_1 and v_2. This is defined to be the total length (number of edges) of the shortest path between v_1 and v_2.

9.2.1 Hamiltonian Paths

A *Hamiltonian path*[3] in a graph G = (V, E) is a path that visits every vertex once and once only. In other words, the length of a Hamiltonian path is $|V| - 1$. A graph is Hamiltonian-connected if for every pair of vertices, there is a Hamiltonian path between the two vertices.

Hamiltonian paths are applicable to the travelling salesman problem, where a salesman[4] wishes to travel to k cities in the country without visiting any city more than once. In principle, this problem may be solved by looking at all of the possible routes between the various cities, and choosing the route with the minimal distance.

For example, Fig. 9.7 shows five cities and the connections (including distance) between them. Then, a travelling salesman starting at A would visit the cities in the order AEDCBA (or in reverse order ABCDEA) covering a total distance of 14.

However, the problem becomes much more difficult to solve as the number of cities increase, and there is no general algorithm for its solution. For example, for

[3]These are named after Sir William Rowan Hamilton, a nineteenth-century Irish mathematician and astronomer, who is famous for discovering quaternions discussed in Chap. 26.
[4]We use the term 'salesman' to stand for 'salesman' or 'saleswoman'.

Fig. 9.7 Travelling salesman
problem

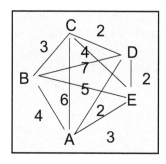

the case of ten cities, the total number of possible routes is given by 9! = 362,880,
and an exhaustive search by a computer is feasible and the solution may be
determined quite quickly. However, for 20 cities, the total number of routes is given
by 19! = 1.2×10^{17}, and, in this case, it is no longer feasible to do an exhaustive
search by a computer.

There are several sufficient conditions for the existence of a Hamiltonian path,
and Theorem 9.4 describes one condition that is sufficient for its existence.

Theorem 9.4 Let G = (V, E) be a graph with $|V| = n$ and such that deg v + deg
$w \geq n - 1$ for all non-adjacent vertices v and w. Then G possesses a Hamiltonian
path.

Proof The first part of the proof involves showing that G is connected, and the
second part involves considering the largest path in G of length $k - 1$ and assuming
that $k < n$. A contradiction is then derived and it is deduced that $k = n$.

We assume that G' = (V', E') and G'' = (V'', E'') are two connected components
of G, then $|V'| + |V''| \leq n$ and so if $v \in V'$ and $w \in V''$ then $n - 1 \leq$ deg v + deg
$w \leq |V'| - 1 + |V''| - 1 = |V'| + |V''| - 2 \leq n - 2$ which is a contradiction, and
so G must be connected.

Let P = $v_1, v_2, ..., v_k$ be the largest path in G and suppose $k < n$. From this, a
contradiction is derived, and the details are in (Piff 1991).

9.3 Trees

An acyclic graph is termed a *forest* and a connected forest is termed a *tree*. A graph
G is a tree if and only if for each pair of vertices in G there exists a unique path in G
joining these vertices. This is since G is connected and acyclic, with the connected
property giving the existence of at least one path and the acyclic property giving
uniqueness.

A *spanning tree* T = (V, E') for the connected graph G = (V, E) is a tree with the
same vertex set V. It is formed from the graph by removing edges from it until it is
acyclic (while ensuring that the graph remains connected).

Theorem 9.5 Let G = (V, E) be a tree and let $e \in$ E then G′ = (V, E\{e}) is disconnected and has two components.

Proof Let $e = uv$ then since G is connected and acyclic uv is the unique path from u to v, and thus G′ is disconnected since there is no path from u to v in G′.

It is thus clear that there are at least two components in G′ with u and v in different components. We show that any other vertex w is connected to u or to v in G′. Since G is connected there is a path from w to u in G, and if this path does not use e then it is in G′ as well, and therefore u and w are in the same component of G′.

If it does use e then e is the last edge of the graph since u cannot appear twice in the path, and so the path is of the form w, …, v, u in G. Therefore, there is a path from w to v in G′, and so w and v are in the same component in G′. Therefore, there are only two components in G′.

Theorem 9.6 Any connected graph G = (V, E) possesses a spanning tree.

Proof This result is proved by considering all connected subgraphs of (G = V, E) and choosing a subgraph T with |E′| as small as possible. The final step is to show that T is the desired spanning tree, and this involves showing that T is acyclic. The details of the proof are left to the reader.

Theorem 9.7 Let G = (V, E) be a connected graph, then G is a tree if and only if | E| = |V| − 1.

Proof This result may be proved by induction on the number of vertices |V| and the applications of Theorems 9.5 and 9.6.

9.3.1 Binary Trees

A *binary tree* (Fig. 9.8) is a tree in which each node has at most two child nodes (termed left and right child nodes). A node with children is termed a *parent node*, and the top node of the tree is termed the root node. Any node in the tree can be reached by starting from the root node, and by repeatedly taking either the left branch (left child) or right branch (right child) until the node is reached. Binary trees are used in computing to implement efficient searching algorithms. (We gave an alternative recursive definition of a binary tree in Chap. 8.)

The *depth* of a node is the length of the path (i.e. the number of edges) from the root to the node. The depth of a tree is the length of the path from the root to the deepest node in the tree. A *balanced* binary tree is a binary tree in which the depth of the two subtrees of any node never differs by more than one. The root of the binary tree in Fig. 9.8 is A and its depth is 4. The tree is unbalanced and unsorted.

Fig. 9.8 Binary tree

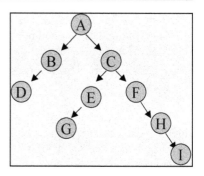

Tree traversal is a systematic way of visiting each node in the tree exactly once, and we distinguish between *breadth-first search* in which every node on a particular level is visited before going to a lower level, and *depth-first search* where one starts at the root and explores as far as possible along each branch before backtracking. The traversal in depth-first search may be in preorder, inorder or postorder.

9.4 Graph Algorithms

Graph algorithms are employed to solve various problems in graph theory including network cost minimization problems; construction of spanning trees; shortest path algorithms; longest path algorithms and timetable construction problems.

A length function l: $E \rightarrow \mathbb{R}$ may be defined on the edges of a connected graph $G = (V, E)$, and a shortest path from u to v in G is a path P with edge set E' such that $l(E')$ is minimal.

The reader may consult the many texts on graph theory to explore many well-known graph algorithms. These include Dijkstra's shortest path algorithm and longest path algorithm, Kruskal's minimal spanning tree algorithm and Prim's minimal spanning tree algorithms which are all described in (Piff 1991). Next, we briefly discuss graph colouring.

9.5 Graph Colouring and Four-Colour Problem

It is very common for maps to be coloured in such a way that neighbouring states or countries are coloured differently. This allows different states or countries to be easily distinguished as well as the borders between them. The question naturally arises as to how many colours are needed (or determining the least number of colours needed) to colour the entire map, as it might be expected that a large number of colours would be needed to colour a large complicated map.

However, it may come as a surprise that, in fact, very few colours are required to colour any map. A former student of the British logician, Augustus De Morgan, had noticed this in the mid-1800s, and he proposed the conjecture of the four-colour theorem. There were various attempts to prove that four colours were sufficient from the mid-1800s onwards, and it remained a famous unsolved problem in mathematics until the late twentieth century.

Kempe gave an erroneous proof of the four-colour problem in 1879, but his attempt led to the proof that five colours are sufficient (which was proved by Heawood in the late 1800s). Appel and Haken of the University of Illinois finally provided the proof that four colours are sufficient in the mid-1970s (using over 1000 h of computer time in their proof).

Each map in the plane can be represented by a graph, with each region of the graph represented by a vertex. Edges connect two vertices if the regions have a common border. The colouring of a graph is the assignment of a colour to each vertex of the graph so that no two adjacent vertices in this graph have the same colour.

Definition Let G = (V, E) be a graph and let C be a finite set called the colours. Then, a colouring of G is a mapping $\kappa{:}V \rightarrow C$ such that if $uv \in E$ then $\kappa(u) \neq \kappa(v)$.

That is, the colouring of a simple graph is the assignment of a colour to each vertex of the graph such that if two vertices are adjacent then they are assigned a different colour. The chromatic number of a graph is the least number of colours needed for colouring the graph. It is denoted by $\chi(G)$.

Example 9.2 Show that the chromatic colour of the following graph G is 3 (this example is adapted from [Ros:12]) (Fig. 9.9).

Solution The chromatic colour of G must be at least three since vertices p, q and r must have different colours, and so we need to show that three colours are, in fact, sufficient to colour G. We assign the colours red, blue and green to p, q and r, respectively. We immediately deduce that the colour of s must be red (as adjacent to q and r). From this, we deduce that t is coloured green (as adjacent to q and s) and u is coloured blue (as adjacent to s and t). Finally, v must be coloured red (as adjacent to u and t). This leads to the colouring of the graph G in Fig. 9.10.

Theorem 9.8 (Four-Colour Theorem) The chromatic number of a planar graph G is less than or equal to 4.

Fig. 9.9 Determining the chromatic colour of G

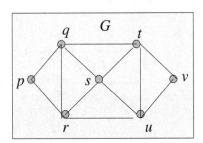

Fig. 9.10 Chromatic
colouring of G

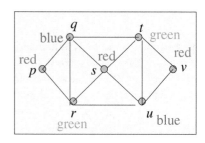

9.6 Review Questions

1. What is a graph and explain the difference between an adirected graph and a directed graph.
2. Determine the adjacency and incidence matrices of the following graph where $V = \{a, b, c, d, e\}$ and $E = \{ab, bc, ae, cd, bd\}$.
3. Determine if the two graphs G and G' defined below are isomorphic.
   $G = (V, E), V = \{a, b, c, d, e, f, g\}$ and $E = \{ab, ad, ae, bd, ce, cf, dg, fg, bf\}$
   $G' = (V', E'), V' = \{a, b, c, d, e, f, g\}$ and $E' = \{ab, bc, cd, de, ef, fg, ga, ac, be\}$.
4. What is a binary tree? Describe applications of binary trees.
5. Describe the travelling salesman problem and its applications.
6. Explain the difference between a walk, trail and path.
7. What is a connected graph?
8. Explain the difference between an incidence matrix and an adjacency matrix.
9. Complete the details of Theorems 9.6 and 9.7.
10. Describe the four-colour problem and its applications.

9.7 Summary

This chapter provided a brief introduction to graph theory, which is a practical branch of mathematics that deals with the arrangements of vertices and the edges between them. It has been applied to practical problems such as the modelling of computer networks, determining the shortest driving route between two cities and the travelling salesman problem.

The seven bridges of Königsberg is one of the earliest problems in graph theory, and it was concerned with the problem was of finding a walk through the city that would cross each bridge once and once only. Euler showed that the problem had no solution.

An undirected graph G is a pair of finite sets (V, E) such that E is a binary symmetric relation on V, whereas a directed graph is a binary relation that is not symmetric. An adjacency matrix is used to represent whether two vertices are adjacent to each other, whereas an incidence matrix indicates whether a vertex is part of a particular edge.

A Hamiltonian path in a graph is a path that visits every vertex once and once only. Hamiltonian paths are applicable to the travelling salesman problem, where a salesman wishes to travel to k cities in the country without visiting any city more than once.

Graph colouring arose to answer the question as to how many colours are needed to colour an entire map. It may be expected that many colours would be required, but the four-colour theorem demonstrates that, in fact, four colours are sufficient to colour a planar graph.

A tree is a connected and acyclic graph, and a binary tree is a tree in which each node has at most two child nodes.

Reference

Piff M (1991) Discrete mathematics: an introduction for software engineers. Cambridge University Press

Cryptography

10

Key Topics

Caesar cipher
Enigma codes
Bletchley Park
Turing
Public and private keys
Symmetric keys
Block ciphers
RSA

10.1 Introduction

Cryptography was originally employed to protect communication of private information between individuals. Today, it consists of mathematical techniques that provide secrecy in the transmission of messages between computers, and its objective is to solve security problems such as privacy and authentication over a communication channel.

It involves enciphering and deciphering messages, and it employs theoretical results from number theory (see Chap. 5) to convert the original message (or plaintext) into ciphertext that is then transmitted over a secure channel to the intended recipient. The ciphertext is meaningless to anyone other than the intended recipient, and the recipient uses a key to decrypt the received ciphertext and to read the original message.

© Springer Nature Switzerland AG 2020
G. O'Regan, *Mathematics in Computing*, Undergraduate Topics
in Computer Science, https://doi.org/10.1007/978-3-030-34209-8_10

The origin of the word 'cryptography' is from the Greek words '*kryptos*', meaning hidden and '*graphein*', meaning to write. The field of cryptography is concerned with techniques by which information may be concealed in ciphertexts and made unintelligible to all but the intended recipient. This ensures the privacy of the information sent, as any information intercepted will be meaningless to anyone other than the authorized recipient.

Julius Caesar developed one of the earliest ciphers on his military campaigns in Gaul (see Chap. 4). His objective was to communicate important messages safely to his generals. His solution is one of the simplest and widely known encryption techniques, and it involves the substitution of each letter in the plaintext (i.e. the original message) by a letter a fixed number of positions down in the alphabet. The Caesar cipher involves a shift of three positions and this leads to the letter B being replaced by E, the letter C by F and so on.

The Caesar cipher is easily broken, as the frequency distribution of letters may be employed to determine the mapping. However, the Gaulish tribes were mainly illiterate, and so it is highly likely that the cipher provided good security. The translation of the Roman letters by the Caesar cipher (with a shift key of 3) can be seen in Fig. 4.3.

The process of enciphering a message (i.e. the plaintext) simply involves going through each letter in the plaintext and writing down the corresponding cipher letter. The enciphering of the plaintext message 'summer' involves the following:

Plaintext: summer
Cipher Text: vxpphu

The process of deciphering a cipher message involves doing the reverse operation, i.e. for each cipher letter, the corresponding plaintext letter is identified from the table.

Cipher Text: vxpphu
Plaintext: summer

The encryption may also be represented using modular arithmetic. This involves using the numbers 0–25 to represent the alphabet letters, and the encryption of a letter is given by a shift transformation of three (modulo 26). This is simply addition (modulo 26), i.e. the encoding of the plaintext letter x is given by

$$x + 3 (\mathrm{mod}\, 26) = a$$

Similarly, the decoding of the cipher letter a is given by

$$a - 3 (\mathrm{mod}\, 26) = x$$

The Caesar cipher was still in use up to the early twentieth century. However, by then frequency analysis techniques were available to break the cipher. The Vigenère cipher uses a Caesar cipher with a different shift at each position in the text. The value of the shift to be employed with each plaintext letter is defined using a repeating keyword.

10.2 Breaking the Enigma Codes

The Enigma codes were used by the Germans during the Second World War for the secure transmission of naval messages to their submarines. These messages contained top-secret information on German submarine and naval activities in the Atlantic, and the threat that they posed to British and Allied shipping.

The codes allowed messages to be passed secretly using encryption, and this meant that any unauthorized interception was meaningless to the Allies. The plaintext (i.e. the original message) was converted by the Enigma machine (Fig. 10.1) into the encrypted text, and these messages were then transmitted by the German military to their submarines in the Atlantic, or to their bases throughout Europe.

The Enigma cipher was invented in 1918 and the Germans believed it to be unbreakable. A letter was typed in German into the machine, and electrical impulses through a series of rotating wheels and wires produced the encrypted letter which was lit up on a panel above the keyboard. The recipient typed the received message into his machine and the decrypted message was lit up letter by letter above the keyboard. The rotors and wires of the machine could be configured in many different ways, and during the war, the cipher settings were changed at least once a day. The odds against anyone breaking the Enigma machine without knowing the setting were 150×10^{18} to 1.

The British code and cipher school was relocated from London to Bletchley Park at the start of the Second World War (Fig. 10.2). It was located in the town of Bletchley (about 50 miles northwest of London). It was commanded by Alistair

Fig. 10.1 The Enigma machine

Fig. 10.2 Bletchley Park

Dennison and was known as Station X, and several thousands were working there during the Second World War. The team at Bletchley Park broke the Enigma codes, and therefore made vital contributions to the British and Allied war effort.

Polish cryptanalysts did important work in breaking the Enigma machine in the early 1930s, and they constructed a replica of the machine. They passed their knowledge on to the British and gave them the replica just prior to the German invasion of Poland. The team at Bletchley built upon the Polish work, and the team included Alan Turing[1] (Fig. 10.3) and other mathematicians.

The codebreaking teams worked in various huts in Bletchley Park. Hut 6 focused on air force and army ciphers and hut 8 focused on naval ciphers. The deciphered messages were then converted into intelligence reports, with air force and army intelligence reports produced by the team in hut 3 and naval intelligence reports produced by the team in hut 4. The raw material (i.e. the encrypted messages) to be deciphered came from wireless intercept stations dotted around Britain, and from various countries overseas. These stations listened to German radio messages, and sent them to Bletchley Park to be deciphered and analysed.

Turing devised a machine to assist with breaking the codes (an idea that was originally proposed by the Polish cryptanalysts). This electromechanical machine was known as the bombe (Fig. 10.4), and its goal was to find the right settings of the Enigma machine for that particular day. The machine greatly reduced the odds and the time required to determine the settings on the Enigma machine, and it became the main tool for reading the Enigma traffic during the war. The bombe was first installed in early 1940 and it weighed over a tonne. It was named after a cryptological device designed in 1938 by the Polish cryptologist, Marian Rejewski.

A standard Enigma machine employed a set of rotors, and each rotor could be in any of 26 positions. The bombe tried each possible rotor position and applied a test. The test eliminated almost all of the positions and left a smaller number of cases to be dealt with. The test required the cryptologist to have a suitable '*crib*', i.e. a section of ciphertext for which he could guess the corresponding plaintext.

---

[1]Turing made fundamental contributions to computing, including the theoretical Turing machine.

Fig. 10.3 Alan Turing

Fig. 10.4 Replica of Bombe

For each possible setting of the rotors, the bombe employed the crib to perform a chain of logical deductions. The bombe detected when a contradiction had occurred and it then ruled out that setting and moved on to the next. Most of the possible settings would lead to contradictions and could then be discarded. This would leave only a few settings to be investigated in detail.

The Government Communication Headquarters (GCHQ) was the successor of Bletchley Park, and it relocated to Cheltenham after the war. The site at Bletchley Park was then used for training purposes.

The codebreakers who worked at Bletchley Park were required to remain silent about their achievements until the mid-1970s when the wartime information was declassified. The link between British Intelligence and Bletchley Park came to an end in the mid-1980s.

It was decided in the mid-1990s to restore Bletchley Park, and today it is run as a museum by the Bletchley Park Trust.

10.3 Cryptographic Systems

A cryptographic system is a computer system that is concerned with the secure transmission of messages. The message is encrypted prior to its transmission, which ensures that any unauthorized interception and viewing of the message is meaningless to anyone other than the intended recipient. The recipient uses a key to decrypt the ciphertext, and to retrieve the original message.

There are essentially two different types of cryptographic systems employed, and these are public-key cryptosystems and secret-key cryptosystems. A *public-key cryptosystem* is an asymmetric cryptosystem where two different keys are employed: one for encryption and one for decryption. The fact that a person is able to encrypt a message does not mean that the person is able to decrypt a message.

In a *secret-key cryptosystem*, the same key is used for both encryption and decryption. Anyone who has knowledge of how to encrypt messages has sufficient knowledge to decrypt messages, and the sender and receiver need to agree on a shared key prior to any communication. The following notation is employed (Table 10.1).

Table 10.1 Notation in cryptography

Symbol	Description
M	Represents the message (plaintext)
C	Represents the encrypted message (ciphertext)
e_k	Represents the encryption key
d_k	Represents the decryption key
E	Represents the encryption process
D	Represents the decryption process

The encryption and decryption algorithms satisfy the following equation:

$$Dd_k(C) = Dd_k(Ee_k(M)) = M$$

There are two different keys employed in a public-key cryptosystem. These are the encryption key e_k and the decryption key d_k with $e_k \neq d_k$. It is called asymmetric since the encryption key differs from the decryption key.

There is just one key employed in a secret-key cryptosystem, with the same key e_k is used for both encryption and decryption. It is called *symmetric* since the encryption key is the same as the decryption key, i.e. $e_k = d_k$.

10.4   Symmetric-Key Systems

A symmetric-key cryptosystem (Fig. 10.5) uses the same secret key for encryption and decryption. The sender and the receiver first need to agree a shared key prior to communication. This needs to be done over a secure channel to ensure that the shared key remains secret. Once this has been done, they can begin to encrypt and decrypt messages using the secret key. Anyone who is able to encrypt a message has sufficient information to decrypt the message.

The encryption of a message is in effect a transformation from the space of messages M to the space of cryptosystems $\mathbb{C}$. That is, the encryption of a message with key k is an invertible transformation f, such that

$$f : M \xrightarrow{k} \mathbb{C}$$

The ciphertext is given by $C = E_k(M)$, where $M \in M$ and $C \in \mathbb{C}$. The legitimate receiver of the message knows the secret key k (as it will have been transmitted

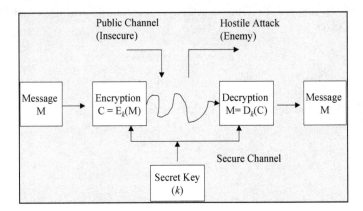

Fig. 10.5 Symmetric-key cryptosystem

Table 10.2 Advantages and disadvantages of symmetric-key systems

Advantages	Disadvantages
Encryption process is simple (as the same key is used for encryption and decryption)	A shared key must be agreed between two parties
It is faster than public-key systems	Key exchange is difficult as there needs to be a secure channel between the two parties (to ensure that the key remains secret)
It uses less computer resources than public-key systems	If a user has n trading partners then n secret keys must be maintained (one for each partner)
It uses a different key for communication with every different party	There are problems with the management and security of all of these keys (due to volume of keys that need to be maintained)
	Authenticity of origin or receipt cannot be proved (as key is shared)

previously over a secure channel), and so the ciphertext C can be decrypted by the inverse transformation f^{-1} defined by

$$f^{-1} : \mathbb{C} \xrightarrow{k} \mathsf{M}$$

Therefore, we have that $D_k(C) = D_k (E_k(M)) = M$, the original plaintext message.

There are advantages and disadvantages to symmetric-key systems (Table 10.2), and these include the following.

Examples of Symmetric-Key Systems

(i)   *Caesar Cipher*

The Caesar cipher may be defined using modular arithmetic. It involves a shift of three places for each letter in the plaintext, and the alphabetic letters are represented by the numbers 0–25. The encryption is carried out by addition (modulo 26). The encryption of a plaintext letter x to a cipher letter c is given by[2]

$$c = x + 3 (\mathrm{mod}\, 26)$$

Similarly, the decryption of a cipher letter c is given by

$$x = c - 3 (\mathrm{mod}\, 26)$$

---

[2]Here x and c are variables rather than the alphabetic characters '*x*' and '*c*'.

(ii)   *Generalized Caesar Cipher*
This is a generalization of the Caesar cipher to a shift of k (the Caesar cipher involves a shift of three). This is given by

$$f_k = E_k(x) \equiv x + k \pmod{26} \quad 0 \le k \le 25$$
$$f_k^{-1} = D_k(c) \equiv c - k \pmod{26} \quad 0 \le k \le 25$$

(iii) *Affine Transformation*
This is a more general transformation and is defined by

$$f_{(a,b)} = E_{(a,b)}(x) \equiv ax + b \pmod{26} \quad 0 \le a,b,x \le 25 \text{ and } \gcd(a,26) = 1$$
$$f_{(a,b)}^{-1} = D_{(a,b)}(c) \equiv a^{-1}(c - b) \pmod{26} \quad a^{-1} \text{ is the inverse of } a \bmod 26$$

(iv)   *Block Ciphers*
Stream ciphers encrypt a single letter at a time and are easy to break. Block ciphers offer greater security, and the plaintext is split into groups of letters, and the encryption is performed on the block of letters rather than on a single letter.

The message is split into blocks of n-letters: $M_1, M_2, \ldots, M_k$, where each M_i $(1 \le i \le k)$ is a block n letters. The letters in the message are translated into their numerical equivalents, and the ciphertext is formed as follows:

$$C_i \equiv AM_i + B \pmod{N} \quad i = 1, 2, \ldots k$$

$$\begin{pmatrix} a_{11} & a_{12} & a_{13} & \cdots & a_{1n} \\ a_{21} & a_{22} & a_{23} & \cdots & a_{2n} \\ a_{31} & a_{32} & a_{33} & \cdots & a_{3n} \\ \cdots & \cdots & \cdots & \cdots & \cdots \\ \cdots & \cdots & \cdots & \cdots & \cdots \\ a_{n1} & a_{n2} & a_{n3} & \cdots & a_m \end{pmatrix} \begin{pmatrix} m_1 \\ m_2 \\ m_3 \\ \cdots \\ \cdots \\ m_n \end{pmatrix} + \begin{pmatrix} b_1 \\ b_2 \\ b_3 \\ \cdots \\ \cdots \\ b_n \end{pmatrix} = \begin{bmatrix} c_1 \\ c_2 \\ c_3 \\ \cdots \\ \cdots \\ c_n \end{bmatrix}$$

where (A, B) is the key, A is an invertible $n \times n$ matrix with $\gcd(\det(A), N) = 1$,[3] $M_i = (m_1, m_2, \ldots, m_n)^T$, $B = (b_1, b_2, \ldots, b_n)^T$, $C_i = (c_1, c_2, \ldots, c_n)^T$. The decryption is performed by

$$M_i \equiv A^{-1}(C_i - B) \pmod{N} \quad i = 1, 2, \ldots, k$$

---

[3] This requirement is to ensure that the matrix A is invertible. Matrices are discussed in Chap. 14.

$$\begin{pmatrix} m_1 \\ m_2 \\ m_3 \\ \cdots \\ \cdots \\ m_n \end{pmatrix} = \begin{pmatrix} a_{11} & a_{12} & a_{13} & \cdots & a_{1n} \\ a_{21} & a_{22} & a_{23} & \cdots & a_{2n} \\ a_{31} & a_{32} & a_{33} & \cdots & a_{3n} \\ \cdots & \cdots & \cdots & \cdots & \cdots \\ \cdots & \cdots & \cdots & \cdots & \cdots \\ a_{n1} & a_{n2} & a_{n3} & \cdots & a_{nn} \end{pmatrix}^{-1} \begin{pmatrix} c_1 - b_1 \\ c_2 - b_2 \\ c_3 - b_3 \\ \cdots \\ \cdots \\ c_n - b_n \end{pmatrix}$$

(v)   *Exponential Ciphers*

Pohlig and Hellman (1978) invented the exponential cipher in 1976. This cipher is less vulnerable to frequency analysis than block ciphers.

Let p be a prime number and let M be the numerical representation of the plaintext, with each letter of the plaintext replaced with its two-digit representation (00–25). That is, A = 00, B = 01, ..., Z = 25.

M is divided into blocks M_i (these are equal size blocks of m letters where the block size is approximately the same number of digits as p). The number of letters m per block is chosen, such that

$$\underbrace{2525...25}_{m\ \text{times}} < p < \underbrace{2525...25}_{m+1\ \text{times}}$$

For example, for the prime 8191, a block size of $m = 2$ letters (4 digits) is chosen, since

$$2525 < 8191 < 252525$$

The secret encryption key is chosen to be an integer k such that $0 < k < p$ and $\gcd(k, p - 1) = 1$. Then the encryption of the block M_i is defined by

$$C_i = E_k(M_i) \equiv M_i^k (\bmod\, p)$$

The ciphertext C_i is an integer, such that $0 \le C_i < p$.

The decryption of C_i involves first determining the inverse k^{-1} of the key $k\ (\bmod\, p - 1)$, i.e. we determine k^{-1} such that $k \cdot k^{-1} \equiv 1\ (\bmod\, p - 1)$. The secret key k was chosen so that $(k, p - 1) = 1$, and this means that there are integers d and n such that $kd = 1 + n(p - 1)$, and so k^{-1} is d and $kk^{-1} = 1 + n(p - 1)$. Therefore,

$$D_{k^{-1}}(C_i) \equiv C_i^{k^{-1}} \equiv (M_i^k)^{k^{-1}} \equiv M_i^{1+n(p-1)} \equiv M_i (\bmod\, p)$$

The fact that $M_i^{1+n(p-1)} \equiv M_i\ (\bmod\ p)$ follows from Euler's theorem and Fermat's little theorem (Theorems 5.7 and 5.8), which were discussed in

Chap. 5. Euler's theorem states that for two positive integers a and n with $\gcd(a, n) = 1$ that $a^{\phi(n)} \equiv 1 \pmod{n}$.

Clearly, for a prime p, we have that $\phi(p) = p - 1$. This allows us to deduce that

$$M_i^{1+n(p-1)} \equiv M_i^1 M_i^{n(p-1)} \equiv M_i\left(M_i^{(p-1)}\right)^n \equiv M_i(1)^n \equiv M_i \pmod{p}$$

(vi)   *Data Encryption Standard (DES)*

DES is a popular cryptographic system (Data Encryption Standard 1977) used by governments and private companies around the world. It is based on a symmetric-key algorithm and uses a shared secret key that is known only to the sender and receiver. It was designed by IBM and approved by the National Bureau of Standards (NBS[4]) in 1976. It is a block cipher and a message is split into 64-bit message blocks. The algorithm is employed in reverse to decrypt each ciphertext block.

Today, DES is considered to be insecure for many applications as its key size (56 bits) is viewed as being too small, and the cipher has been broken in less than 24 h. This has led to it being withdrawn as a standard and replaced by the Advanced Encryption Standard (AES), which uses a larger key of 128 bits or 256 bits.

The DES algorithm uses the same secret 56-bit key for encryption and decryption. The key consists of 56 bits taken from a 64-bit key that includes 8 parity bits. The parity bits are at positions 8, 16, ..., 64, and so every eighth bit of the 64-bit key is discarded leaving behind only the 56-bit key.

The algorithm is then applied to each 64-bit message block and the plaintext message block is converted into a 64-bit ciphertext block. An initial permutation is first applied to M to create M', and M' is divided into a 32-bit left half L_0 and a 32-bit right half R_0. There are then 16 iterations, with the iterations having a left half and a right half:

$$L_i = R_{i-1}$$
$$R_i = L_{i-1} \oplus f(R_{i-1}, K_i)$$

The function f is a function that takes a 32-bit right half and a 48-bit round key K_i (each K_i contains a different subset of the 56-bit key) and produces a 32-bit output. Finally, the pre-ciphertext (R_{16}, L_{16}) is permuted to yield the final ciphertext C. The function f operates on half a message block (Table 10.3) and involves the following:

The decryption of the ciphertext is similar to the encryption and it involves running the algorithm in reverse.

---

[4]The NBS is now known as the National Institute of Standards and Technology (NIST).

Table 10.3 DES encryption

Step	Description
1.	Expansion of the 32-bit half block to 48 bits (by duplicating half of the bits)
2.	The 48-bit result is combined with a 48-bit subkey of the secret key using an XOR operation
3.	The 48-bit result is broken into 8 * 6 bits and passed through 8 substitution boxes to yield 8 * 4 = 32 bits (this is the core part of the encryption algorithm)
4.	The 32-bit output is rearranged according to a fixed permutation

DES has been implemented on a microchip. However, it has been superseded in recent years by AES due to security concerns with its small 56-bit key size. The AES uses a key size of 128 bits or 256 bits.

10.5   Public-Key Systems

A public-key cryptosystem (Fig. 10.6) is an asymmetric-key system where there is a separate key e_k for encryption and d_k decryption with $e_k \neq d_k$. Martin Hellman and Whitfield Diffie invented it in 1976. The fact that a person is able to encrypt a message does not mean that the person has sufficient information to decrypt messages.

The public-key cryptosystem is based on the following (Table 10.4):

The advantages and disadvantages of public-key cryptosystems include (Table 10.5).

The implementation of public-key cryptosystems is based on *trapdoor one-way functions*. A function $f: X \rightarrow Y$ is a trapdoor one-way function if

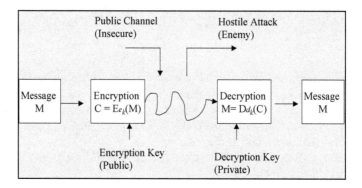

Fig. 10.6 Public-key cryptosystem

Table 10.4 Public-key encryption system

Item	Description
1.	It uses the concept of a key pair (e_k, d_k)
2.	One half of the pair can encrypt messages and the other half can decrypt messages
3.	One key is private and one key is public
4.	The private key is kept secret and the public key is published (but associated with trading partner)
5.	The key pair is associated with exactly one trading partner

Table 10.5 Advantages and disadvantages of public-key cryptosystems

Advantages	Disadvantages
Only the private key needs to be kept secret	Public keys must be authenticated
The distribution of keys for encryption is convenient as everyone publishes their public key and the private key is kept private	It is slow and uses more computer resources
It provides message authentication as it allows the use of digital signatures (which enables the recipient to verify that the message is really from the particular sender)	Security compromise is possible (if private key compromised)
The sender encodes with the private key that is known only to sender. The receiver decodes with the public key and therefore knows that the message is from the sender	Loss of private key may be irreparable (unable to decrypt messages)
Detection of tampering (digital signatures enable the receiver to detect whether message was altered in transit)	
Provides for non-repudiation	

- f is easy to computer,
- f^{-1} is difficult to compute and
- f^{-1} is easy to compute if a trapdoor (secret information associated with the function) becomes available.

A function satisfying just the first two conditions above is termed a *one-way function*.

Examples of Trapdoor and One-way Functions

(i) The function $f\colon pq \to n$ (where p and q are primes) is a one-way function since it is easy to compute. However, the inverse function f^{-1} is difficult to compute for large n since there is no efficient algorithm to factorize a large integer into its prime factors (*integer factorization problem*).

(ii) The function $f_{g,N}\colon x \to g^x \pmod{N}$ is a one-way function since it is easy to compute. However, the inverse function f^{-1} is difficult to compute as there is no efficient method to determine x from the knowledge of g^x (mod N) and g and N (*the discrete logarithm problem*).

(iii)   The function $f_{k,N}: x \rightarrow x^k$ (mod N) (where N = pq and p and q are primes) and $kk' \equiv 1$ (mod $\varphi(n)$) is a trapdoor function. It is easy to compute but the inverse of f (the kth root modulo N) is difficult to compute. However, if the trapdoor k' is given then f can easily be inverted as $(x^k)^{k'} \equiv x$ (mod N).

10.5.1   RSA Public-Key Cryptosystem

Rivest, Shamir and Adleman proposed a practical public-key cryptosystem (RSA) based on primality testing and integer factorization in the late 1970s. The RSA algorithm was filed as a patent (Patent No. 4,405,829) at the U.S. Patent Office in December 1977. The RSA public-key cryptosystem is based on the following assumptions:

–   It is straightforward to find two large prime numbers.
–   The integer factorization problem is infeasible for large numbers.

The algorithm is based on mod n arithmetic where n is a product of two large prime numbers.

The encryption of a plaintext message M to produce the ciphertext C is given by

$$C \equiv M^e (\mathrm{mod}\, n)$$

where e is the public encryption key, M is the plaintext, **C** is the ciphertext and n is the product of two large primes p and q. Both e and n are made public, and e is chosen such that $1 < e < \phi(n)$, where $\phi(n)$ is a number of positive integers that are relatively prime to n.

The ciphertext C is decrypted by

$$M \equiv C^d (\mathrm{mod}\, n)$$

where d is the private decryption key that is known only to the receiver, and $ed \equiv 1$ (mod $\phi(n)$) and d and $\phi(n)$ are kept private.

The calculation of $\phi(n)$ is easy if both p and q are known, as it is given by $\phi(n) = (p - 1)(q - 1)$. However, its calculation for large n is infeasible if p and q are unknown.

$$ed \equiv 1(\mathrm{mod}\, \phi(n))$$
$$\Rightarrow ed = 1 + k\,\phi(n) \text{ for some } k \in \mathbb{Z}$$

Table 10.6 Steps for A to send secure message and signature to B

Step	Description
1.	A uses B's public key to encrypt the message
2.	A uses its private key to encrypt its signature
3.	A sends the message and signature to B
4.	B uses A's public key to decrypt A's signature
5.	B uses its private key to decrypt A's message

We discussed Euler's theorem in Chap. 5, and this result states that if a and n are positive integers with $\gcd(a, n) = 1$ then $a^{\phi(n)} \equiv 1(\mathrm{mod}\ n)$. Therefore, $\mathrm{M}^{\phi(n)} \equiv 1$ $(\mathrm{mod}\ n)$ and $\mathrm{M}^{k\phi(n)} \equiv 1(\mathrm{mod}\ n)$. The decryption of the ciphertext is given by

$$C^d(\mathrm{mod}\ n) \equiv \mathrm{M}^{ed}(\mathrm{mod}\ n)$$
$$\equiv \mathrm{M}^{1+k\phi(n)}(\mathrm{mod}\ n)$$
$$\equiv \mathrm{M}^1 \mathrm{M}^{k\phi(n)}(\mathrm{mod}\ n)$$
$$\equiv \mathrm{M}.1(\mathrm{mod}\ n)$$
$$\equiv \mathrm{M}(\mathrm{mod}\ n)$$

10.5.2   Digital Signatures

The RSA public-key cryptography may also be employed to obtain digital signatures. Suppose A wishes to send a secure message to B as well as a digital signature. This involves signature generation using the private key and signature verification using the public key. The steps involved are given in Table 10.6.

The National Institute of Standards and Technology (NIST) proposed an algorithm for digital signatures in 1991. The algorithm is known as the Digital Signature Algorithm (DSA) and later became the Digital Signature Standard (DSS).

10.6   Review Questions

1. Discuss the early cipher developed by Julius Caesar. How effective was it at that period in history, and what are its weaknesses today?
2. Describe how the team at Bletchley Park cracked the German Enigma codes.
3. Explain the differences between a public-key cryptosystem and a private-key cryptosystem.

4. What are the advantages and disadvantages of private-key (symmetric) cryptosystems?
5. Describe the various types of symmetric-key systems.
6. What are the advantages and disadvantages of public-key cryptosystems?
7. Describe public-key cryptosystems including the RSA public-key cryptosystem.
8. Describe how digital signatures may be generated.

10.7 Summary

This chapter provided a brief introduction to cryptography, which is the study of mathematical techniques that provide secrecy in the transmission of messages between computers. It was originally employed to protect communication between individuals, and today it is employed to solve security problems such as privacy and authentication over a communications channel.

It involves enciphering and deciphering messages, and theoretical results from number theory are employed to convert the original messages (or plaintext) into ciphertext that is then transmitted over a secure channel to the intended recipient. The ciphertext is meaningless to anyone other than the intended recipient, and the received ciphertext is then decrypted to allow the recipient to read the message.

A public-key cryptosystem is an asymmetric cryptosystem. It has two different encryption and decryption keys, and the fact that a person has knowledge of how to encrypt messages does not mean that the person has sufficient information to decrypt messages.

In a secret-key cryptosystem, the same key is used for both encryption and decryption. Anyone who has knowledge on how to encrypt messages has sufficient knowledge to decrypt messages, and it is essential that the key is kept secret between the two parties.

References

Data Encryption Standard (1977) FIPS-Pub 46. National Bureau of Standards. U.S. Department of Commerce. January 1977

Pohlig S, Hellman M (1978) An improved algorithm for computing algorithms over GF(p) and its cryptographic significance. IEEE Trans Inf Theory 24:106–110

Coding Theory

11

Key Topics

Groups, rings and fields
Block codes
Error detection and correction
Generation matrix
Hamming codes

11.1 Introduction

Coding theory is a practical branch of mathematics concerned with the reliable transmission of information over communication channels. It allows errors to be detected and corrected, which is essential when messages are transmitted through a noisy communication channel. The channel could be a telephone line, radio link or satellite link, and coding theory is applicable to mobile communications and satellite communications. It is also applicable to storing information on storage systems such as the compact disc.

It includes theory and practical algorithms for error detection and correction, and it plays an important role in modern communication systems that require reliable and efficient transmission of information.

An error-correcting code encodes the data by adding a certain amount of redundancy to the message. This enables the original message to be recovered if a small number of errors have occurred. The extra symbols added are also subject to errors, as accurate transmission cannot be guaranteed in a noisy channel.

© Springer Nature Switzerland AG 2020
G. O'Regan, *Mathematics in Computing*, Undergraduate Topics
in Computer Science, https://doi.org/10.1007/978-3-030-34209-8_11

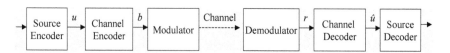

Fig. 11.1 Basic digital communication

The basic structure of a digital communication system is shown in Fig. 11.1 . It includes transmission tasks such as source encoding, channel encoding and modulation; and receiving tasks such as demodulation, channel decoding and source decoding.

The modulator generates the signal that is used to transmit the sequence of symbols b across the channel. The transmitted signal may be altered due to the fact that there is noise in the channel, and the signal received is demodulated to yield the sequence of received symbols r.

The received symbol sequence r may differ from the transmitted symbol sequence b due to the noise in the channel, and therefore a channel code is employed to enable errors to be detected and corrected. The channel encoder introduces redundancy into the information sequence u, and the channel decoder uses the redundancy for error detection and correction. This enables the transmitted symbol sequence $\hat{u}$ to be estimated.

Shannon (1948) showed that it is theoretically possible to produce an information transmission system with an error probability as small as required provided that the information rate is smaller than the channel capacity.

Coding theory uses several results from pure mathematics, and we briefly discuss its mathematical foundations. We discussed mathematical structures used in coding theory in Chap. 6.

11.2 Mathematical Foundations

Coding theory is built from the results of modern algebra, and it uses abstract algebraic structures such as groups, rings, fields and vector spaces. These abstract structures provide a solid foundation for the discipline, and the main structures used include vector spaces and fields. A *group* is a non-empty set with a single binary operation, whereas *rings* and *fields* are algebraic structures with two binary operations satisfying various laws. A *vector space* consists of vectors over a field.

These mathematical structures were discussed in Chap. 6, and examples presented. The representation of codewords is by n-dimensional vectors over the finite field F_q. A codeword vector v is represented as the n-tuple:

$$v = (a_0, a_1, \ldots a_{n-1})$$

where each $a_i \in F_q$. The set of all n-dimensional vectors is the n-dimensional vector space $\mathbf{F}_q^n$ with q^n elements. The addition of two vectors v and w, where $v = (a_0, a_1, \ldots a_{n-1})$ and $w = (b_0, b_1, \ldots b_{n-1})$ is given by

$$v + w = (a_0 + b_0, a_1 + b_1, \ldots a_{n-1} + b_{n-1})$$

The scalar multiplication of a vector $v = (a_0, a_1, \ldots a_{n-1}) \in \mathbf{F}_q^n$ by a scalar $\beta \in F_q$ is given by

$$\beta v = (\beta a_0, \beta a_1, \ldots . \beta a_{n-1})$$

The set $\mathbf{F}_q^n$ is called the vector space over the finite field F_q.if the vector space properties above hold. A finite set of vectors $v_1, v_2, \ldots v_k$ is said to be *linearly independent* if

$$\beta_1 v_1 + \beta_2 v_2 + \cdots + \beta_k v_k = 0 \Rightarrow \beta_1 = \beta_2 = \cdots \beta_k = 0$$

Otherwise, the set of vectors $v_1, v_2, \ldots v_k$ is said to be *linearly dependent*.

The *dimension* (dim W) of a subspace $W \subseteq V$ is k if there are k linearly independent vectors in W but every $k + 1$ vector is linearly dependent. A subset of a vector space is a *basis* for V if it consists linearly independent vectors, and its linear span is V (i.e. the basis generates V). We shall employ the basis of the vector space of codewords to create the generator matrix to simplify the encoding of the information words. The linear span of a set of vectors $v_1, v_2, \ldots, v_k$ is defined as $\beta_1 v_1 + \beta_2 v_2 + \cdots + \beta_k v_k$.

11.3   Simple Channel Code

This section presents a simple example to illustrate the concept of an error-correcting code. The example code presented is able to correct a single transmitted error only.

We consider the transmission of binary information over a noisy channel that leads to differences between the transmitted sequence and the received sequence. The differences between the transmitted and received sequence are illustrated by underlining the relevant digits in the example.

$$
\begin{aligned}
&\text{Sent} \qquad\quad 0010\underline{1}1110 \\
&\text{Received} \quad\ 000\underline{00}110
\end{aligned}
$$

Initially, it is assumed that the transmission is done without channel codes as follows:

$$00101110 \xrightarrow[]{\text{Channel}} 00000110$$

Next, the use of an encoder is considered and a triple repetition-encoding scheme is employed. That is, the binary symbol 0 is represented by the codeword 000 and the binary symbol 1 is represented by the codeword 111.

$$00101110 \rightarrow \boxed{\text{Encoder}} \rightarrow 000000111000111111111000$$

In other words, if the symbol 0 is to be transmitted then the encoder emits the codeword 000, and similarly the encoder emits 111 if the symbol 1 is to be transmitted. Assuming that on average one symbol in four is incorrectly transmitted, then transmission with binary triple repetition may result in a received sequence such as

$$000000111000111111111000 \rightarrow \boxed{\text{Channel}} \rightarrow 0\underline{1}00000\underline{1}10\underline{1}0111\underline{0}10\underline{0}1110\underline{1}0$$

The decoder tries to estimate the original sequence by using a *majority decision* on each 3-bit word. Any 3-bit word that contains more zeros than ones is decoded to 0, and similarly if it contains more ones than zero it is decoded to 1. The decoding algorithm yields

$$0\underline{1}00000\underline{1}10\underline{1}0111\underline{0}10\underline{0}1110\underline{1}0 \rightarrow \boxed{\text{Decoder}} \rightarrow 00101\underline{0}10$$

In this example, the binary triple repetition code is able to correct a single error within a codeword (as the majority decision is two to one). This helps to reduce the number of errors transmitted compared to unprotected transmission. In the first case, where an encoder is not employed there are two errors, whereas there is just one error when the encoder is used.

However, there are disadvantages with this approach in that the transmission bandwidth has been significantly reduced. It now takes three times as long to transmit an information symbol with the triple replication code than with standard transmission. Therefore, it is desirable to find more efficient coding schemes.

11.4 Block Codes

There were two codewords employed in the simple example above, namely, 000 and 111. This is an example of a (n,k) code where the codewords are of length $n = 3$, and the information words are of length $k = 1$ (as we were just encoding a single symbol 0 or 1). This is an example of a (3,1) block code, and the objective of this section is to generalize the simple coding scheme to more efficient and powerful channel codes.

The fundamentals of the q-nary (n,k) block codes (where q is the number of elements in the finite field F_q) involve converting an information block of length k to a codeword of length n. Consider an information sequence u_0, u_1, u_2, ... of discrete information symbols where $u_i \in \{0,1,...q{-}1\} = F_q$. The normal class of channel codes is when we are dealing with binary codes, i.e. $q = 2$. The information sequence is then grouped into blocks of length k as follows:

$$\underbrace{u_0u_1u_2...u_{k-1}} \quad \underbrace{u_ku_{k+1}u_{k+2}...u_{2k-1}} \quad \underbrace{u_{2k}u_{2k+1}u_{2k+2}...u_{3k-1}} ...$$

Each block is of length k (i.e. the information words are of length k), and it is then encoded separately into codewords of length n. For example, the block $u_ku_{k+1}u_{k+2}\cdots u_{2k-1}$ is encoded to the codeword $b_nb_{n+1}b_{n+2}...b_{2n-1}$ of length n where $b_i \in F_q$. Similarly, the information word $u_0u_1u_2...u_{k-1}$ is uniquely mapped to a codeword $b_0b_1b_2...b_{n-1}$ of length n as follows:

$$(u_0u_1u_2...u_{k-1}) \rightarrow \boxed{\text{Encoder}} \rightarrow (b_0b_1b_2...b_{n-1})$$

These codewords are then transmitted across the communication channel and the received words are then decoded. The received word $\mathrm{r} = (r_0r_1r_2...r_{n-1})$ is decoded into the information word $\hat{\mathrm{u}} = (\hat{u}_0\hat{u}_1\hat{u}_2...\hat{u}_{k-1})$.

$$(r_0r_1r_2...r_{n-1}) \rightarrow \boxed{\text{Decoder}} \rightarrow (\hat{u}_0\hat{u}_1\hat{u}_2...\hat{u}_{k-1})$$

Strictly speaking, the decoding is done in two steps with the received n-block word r first decoded to the n-block codeword b^*. This is then decoded into the k-block information word $\hat{u}$. The encoding, transmission and decoding of an (n,k) block may be summarized as follows (Fig. 11.2).

A lookup table may be employed for the encoding to determine the codeword b for each information word u. However, the size of the table grows exponentially with increasing information word length k, and so this is inefficient due to the large memory size required. We shall discuss later how a generator matrix provides an efficient encoding and decoding mechanism.

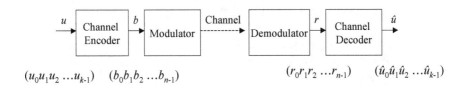

$(u_0u_1u_2...u_{k-1})$ $(b_0b_1b_2...b_{n-1})$ $(r_0r_1r_2...r_{n-1})$ $(\hat{u}_0\hat{u}_1\hat{u}_2...\hat{u}_{k-1})$

Fig. 11.2 Encoding and decoding of an (n,k) block

Notes

(i)   The codeword is of length n.

(ii)   The information word is of length k.

(iii) The codeword length n is larger than the information word length k.

(iv)   A block (n,k) code is a code in which all codewords are of length n and all information words are of length k.

(v)   The number of possible information words is given by $M = q^k$ (where each information symbol can take one of q possible values and the length of the information word is k).

(vi)   The code rate R in which information is transmitted across the channel is given by

$$R = \frac{k}{n}.$$

(vii) The weight of a codeword is $b = (b_0 b_1 b_2 \ldots b_{n-1})$ which is given by the number of non-zero components of b. That is,

$$\mathrm{wt}(b) = \left| \{ i : b_i \neq 0, 0 \leq i < n \} \right|.$$

(viii) The distance between two codewords $b = (b_0 b_1 b_2 \ldots b_{n-1})$ and $b' = (b_0' b_1' b_2' \ldots b_{n-1}')$ measures how close the codewords b and b' are to each other. It is given by the Hamming distance:

$$\mathrm{dist}(b, b') = \left| \{ i : b_i \neq b_i', 0 \leq i < n \} \right|.$$

(ix)   The minimum Hamming distance for a code **B** consisting of M codewords $b_1, \ldots, b_M$ is given by

$$d = \min \{ \mathrm{dist}(b, b') \quad : \quad \text{where } b \neq b' \}.$$

(x)   The (n,k) block code $\mathbf{B} = \{ b_1, \ldots, b_M \}$ with M $(= q^k)$ codewords of length n and minimum Hamming distance d is denoted by $\mathbf{B}(n, k, d)$.

11.4.1 Error Detection and Correction

The minimum Hamming distance offers a way to assess the error detection and correction capability of a channel code. Consider two codewords b and b' of an (n, k) block code $\mathbf{B}(n, k, d)$.

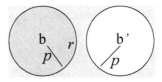

Fig. 11.3 Error-correcting capability sphere

Then, the distance between these two codewords is greater than or equal to the minimum Hamming distance d, and so errors can be detected as long as the erroneously received word is not equal to a codeword different from the transmitted codeword.

That is, the *error detection capability* is guaranteed as long as the number of errors is less than the minimum Hamming distance d, and so the number of detectable errors is $d-1$.

Any two codewords are of distance at least d and so if the number of errors is less than $^d/_2$ then the received word can be properly decoded to the codeword b. That is, the *error correction capability* is given by

$$E_{cor} = \frac{d-1}{2}$$

An error-correcting sphere (Fig. 11.3) may be employed to illustrate the error correction of a received word to the correct codeword b. This may be done when all received words are within the error-correcting sphere with radius p ($< {^d/_2}$).

If the received word r is different from b in less than $^d/_2$ positions, then it is decoded to b as it is more than $^d/_2$ positions from the next closest codeword. That is, b is the closest codeword to the received word r (provided that the error-correcting radius is less than $^d/_2$).

11.5 Linear Block Codes

Linear block codes have nice algebraic properties and the codewords in a linear block code are considered to be vectors in the finite vector space $\mathbf{F}_q^n$. The representation of codewords by vectors allows the nice algebraic properties of vector spaces to be used, and this simplifies the encoding of information words as a generator matrix may be employed to create the codewords.

An (n,k) block code $\mathbf{B}(n,k,d)$ with minimum Hamming distance d over the finite field F_q is called *linear* if $\mathbf{B}(n,k,d)$ is a subspace of the vector space $\mathbf{F}_q^n$ of dimension k. The number of codewords is then given by

$$M = q^k$$

The rate of information (R) through the channel is given by

$$R = \frac{k}{n}$$

Clearly, since $\mathbf{B}(n,k,d)$ is a subspace of $\mathbf{F}_q^n$, any linear combination of the codewords (vectors) will be a codeword. That is, for the codewords $b_1, b_2, \ldots, b_r$ we have that

$$\alpha_1 b_1 + \alpha_2 b_2 + \cdots + \alpha_r b_r \in \mathbf{B}(n, k, d)$$

where $\alpha_1, \alpha_2, \ldots, \alpha_r \in \mathbf{F}_q$ and $b_1, b_2, \ldots, b_r \in \mathbf{B}(n, k, d)$.

Clearly, the n-dimensional zero row vector $(0, 0, \ldots, 0)$ is always a codeword, and so $(0, 0, \ldots, 0) \in \mathbf{B}(n, k, d)$. The minimum Hamming distance of a linear block code $\mathbf{B}(n, k, d)$ is equal to the minimum weight of the non-zero codewords: That is,

$$d = \min_{\forall b \neq b'} \{\mathrm{dist}(b, b')\} = \min_{\forall b \neq 0} \mathrm{wt}(b)$$

In summary, an (n, k) linear block code $\mathbf{B}(n, k, d)$ is

1. A subspace of $\mathbf{F}_q^n$.]]
2. The number of codewords is $M = q^k$.
3. The minimum Hamming distance d is the minimum weight of the non-zero codewords.

The encoding of a specific k-dimensional information word $u = (u_0, u_1, \ldots u_{k-1})$ to an n-dimensional codeword $b = (b_0, b_1, \ldots, b_{n-1})$ may be done efficiently with a generator matrix (matrices are discussed in Chap. 14). First, a basis $\{g_0, g_1, \ldots g_{k-1}\}$ of the k-dimensional subspace spanned by the linear block code is chosen, and this consists of k linearly independent n-dimensional vectors. Each basis element g_i (where $0 \leq i \leq k-1$) is an n-dimensional vector:

$$g_i = (g_{i,0}, g_{i,1}, \ldots, g_{i,n-1})$$

The corresponding codeword $b = (b_0, b_1, \ldots, b_{n-1})$ is then a linear combination of the information word with the basis elements. That is,

$$b = u_0 g_0 + u_1 g_1 + \cdots + u_{k-1} g_{k-1}$$

where each information symbol $u_i \in \mathbf{F}_q$. The generator matrix G is then constructed from the k linearly independent basis vectors as follows (Fig. 11.4).

The encoding of the k-dimensional information word u to the n-dimensional codeword b involves matrix multiplication (Fig. 11.5).

This may also be written as

$$G = \begin{pmatrix} g_{0,} \\ g_{1,} \\ g_{2,} \\ .... \\ .... \\ .... \\ g_{k-1} \end{pmatrix} = \begin{pmatrix} g_{0,0} & g_{0,1} & g_{0,2} & .... & g_{0,n-1} \\ g_{1,0} & g_{1,1} & g_{1,2} & .... & g_{1,n-1} \\ g_{2,0} & g_{2,1} & g_{2,2} & .... & g_{2,n-1} \\ ... & .... & .... & .... & .... \\ ... & .... & .... & .... & .... \\ ... & .... & .... & .... & .... \\ g_{k-1,0} & g_{k-1,1} & g_{k-1,2} & .... & g_{k-1,n-1} \end{pmatrix}$$

Fig. 11.4 Generator matrix

$$(u_0, u_1, ..., u_{k-1}) \begin{pmatrix} g_{0,0} & g_{0,1} & g_{0,2} & .... & g_{0,n-1} \\ g_{1,0} & g_{1,1} & g_{1,2} & .... & g_{1,n-1} \\ g_{2,0} & g_{2,1} & g_{2,2} & .... & g_{2,n-1} \\ ... & .... & .... & .... & .... \\ ... & .... & .... & .... & .... \\ ... & .... & .... & .... & .... \\ g_{k-1,0} & g_{k-1,1} & g_{k-1,2} & .... & g_{k-1,n-1} \end{pmatrix} = (b_0, b_1, ..., b_{n-1})$$

Fig. 11.5 Generation of codewords

$$b = uG$$

Clearly, all $M = q^k$ codewords $b \in \mathbf{B}(n, k, d)$ can be generated according to this rule, and so the matrix G is called the generator matrix. The generator matrix defines the linear block code $\mathbf{B}(n, k, d)$. There is an equivalent $k \times n$ generator matrix for $\mathbf{B}(n, k, d)$ defined as

$$G = I_k | A_{k,n-k}$$

where I_k is the $k \times k$ identity matrix (Fig. 11.6).

The encoding of the information word u yields the codeword b such that the first k symbols b_i of b are the same as the information symbols u_i, $0 \le i \le k$.

$$b = uG = \left(u | u\, A_{k,n-k} \right)$$

The remaining $m = n - k$ symbols are generated from $uA_{k,n-k}$ and the last m symbols are the m parity check symbols. These are attached to the information vector u for the purpose of error detection and correction.

Fig. 11.6 Identity matrix
$(k \times k)$

$$I_k = \begin{pmatrix} 1 & 0 & 0 & \ldots & 0 \\ 0 & 1 & 0 & \ldots & 0 \\ 0 & 0 & 1 & \ldots & 0 \\ \ldots & \ldots & \ldots & \ldots & \ldots \\ \ldots & \ldots & \ldots & \ldots & \ldots \\ \ldots & \ldots & \ldots & \ldots & \ldots \\ 0 & 0 & 0 & \ldots & 1 \end{pmatrix}$$

11.5.1 Parity Check Matrix

The linear block code $\mathbf{B}(n,\ k,\ d)$ with generator matrix $G = (I_k,| \ A_{k,n-k})$ may be defined equivalently by the $(n-k) \times n$ parity check matrix H, where this matrix is defined as

$$H = \left(-A_{k,\,n-k}^{T} | I_{n-k} \right)$$

The generator matrix G and the parity check matrix H are orthogonal, i.e.

$$HG^{T} = 0_{n-k,\,k}$$

The parity check orthogonality property holds if and only if the vector belongs to the linear block code. That is, for each code vector in $b \in \mathbf{B}(n,k,d)$, we have

$$Hb^{T} = 0_{n-k,\,1}$$

and vice versa whenever the property holds for a vector r, then r is a valid codeword in $\mathbf{B}(n,k,d)$. We present an example of a parity check matrix in Example 11.5.

11.5.2 Binary Hamming Code

The Hamming code is a linear code that has been employed in dynamic random access memory to detect and correct deteriorated data in memory. The generator matrix for the $\mathbf{B}$ (7, 4, 3) binary Hamming code is given by (Fig. 11.7).

The information words are of length $k = 4$ and the codewords are of length $n = 7$. For example, it can be verified by matrix multiplication that the information word (0, 0, 1, 1) is encoded into the codeword (0, 0, 1, 1, 0, 0, 1).

That is, three parity bits 001 have been added to the information word (0,0,1,1) to yield the codeword (0, 0, 1, 1, 0, 0, 1).

$$G = \begin{pmatrix} 1 & 0 & 0 & 0 & 0 & 1 & 1 \\ 0 & 1 & 0 & 0 & 1 & 0 & 1 \\ 0 & 0 & 1 & 0 & 1 & 1 & 0 \\ 0 & 0 & 0 & 1 & 1 & 1 & 1 \end{pmatrix}$$

Fig. 11.7 Hamming code B (7, 4, 3) generator matrix

The minimum Hamming distance is $d = 3$, and the Hamming code can detect up to two errors, and it can correct one error.

Example 11.5 (*Parity Check Matrix—Hamming Code*) The objective of this example is to construct the Parity Check Matrix of the Binary Hamming Code (7, 4, 3), and to show an example of the parity check orthogonality property.

First, we construct the parity check matrix H which is given by $H = \left(-A_{k,n-k}^T | I_{n-k}\right)$ or, in other words, $H = \left(-A_{4,3}^T | I_3\right)$. We first note that

$$A_{4,3} = \begin{bmatrix} 0 & 1 & 1 \\ 1 & 0 & 1 \\ 1 & 1 & 0 \\ 1 & 1 & 1 \end{bmatrix} \quad A_{4,3}^T = \begin{bmatrix} 0 & 1 & 1 & 1 \\ 1 & 0 & 1 & 1 \\ 1 & 1 & 0 & 1 \end{bmatrix}$$

Therefore, H is given by

$$H = \begin{bmatrix} 0 & -1 & -1 & -1 & 1 & 0 & 0 \\ -1 & 0 & -1 & -1 & 0 & 1 & 0 \\ -1 & -1 & 0 & -1 & 0 & 0 & 1 \end{bmatrix}$$

We noted that the encoding of the information word u = (0011) yields the codeword b = (0011001). Therefore, the calculation of Hb^T yields (recalling that addition is modulo two)

$$Hb^T = \begin{bmatrix} 0 & -1 & -1 & -1 & 1 & 0 & 0 \\ -1 & 0 & -1 & -1 & 0 & 1 & 0 \\ -1 & -1 & 0 & -1 & 0 & 0 & 1 \end{bmatrix} \begin{bmatrix} 0 \\ 0 \\ 1 \\ 1 \\ 0 \\ 0 \\ 1 \end{bmatrix} = \begin{bmatrix} 0 \\ 0 \\ 0 \end{bmatrix}$$

11.5.3 Binary Parity Check Code

The binary parity check code is a linear block code over the finite field F_2. The code takes a k-dimensional information word $u = (u_0, u_1,\ldots u_{k-1})$ and generates the codeword $b = (b_0, b_1,\ldots, b_{k-1}, b_k)$, where $u_i = b_i$ $(0 \leq i \leq k-1)$ and b_k is the parity bit chosen so that the resulting codeword is of even parity. That is,

$$b_k = u_0 + u_1 + \cdots + u_{k-1} = \sum_{i=0}^{k-1} u_i$$

11.6 Miscellaneous Codes in Use

There are many examples of codes in use such as repetition codes (such as the triple replication code considered earlier in Sect. 11.3); parity check codes, where a parity symbol is attached such as the binary parity check code and Hamming codes such as the (7,4) code that was discussed in Sect. 11.5.2, and which has been applied for error correction of faulty memory.

The Reed–Muller codes form a class of error-correcting codes that can correct more than one error. Cyclic codes are special linear block codes with efficient algebraic decoding algorithms. The BCH codes are an important class of cyclic codes, and the Reed–Solomon codes are an example of a BCH code.

Convolution codes have been applied in the telecommunications field, for example, in GSM, UMTS and in satellite communications. They belong to the class of linear codes, but also employ a memory so that the output depends on the current input symbols and previous input. For more detailed information on coding theory, see Neubauer (2007).

11.7 Review Questions

1. Describe the basic structure of a digital communication system.
2. Describe the mathematical structure known as the field. Give examples of fields.
3. Describe the mathematical structure known as the ring and give examples of rings. Give examples of zero divisors in rings.
4. Describe the mathematical structure known as the vector space and give examples
5. Explain the terms linear independence and linear dependence and a basis.
6. Describe the encoding and decoding of an (n,k) block code where an information word of length k is converted to a codeword of length n.

7. Show how the minimum Hamming distance may be employed for error detection and error correction.
8. Describe linear block codes and show how a generator matrix may be employed to generate the codewords from the information words.

11.8 Summary

Coding theory is the branch of mathematics that is concerned with the reliable transmission of information over communication channels. It allows errors to be detected and corrected, and this is extremely useful when messages are transmitted through a noisy communication channel. This branch of mathematics includes theory and practical algorithms for error detection and correction.

The theoretical foundations of coding theory were considered, and its foundations lie in abstract algebra including group theory, ring theory, fields and vector spaces. The codewords are represented by n-dimensional vectors over a finite field F_q.

An error-correcting code encodes the data by adding a certain amount of redundancy to the message so that the original message can be recovered if a small number of errors have occurred.

The fundamentals of block codes were discussed where an information word is of length k and a codeword is of length n. This led to the linear block codes $\mathbf{B}(n,k,d)$ and a discussion on error detection and error correction capabilities of the codes.

The goal of this chapter was to give a flavour of coding theory, and the reader is referred to more specialized texts (e.g. Neubauer et al. 2007) for more detailed information.

References

Neubauer A, Freunderberger J, Kühn V (2007) Coding theory algorithms, architectures and applications. Wiley, New York
Shannon C (1948) A mathematical theory of communication. Bell Syst Tech J 27:379–423

Language Theory and Semantics

12

12.1 Introduction

There are two key parts to any programming language, and these are its syntax and semantics. The syntax is the grammar of the language and a program needs to be syntactically correct with respect to its grammar. The semantics of the language is deeper, and determines the meaning of what has been written by the programmer.

The difference between syntax and semantics may be illustrated by an example in a natural language. A sentence may be syntactically correct but semantically meaningless, or a sentence may have semantic meaning but be syntactically incorrect. For example, consider the sentence:

I will go to Dublin yesterday

© Springer Nature Switzerland AG 2020
G. O'Regan, *Mathematics in Computing*, Undergraduate Topics
in Computer Science, https://doi.org/10.1007/978-3-030-34209-8_12

Then this sentence is syntactically valid but semantically meaningless. Similarly, if a speaker utters the sentence 'Me Dublin yesterday' we would deduce that the speaker had visited Dublin the previous day even though the sentence is syntactically incorrect.

The semantics of a programming language determines what a syntactically valid program will compute. A programming language is therefore given by

$$\text{Programming Language} = \text{Syntax} + \text{Semantics}$$

Many programming languages have been developed since the birth of digital computing including Plankalkül which was developed by Zuse in the 1940s, Fortran developed by IBM in the 1950s, COBOL was developed by a committee in the late 1950s, Algol 60 and Algol 68 were developed by an international committee in the 1960s, Pascal was developed by Wirth in the early 1970s, Ada was developed for the US military in the late 1970s, the C language was developed by Richie and Thompson at Bell Labs in the early 1970s, C ++ was developed by Stroustrup at Bell Labs in the early 1980s and Java developed by Gosling at Sun Microsystems in the mid-1990s. A short description of a selection of programming languages in use is in O'Regan (2016).

A programming language needs to have a well-defined syntax and semantics, and the compiler preserves the semantics of the language (rather than giving the semantics of a language). Compilers are programs that translate a program that is written in some programming language into another form. It involves syntax analysis and parsing to check the syntactic validity of the program; semantic analysis to determine what the program should do; optimization to improve the speed and performance and code generation in some target language.

Alphabets are a fundamental building block in language theory, as words and language are generated from alphabets. They are discussed in the next section.

12.2 Alphabets and Words

An *alphabet* is a finite non-empty set A, and the elements of A are called letters. For example, consider the set A which consists of the letters a to z.

Words are finite strings of letters, and a set of words is generated from the alphabet. For example, the alphabet $A = \{a,b\}$ generates the following set of words:

$$\{\varepsilon, a, b, aa, ab, bb, ba, aaa, bbb, \ldots\}$$

[1]Each word consists of an ordered list of one or more letters and the set of words of length two consists of all ordered lists of two letters. It is given by

$$A^2 = \{aa, ab, bb, ba\}$$

Similarly, the set of words of length three is given by

---

[1]ε denotes the empty word.

$$A^3 = \{aaa, aab, abb, aba, baa, bab, bbb, bba\}$$

The set of all words over the alphabet A is given by the positive closure A^+, and it is defined by

$$A^+ = A \cup A^2 \cup A^3 \cup \ldots = \bigcup_{n=1}^{\infty} A^n$$

Given any two words $w_1 = a_1 a_2 \ldots a_k$ and $w_2 = b_1 b_2 \ldots b_r$ then the concatenation of w_1 and w_2 is given by

$$w = w_1 w_2 = a_1 a_2 \ldots a_k b_1 b_2 \ldots b_r$$

The empty word is a word of length zero and is denoted by ε. Clearly, $\varepsilon w = w \varepsilon = w$ for all w and so ε is the identity element under the concatenation operation, A^0 is used to denote the set containing the empty word $\{\varepsilon\}$ and the closure A^* ($= A^+ \cup \{\varepsilon\}$) denotes the infinite set of all words over A (including empty words). It is defined as

$$A^* = \bigcup_{n=0}^{\infty} A^n$$

The mathematical structure $(A^*, \wedge, \varepsilon)$ forms a monoid[2], where $\wedge$ is the concatenation operator for words and the identity element is ε. The length of a word w is denoted by $|w|$ and the length of the empty word is zero, i.e. $|\varepsilon| = 0$.

A subset L of A^* is termed a formal language over A. Given two languages L_1, L_2 then the concatenation (or product) of L_1 and L_2 is defined by

$$L_1 L_2 = \{w | w = w_1 w_2 \text{ where } w_1 \in L_1 \text{ and } w_2 \in L_2\}$$

The positive closure of L and the closure of L may also be defined as

$$L^+ = \bigcup_{n=1}^{\infty} L^n \quad L^* = \bigcup_{n=0}^{\infty} L^n$$

12.3 Grammars

A formal grammar describes the syntax of a language, and we distinguish between *concrete* and *abstract syntax*. Concrete syntax describes the external appearance of programs as seen by the programmer, whereas abstract syntax aims to describe the

---

[2]Recall from Chap. 6 that a monoid (M, *, e) is a structure that is closed and associative under the binary operation '*', and it has an identity element 'e'.

essential structure of programs rather than its external form. In other words, abstract syntax aims to give the components of each language structure while leaving out the representation details (e.g. syntactic sugar). Backus–Naur Form (BNF) notation is often used to specify the concrete syntax of a language. A grammar consists of

- A finite set of terminal symbols,
- A finite set of non-terminal symbols,
- A set of production rules and
- A start symbol.

A formal grammar generates a formal language, which is set of finite length sequences of symbols created by applying the production rules of the grammar. The application of a production rule involves replacing symbols at the left-hand side of the rule with the symbols on the right-hand side of the rule. The formal language then consists of all words consisting of terminal symbols that are reached by a derivation (i.e. the application of production rules) starting from the start symbol of the grammar.

A construct that appears on the left-hand side of a production rule is termed a *non-terminal*, whereas a construct that only appears on the right-hand side of a production rule is termed a *terminal*. The set of non-terminals N is disjoint from the set of terminals A.

The theory of the syntax of programming languages is well established, and programming languages have a well-defined grammar that allows syntactically valid programs to be derived from the grammars.

Chomsky[3] (Figure 12.1) is a famous linguist who classified a number of different types of grammar that occur. The Chomsky hierarchy (Table 12.1) consists of four levels including regular grammars; context-free grammars; context-sensitive grammars and unrestricted grammars. The grammars are distinguished by the production rules, which determine the type of language that is generated.

Regular grammars are used to generate the words that may appear in a programming language. This includes the identifiers (e.g. names for variables, functions and procedures); special symbols (e.g. addition, multiplication, etc.) and the reserved words of the language.

A rewriting system for context-free grammars is a finite relation between N and $(A \cup N)^*$, i.e. a subset of $N \times (A \cup N)^*$: A production rule $< N> \to w$ is one element of this relation, and is an ordered pair $(< N>, w)$ where w is a word consisting of zero or more terminal and non-terminal letters. This production rule means that $< N>$ may be replaced by w.

---

[3]Chomsky made important contributions to linguistics and the theory of grammars. He is more widely known today as a critic of United States foreign policy.

Fig. 12.1 Noam Chomsky.
public domain

Table 12.1 Chomsky hierarchy of grammars

Grammar type	Description
Type 0 grammar	Type 0 grammars include all formal grammars. They have production rules of the form $\alpha \rightarrow \beta$ where α and β are strings of terminals and non-terminals. They generate all languages that can be recognized by a Turing machine (see Chap. 23)
Type 1 grammar (Context sensitive)	These grammars generate the context-sensitive languages. They have production rules of the form $\alpha A \beta \rightarrow \alpha \gamma \beta$ where A is a non-terminal and α, β and γ are strings of terminals and non-terminals. They generate all languages that can be recognized by a linear bounded automaton[a]
Type 2 grammar (Context free)	These grammars generate the context-free languages. These are defined by rules of the form $A \rightarrow \gamma$ where A is a non-terminal and γ is a string of terminals and non-terminals. These languages are recognized by a pushdown automaton[b] and are used to define the syntax of most programming languages
Type 3 grammar (Regular grammars)	These grammars generate the regular languages (or regular expressions). These are defined by rules of the form $A \rightarrow a$ or $A \rightarrow aB$ where A and B are non-terminals and a is a single terminal. A finite-state automaton recognizes these languages (see Chap. 23), and regular expressions are used to define the lexical structure of programming languages

[a]A linear bounded automaton is a restricted form of a non-deterministic Turing machine in which a limited finite portion of the tape (a function of the length of the input) may be accessed
[b]A pushdown automaton is a finite automaton that can make use of a stack containing data, and it is discussed in Chap. 23

12.3.1 Backus–Naur Form

Backus–Naur form[4] (BNF) provides an elegant means of specifying the syntax of programming languages. It was originally employed to define the grammar for the Algol 60 programming language (Naur 1960), and a variant was used by Wirth to specify the syntax of the Pascal programming language. BNF is widely used and accepted today as the way to specify the syntax of programming languages.

BNF specifications essentially describe the external appearance of programs as seen by the programmer. The grammar of a context-free grammar may then be input into a parser (e.g. Yacc), and the parser is used to determine if a program is syntactically correct or not.

A BNF specification consists of a set of production rules with each production rule describing the form of a class of language elements such expressions, statements and so on. A production rule is of the following form:

$$<symbol> ::= <expression\ with\ symbols>$$

where < symbol > is a *non-terminal*, and the expression consists of a sequence of terminal and non-terminal symbols. A construct that has alternate forms appears more than once, and this is expressed by sequences separated by the vertical bar '|' (which indicates a choice). In other words, there is more than one possible substitution for the symbol on the left-hand side of the rule. Symbols that never appear on the left-hand side of a production rule are called *terminals*.

The following example defines the syntax of various statements in a sample programming language:

$$<loop\ statement> ::= <while\ loop> \ | <for\ loop>$$
$$<while\ loop> ::= while(<condition>)<statement>$$
$$<for\ loop> ::= for(<expression>)<statement>$$
$$<statement> ::= <assignment\ statement> \ | <loop\ statement>$$
$$<assignment\ statement> ::= <variable> := <expression>$$

This is a partial definition of the syntax of various statements in the language. It includes various non-terminals such as < loop statement > , < while loop > and so on. The terminals include 'while', 'for', ':=', '(' and ')'. The production rules for < condition > and < expression > are not included.

The grammar of a context-free language (e.g. LL(1), LL(k), LR(1), LR(k)) grammar expressed in BNF notation) may be translated by a parser into a parse table. The parse table may then be employed to determine whether a particular program is valid with respect to its grammar.

---

[4]Backus–Naur form is named after John Backus and Peter Naur. It was created as part of the design of the Algol 60 programming language, and is used to define the syntax rules of the language.

Example 12.1 (*Context-Free Grammar*) The example considered is that of parenthesis matching in which there are two terminal symbols and one non-terminal symbol

$$S \rightarrow SS$$
$$S \rightarrow (S)$$
$$S \rightarrow ()$$

Then, by starting with S and applying the rules we can construct

$$S \rightarrow SS \rightarrow (S)S \rightarrow (())S \rightarrow (())()$$

Example 12.2 (*Context-Free Grammar*) The example considered is that of expressions in a programming language. The definition is ambiguous as there is more than one derivation tree for some expressions (e.g. there are two parse trees for the expression $5 \times 3 + 1$ discussed below).

<expr> ::= <numeral> | (<expr>)
           | (<expr> <operator> <expr>)
<operator> ::= + | - | × | /
    <digit> ::= 0 | 1 | .... | 9
<numeral>::=<digit> |<digit><numeral>

Example 12.3 (*Regular Grammar*) The definition of an identifier in most programming languages is similar to:

<identifier> ::= <let> <letdig>
    <letdig> ::= <let> | <dig> | ε
    <letdig> ::= <let> <letdig> | <dig> <letdig>
        <let> ::= a | b | c | .... | z
        <dig> ::= 0 | 1 | .... | 9

12.3.2 Parse Trees and Derivations

Let A and N be the terminal and non-terminal alphabet of a rewriting system and let $< X > \rightarrow w$ be a production. Let x be a word in $(A \cup N)^*$ with $x = u < X>v$ for some words $u,v \in (A \cup N)^*$. Then x is said to directly yield uwv and this is written as $x \Rightarrow uwv$.

This single substitution ($\Rightarrow$) can be extended by a finite number of productions ($\Rightarrow^*$), and this gives the set of words that can be obtained from a given word.

Fig. 12.2 Parse tree 5 ×
3 + 1

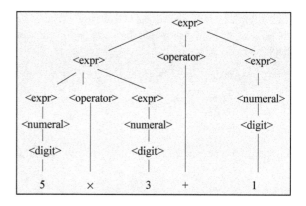

This derivation is achieved by applying several production rules (one production rule is applied at a time) in the grammar.

That is, given $x, y \in (A \cup N)^*$ then x yields y (or y is a derivation of x) if $x = y$, or there exists a sequence of words $w_1, w_2,....w_n \in (A \cup N)^*$ such that $x = w_1$, $y = w_n$ and $w_i \Rightarrow w_{i+1}$ for $1 \leq i \leq n\text{-}1$. This is written as $x \Rightarrow^* y$.

The expression grammar presented in Example 12.2 is ambiguous, and this means that an expression such as $5 \times 3 + 1$ has more than one interpretation. (Figure 12.2 and Fig. 12.3). It is not clear from the grammar whether multiplication is performed first and then addition, or whether addition is performed first and then multiplication.

The interpretation of the parse tree in Fig. 12.2 is that multiplication is performed first and then addition (this is the normal interpretation of such expressions in programming languages as multiplication is a higher precedence operator than addition).

The interpretation of the second parse tree is that addition is performed first and then multiplication (Fig. 12.3). It may seem a little strange that one expression has two parse trees and it shows that the grammar is ambiguous. This means that there is a choice for the compiler in evaluating the expression, and the compiler needs to assign the right meaning to the expression. For the expression grammar one solution would be for the language designer to alter the definition of the grammar to remove the ambiguity.

Fig. 12.3 Parse tree 5 ×
3 + 1

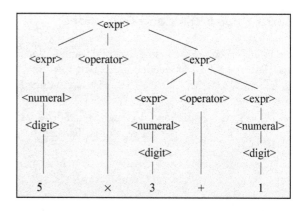

12.4 Programming Language Semantics

The formal semantics of a programming language is concerned with defining the actual meaning of a language. Language semantics is deeper than syntax, and the theory of the syntax of programming languages is well established. A programmer writes a program according to the rules of the language. The compiler first checks the program for syntactic correctness, i.e. it determines whether the program as written is valid according to the rules of the grammar of the language. If the program is syntactically correct, then the compiler determines the meaning of what has been written and generates the corresponding machine code.[5]

The compiler must preserve the semantics of the language, i.e. the semantics are not defined by the compiler, but rather the function of the compiler is to preserve the semantics of the language. Therefore, there is a need to have an unambiguous definition of the meaning of the language independently of the compiler, and the meaning is then preserved by the compiler.

A program's syntax[6] gives no information as to the meaning of the program, and therefore there is a need to supplement the syntactic description of the language with a formal unambiguous definition of its semantics.

We mentioned that it is possible to utter syntactically correct but semantically meaningless sentences in a natural language. Similarly, it is possible to write syntactically correct programs that behave in quite a different way from the intention of the programmer.

The formal semantics of a language is given by a mathematical model that describes the possible computations described by the language. There are three main approaches to programming language semantics, namely, axiomatic semantics, operational semantics and denotational semantics (Table 12.2).

There are several applications of programming language semantics including language design, program verification, compiler writing and language standardization. The three main approaches to semantics are described in more detail below.

12.4.1 Axiomatic Semantics

Axiomatic semantics gives meaning to phrases of the language by describing the logical axioms that apply to them. It was developed by C.A.R. Hoare[7] in a famous paper *'An axiomatic basis for computer programming'* (Hoare 1969). His axiomatic theory consists of *syntactic elements*, *axioms* and *rules of inference*.

The well-formed formulae that are of interest in axiomatic semantics are pre-post assertion formulae of the form $P\{a\}Q$, where a is an instruction in the language and

---

[5]It is possible that what the programmer has written is not be what the programmer had intended.
[6]There are attribute (or affix) grammars that extend the syntactic description of the language with supplementary elements covering the semantics. The process of adding semantics to the syntactic description is termed decoration.
[7]Hoare was influenced by earlier work by Floyd on assigning meanings to programs using flowcharts (Floyd 1967).

Table 12.2 Programming language semantics

Approach	Description
Axiomatic semantics	This involves giving meaning to phrases of the language using logical axioms
	It employs *pre-* and *postcondition assertions* to specify what happens when the statement executes. The relationship between the initial assertion and the final assertion essentially gives the semantics of the code
Operational semantics	This approach describes how a valid program is interpreted as sequences of computational steps. These sequences then define the meaning of the program
	An abstract machine (SECD machine) may be defined to give meaning to phrases, and this is done by describing the transitions they induce on states of the machine
Denotational semantics	This approach provides meaning to programs in terms of mathematical objects such as integers, tuples and functions
	Each phrase in the language is translated into a mathematical object that is the *denotation* of the phrase

P and Q are assertions, i.e. properties of the program objects that may be true or false.

An *assertion* is essentially a predicate that may be true in some states and false in other states. For example, the assertion $(x - y > 5)$ is true in the state in which the values of x and y are 7 and 1, respectively, and false in the state where x and y have values 4 and 2.

The pre- and postcondition assertions are employed to specify what happens when the statement executes. The relationship between the initial assertion and the final assertion gives the semantics of the code statement. The *pre-* and *postcondition* assertions are of the following form:

$$P\{a\}Q$$

The precondition P is a predicate (input assertion), and the postcondition Q is a predicate (output assertion). The braces separate the assertions from the program fragment. The well-formed formula $P\{a\}Q$ is itself a predicate that is either true or false.

This notation expresses the *partial correctness*[8] of a with respect to P and Q, and its meaning is that if statement a is executed in a state in which the predicate P is true and execution terminates, then it will result in a state in which assertion Q is satisfied.

The axiomatic semantics approach is described in more detail in O' Regan (2006), and the axiomatic semantics of a selection of statements is presented below.

---

[8]Total correctness is expressed using $\{P\}a\{Q\}$ and program fragment a is totally correct for precondition P and postcondition Q if and only if whenever a is executed in any state in which P is satisfied then execution terminates, and the resulting state satisfies Q.

• Skip

The skip statement does nothing and whatever condition is true on entry to the command is true on exit from the command. Its meaning is given by

$$P\{skip\}P$$

• Assignment

The meaning of the assignment statement is given by the axiom:

$$P_e^x\{x := e\}P$$

The meaning of the assignment statement is that P will be true after execution of the assignment statement if and only if the predicate P_e^x with the value of x replaced by e in P is true before execution (since x will contain the value of e after execution).

The notation P_e^x denotes the expression obtained by substituting e for all free occurrences of x in P.

• Compound

The meaning of the conditional command is

$$\frac{P\{S_1\}Q, Q\{S_2\}R}{P\{S_1; S_2\}R}$$

The compound statement involves the execution of S_1 followed by the execution of S_2. The meaning of the compound statement is that R will be true after the execution of the compound statement $S_1; S_2$ provided that P is true, if it is established that Q will be true after the execution of S_1 provided that P is true, and that R is true after the execution of S_2 provided Q is true.

There needs to be at least one rule associated with every construct in the language in order to give its axiomatic semantics. The semantics of other programming language statements such as the 'while' statement and the 'if' statement are described in O'Regan (2006).

12.4.2 Operational Semantics

The operational semantics definition is similar to that of an interpreter, where the semantics of the programming language are expressed using a mechanism that makes it possible to determine the effect of any program written in the language. The meaning of a program is given by the evaluation history that an interpreter produces when it interprets the program. The interpreter may be close to an executable programming language or it may be a mathematical language.

The operational semantics for a programming language describes how a valid program is interpreted as sequences of computational steps. The evaluation history defines the meaning of the program, and this is a sequence of internal interpreter configurations.

John McCarthy did early use of operational semantics in the late 1950s in his work on the semantics of LISP in terms of the lambda calculus. The use of lambda calculus allows the meaning of a program to be expressed using a mathematical interpreter, which gives precision through the use of mathematics.

The meaning of a program may be given in terms of a hypothetical or virtual machine that performs the set of actions that corresponds to the program. An abstract machine (SECD machine[9]) may be defined to give meaning to phrases in the language, and this is done by describing the transitions that they induce on states of the machine.

Operational semantics give an intuitive description of the programming language being studied, and its descriptions are close to real programs. It can play a useful role as a testing tool during the design of new languages, as it is relatively easy to design an interpreter to execute the description of example programs. This allows the effects of new languages or new language features to be simulated and studied through actual execution of the semantic descriptions prior to writing a compiler for the language. In other words, operational semantics can play a role in rapid prototyping during language design, and to get early feedback on the suitability of the language.

One disadvantage of the operational approach is that the meaning of the language is understood in terms of execution, i.e. in terms of interpreter configurations, rather than in an explicit *machine-independent specification*. An operational description is just one way to execute programs. Another disadvantage is that the interpreters for non-trivial languages often tend to be large and complex. A more detailed account of operational semantics is in Plotkin (1981), Meyer (1990).

12.4.3   Denotational Semantics

Denotational semantics expresses the semantics of a programming language by a translation schema that associates a meaning (denotation) with each program in the language (Meyer 1990). It maps a program directly to its meaning, and it was originally called mathematical semantics as it provides meaning to programs in terms of mathematical values such as integers, tuples and functions. That is, the meaning of a program is a mathematical object, and an interpreter is not employed. Instead, a valuation function is employed to map a program directly to its meaning,

---

[9]This virtual stack-based machine was originally designed by Peter Landin (a British computer scientist) to evaluate lambda calculus expressions, and it has since been used as a target for several compilers. Landin was influenced by McCarthy's LISP.

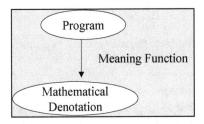

Fig. 12.4 Denotational semantics

and the denotational description of a programming language is given by a set of *meaning functions* M associated with the constructs of the language (Fig. 12.4).

Each meaning function is of the form $M_T: T \to D_T$ where T is some construct in the language and D_T is some semantic domain. Many of the meaning functions will be '*higher order*', i.e. functions that yield functions as results. The signature of the meaning function is from syntactic domains (i.e. T) to semantic domains (i.e. D_T). A valuation map $V_T: T \to \mathbf{B}$ may be employed to check the static semantics prior to giving a meaning of the language construct.[10]

A denotational definition is more abstract than an operational definition. It does not specify the computational steps and its exclusive focus is on the programs to the exclusion of the state and other data elements. The state is less visible in denotational specifications.

It was developed by Christopher Strachey and Dana Scott at the Programming Research Group at Oxford, England in the mid-1960s, and their approach to semantics is known as the Scott–Strachey approach (Stoy 1977). It provides a mathematical foundation for the semantics of programming languages.

Dana Scott's contributions included the formulation of domain theory, which allows programs containing recursive functions and loops to be given a precise semantics. Each phrase in the language is translated into a mathematical object that is the *denotation* of the phrase. Denotational semantics has been applied to language design and implementation.

12.5 Lambda Calculus

Functions are an essential part of mathematics, and they play a key role in specifying the semantics of programming language constructs. We discussed partial and total functions in Chap. 3, and a function was defined as a special type of relation, and simple finite functions may be defined as an explicit set of pairs, e.g.

---

[10]This is similar to what a compiler does in that if errors are found during the compilation phase, the compiler halts and displays the errors and does not continue with code generation.

$$f\underline{\Delta}\{(a, 1), (b, 2), (c, 3)\}$$

However, for more complex functions, there is a need to define the function more abstractly, rather than listing all of its member pairs. This may be done in a similar manner to set comprehension, where a set is defined in terms of a characteristic property of its members.

Functions may be defined (by comprehension) through a powerful abstract notation known as lambda calculus. This notation was introduced by Alonzo Church in the 1930s to study computability (discussed in Chap. 13), and lambda calculus provides an abstract framework for describing mathematical functions and their evaluation. It may be used to study function definition, function application, parameter passing and recursion.

Any computable function can be expressed and evaluated using lambda calculus or Turing machines, as these are equivalent formalisms. Lambda calculus uses a small set of transformation rules, and these include the following:

– Alpha-conversion rule (α-conversion),[11]
– Beta-reduction rule (β-reduction) and[12]
– Eta-conversion rule (η-conversion)[13]

Every expression in the λ-calculus stands for a function with a single argument. The argument of the function is itself a function with a single argument, and so on. The definition of a function is anonymous in the calculus. For example, the function that adds one to its argument is usually defined as $f(x) = x + 1$. However, in λ-calculus, the function is defined as

$$succ\underline{\Delta}\lambda x. x + 1$$

The name of the formal argument x is irrelevant and an equivalent definition of the function is $\lambda z. z + 1$. The evaluation of a function f with respect to an argument (e.g. 3) is usually expressed by $f(3)$. In λ-calculus, this would be written as $(\lambda x. x + 1)$ 3, and this evaluates to $3 + 1 = 4$. Function application is *left associative*, i.e. $f x y = (f x) y$. A function of two variables is expressed in lambda calculus as a function of one argument, which returns a function of one argument. This is known as *currying*, e.g. the function $f(x, y) = x + y$ is written as $\lambda x. \lambda y. x + y$. This is often abbreviated to $\lambda x y. x + y$.

λ-calculus is a simple mathematical system, and its syntax is defined as follows:

---

[11]This essentially expresses that the names of bound variables are unimportant.
[12]This essentially expresses the idea of function application.
[13]This essentially expresses the idea that two functions are equal if and only if they give the same results for all arguments.

```
<exp> ::= <identifier>        |
         λ <identifier>.<exp> | --abstraction
         <exp> <exp>          | --application
         ( <exp> )
```

λ-calculus's four lines of syntax plus *conversion* rules are sufficient to define Booleans, integers, data structures and computations on them. It inspired Lisp and modern functional programming languages. The original calculus was untyped, but typed lambda calculi were later introduced. The typed lambda calculus allows the sets to which the function arguments apply to be specified. For example, the definition of the *plus* function is given as

$$plus \underline{\Delta} \lambda \, a, b : \mathbb{N}. \, a + b$$

The lambda calculus makes it possible to express properties of the function without reference to members of the base sets on which the function operates. It allows functional operations such as function composition to be applied, and one key benefit is that the calculus provides powerful support for higher order functions. This is important in the expression of the denotational semantics of the constructs of programming languages.

12.6 Lattices and Order

This section considers some mathematical structures used in the definition of the semantic domains used in denotational semantics. These mathematical structures may also be employed to give a secure foundation for recursion (discussed in Chap. 8), and it is essential that the conditions in which recursion may be used safely be understood.

It is natural to ask when presented with a recursive definition whether it means anything at all, and in some cases, the answer is negative. Recursive definitions are a powerful and elegant way of giving the denotational semantics of language constructs. The mathematical structures considered in this section include partial orders, total orders, lattices, complete lattices and complete partial orders.

12.6.1 Partially Ordered Sets

A *partial order* $\leq$ on a set P is a binary relation such that for all $x, y, z \in P$ the following properties hold (Fig. 12.5):

(i) $x \leq x$ (*Reflexivity*),
(ii) $x \leq y$ and $y \leq x \Rightarrow x = y$ (*Anti-symmetry*) and
(iii) $x \leq y$ and $y \leq z \Rightarrow x \leq z$ (*Transitivity*).

A set P with an order relation $\leq$ is said to be a *partially ordered* set.

Fig. 12.5 Pictorial
representation of a partial
order

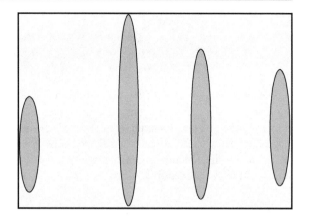

Example 12.4 Consider the power set $\mathbb{P}X$, which consists of all the subsets of the
set X with the ordering defined by set inclusion. That is, $A \leq B$ if and only if $A \subseteq B$ then $\subseteq$ is a partial order on $\mathbb{P}X$.

A partially ordered set is a *totally ordered* set (*also called chain*) if for all $x, y \in P$ then either $x \leq y$ or $y \leq x$. That is, any two elements of P are directly comparable.

A partially ordered set P is an *anti-chain* if for any x, y in P then $x \leq y$ only if $x = y$. That is, the only element in P that is comparable to a particular element is the element itself.

Maps between Ordered Sets

Let P and Q be partially ordered sets then a map ϕ from P to Q may preserve the order in P and Q. We distinguish between order preserving, order embedding and order isomorphism. These terms are defined as follows:

Order Preserving (or *Monotonic Increasing Function*)

A mapping $\phi \colon P \to Q$ is said to be order preserving if

$$x \leq y \Rightarrow \phi(x) \leq \phi(y).$$

Order Embedding

A mapping $\phi \colon P \to Q$ is said to be an order embedding if

$$x \leq y \text{ in } P \text{ if and only if } \phi(x) \leq \phi(y) \text{ in } Q.$$

Order Isomorphism

The mapping $\phi: P \rightarrow Q$ is an order isomorphism if and only if it is an order embedding mapping onto Q.

Dual of a Partially Ordered Set

The dual of a partially ordered set P (denoted P^{∂}) is a new partially ordered set formed from P where $x \leq y$ holds in P^{∂} if and only if $y \leq x$ holds in P (i.e. P^{∂} is obtained by reversing the order on P).

For each statement about P, there is a corresponding statement about P^{∂}. Given any statement Φ about a partially ordered set, then the dual statement Φ^{∂} is obtained by replacing each occurrence of $\leq$ by $\geq$ and vice versa.

Duality Principle

Given that statement Φ is true of a partially ordered set P, then the statement Φ^{∂} is true of P^{∂}.

Maximal and Minimum Elements

Let P be a partially ordered set and let $Q \subseteq P$, then

(i) $a \in Q$ is a *maximal* element of Q if $a \leq x \in Q \Rightarrow a = x$.
(ii) $a \in Q$ is the *greatest* (or *maximum*) element of Q if $a \geq x$ for every $x \in Q$, and in that case we write $a = $max Q.

A *minimal* element of Q and the *least* (or *minimum*) are defined dually by reversing the order. The greatest element (if it exists) is called the top element and is denoted by $\top$. The least element (if it exists) is called the bottom element and is denoted by $\bot$.

Example 12.5 Let X be a set and consider $\mathbb{P}X$ the set of all subsets of X with the ordering defined by set inclusion. The top element $\top$ is given by X and the bottom element $\bot$ is given by $\emptyset$.

A finite totally ordered set always has top and bottom elements, but an infinite chain need not have.

12.6.2   Lattices

Let P be a partially ordered set and let $S \subseteq P$. An element $x \in P$ is an upper bound of S if $s \leq x$ for all $s \in S$. A lower bound is defined similarly.

The set of all upper bounds for S is denoted by S^u, and the set of all lower bounds for S is denoted by S^l.

$$S^u = \{x \in P \mid (\forall s \in S) s \leq x\}$$
$$S^l = \{x \in P \mid (\forall s \in S) s \geq x\}$$

If S^u has a least element x then x is called the *least upper bound* of S. Similarly, if S^l has a greatest element x then x is called the *greatest lower bound* of S.

In other words, x is the least upper bound of S if

(i) x is an upper bound of S.
(ii) $x \leq y$ for all upper bounds y of S.

The least upper bound of S is also called the *supremum* of S denoted (sup S), and the greatest lower bound is also called the infimum of S, and is denoted by inf S.

Join and Meet Operations

The *join* of x and y (denoted by $x \vee y$) is given by sup$\{x,y\}$ when it exists. The *meet* of x and y (denoted by $x \wedge y$) is given by inf$\{x,y\}$ when it exists.

The supremum of S is denoted by $\vee S$, and the infimum of S is denoted by $\wedge S$.

Definition

Let P be a non-empty partially ordered set then

(i)   If $x \vee y$ and $x \wedge y$ exist for all x, $y \in P$ then P is called a *lattice*.
(ii) If $\vee S$ and $\wedge S$ exist for all $S \subseteq P$ then P is called a *complete lattice*.

Every non-empty finite subset of a lattice has a meet and a join (inductive argument can be used), and every finite lattice is a complete lattice. Further, any complete lattice is bounded, i.e. it has top and bottom elements (Fig. 12.6).

Example 12.6 Let X be a set and consider $\mathbb{P}X$ the set of all subsets of X with the ordering defined by set inclusion. Then $\mathbb{P}X$ is a complete lattice in which

$$\vee\{A_i \mid i \in I\} = \cup A_i$$
$$\wedge\{A_i \mid i \in I\} = \cap A_i$$

Consider the set of natural numbers $\mathbb{N}$ and consider the usual ordering of $<$. Then $\mathbb{N}$ is a lattice with the join and meet operations defined as

$$x \vee y = \max(x, y)$$
$$x \wedge y = \min(x, y)$$

Fig. 12.6 Pictorial
representation of a complete
lattice

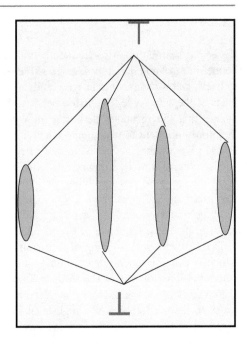

Another possible definition of the meet and join operations is in terms of the greatest common multiple and least common divisor.

$$x \lor y = \text{lcm}(x, y)$$
$$x \land y = \text{gcd}(x, y)$$

12.6.3 Complete Partial Orders

Let S be a non-empty subset of a partially ordered set P. Then

(i) S is said to be a *directed set* if for every finite subset F of S there exists $z \in S$ such that $z \in F^u$.
(ii) S is said to be *consistent* if for every finite subset F of S there exists $z \in P$ such that $z \in F^u$.

A partially ordered set P is a *complete partial order* (CPO) if

(i) P has a bottom element $\bot$.
(ii) $\lor D$ exists for each directed subset D of P.

The simplest example of a directed set is a chain, and we note that any complete lattice is a complete partial order, and that any finite lattice is a complete lattice.

12.6.4 Recursion

Recursive definitions arise frequently in programs and offer an elegant way to define routines and data types. A recursive routine contains a direct or indirect call to itself, and a recursive data type contains a direct or indirect reference to specimens of the same type. Recursion needs to be used with care, as there is always a danger that the recursive definition may be circular (i.e. defines nothing). It is therefore important to investigate when a recursive definition may be used safely, and to give a mathematical definition of recursion.

The control flow in a recursive routine must contain at least one non-recursive branch since if all possible branches included a recursive form the routine could never terminate. The value of at least one argument in the recursive call is different from the initial value of the formal argument as otherwise the recursive call would result in the same sequence of events and therefore would never terminate.

The mathematical meaning of recursion is defined in terms of *fixed-point theory*, which is concerned with determining solutions to equations of the form $x = \tau(x)$, where the function τ is of the form $\tau: X \to X$.

A recursive definition may be interpreted as a fixpoint equation of the form $f = \Phi(f)$, i.e. the fixpoint of a high-level functional Φ that takes a function as an argument. For example, consider the functional Φ defined as follows:

$$\Phi \underline{\Delta} \lambda f \; \lambda n \bullet \text{if } n = 0 \text{ then } 1 \text{ else } n * f(n-1)$$

Then, a fixpoint of Φ is a function f such that $f = \Phi(f)$ or, in other words,

$$f = \lambda n \bullet \text{if } n = 0 \text{ then } 1 \text{ else } n * f(n-1)$$

Clearly, the factorial function is a fixpoint of Φ, and it is the only total function that is a fixpoint. The solution of the equation $f = \Phi(f)$ (where Φ has a fixpoint) is determined as the limit f of the sequence of functions $f_0, f_1, f_2, \ldots$, where f_i are defined inductively as

$$f_0 \underline{\Delta} \varnothing \quad \text{(the empty partial function)}$$
$$f_i \underline{\Delta} \Phi(f_{i-1})$$

Each f_i may be viewed as a successive approximation to the true solution f of the fixpoint equation, with each f_i bringing a little more information on the solution than its predecessor f_{i-1}.

The function f_i is defined for one more value than f_{i-1}, and gives the same result for any value for which they are both defined. The definition of the factorial function is thus built up as follows:

$$f_0 \underline{\Delta} \varnothing \quad \text{(the empty partial function)}$$
$$f_1 \underline{\Delta} \{0 \rightarrow 1\}$$
$$f_2 \underline{\Delta} \{0 \rightarrow 1, 1 \rightarrow 1\}$$
$$f_3 \underline{\Delta} \{0 \rightarrow 1, 1 \rightarrow 1, 2 \rightarrow 2\}$$
$$f_4 \underline{\Delta} \{0 \rightarrow 1, 1 \rightarrow 1, 2 \rightarrow 2, 3 \rightarrow 6\}$$

For every i, the domain of f_i is the interval $1, 2, \ldots\ldots i-1$ and $f_i(n) = n!$ for any n in this interval. In other words, f_i is the factorial function restricted to the interval $1, 2, \ldots.. i-1$. The sequence of f_i may be viewed as successive approximations of the true solution of the fixpoint equation (which is the factorial function), with each f_i bringing defined for one more value than its predecessor f_{i-1}, and defining the same result for any value for which they are both defined.

The candidate fixpoint f_∞ is the limit of the sequence of functions f_i, and is the union of all the elements in the sequence. It may be written as follows:

$$f_\infty \underline{\Delta} \varnothing \cup \Phi(\varnothing) \cup \Phi(\Phi(\varnothing)) \cup \ldots\ldots\ldots = \cup_{i:\mathbb{N}} f_i$$

where the sequence f_i is defined inductively as

$$f_0 \underline{\Delta} \varnothing \qquad \text{(the empty partial function)}$$
$$f_{i+1} \underline{\Delta} f_i \cup \Phi(f_i)$$

This forms a subset chain where each element is a subset of the next, and it follows by induction that

$$f_{i+1} = \cup_{j:0\ldots i} \Phi(f_i)$$

A general technique for solving fixpoint equations of the form $h = \tau(h)$ for some functional τ is to start with the least defined function $\varnothing$ and iterate with τ. The union of all the functions obtained as successive sequence elements is the fixpoint.

The conditions in which f_∞ is a fixpoint of Φ are the requirement for $\Phi(f_\infty) = f_\infty$. This is equivalent to

$$\Phi(\cup_{i:\mathbb{N}} f_i) = \cup_{i:\mathbb{N}} f_i$$
$$\Phi(\cup_{i:\mathbb{N}} f_i) = \cup_{i:\mathbb{N}} \Phi(f_i)$$

A sufficient point for Φ to have a fixpoint is that the property $\Phi(\cup_{i:\mathbb{N}} f_i) = \cup_{i:\mathbb{N}} \Phi(f_i)$ holds for any subset chain f_i.

A more detailed account on the mathematics of recursion is in Chap. 8 of Meyer (1990).

12.7   Review Questions

1. Explain the difference between syntax and semantics.
2. Describe the Chomsky hierarchy of grammars and give examples of each type.
3. Show that a grammar may be ambiguous leading to two difference parse trees. What problems does this create and how should it be dealt with?
4. Describe axiomatic semantics, operation semantics and denotational semantics and explain the differences between them.
5. Explain partial orders, lattices and complete partial orders. Give examples of each.
6. Show how the meaning of recursion is defined with fixpoint theory.

12.8   Summary

This chapter considered the syntax and semantics of programming languages. The syntax of the language is concerned with the production of grammatically correct programs in the language, whereas the semantics of the language is deeper and is concerned with the meaning of what has been written by the programmer.

The semantics of programming languages may be given by axiomatic, operational and denotational semantics. Axiomatic semantics is concerned with defining properties of the language in terms of axioms; operational semantics is concerned with defining the meaning of the language in terms of an interpreter and denotational semantics is concerned with defining the meaning of the phrases in a language by the denotation or mathematical meaning of the phrase.

Compilers are programs that translate a program that is written in some programming language into another form. It involves syntax analysis and parsing to check the syntactic validity of the program; semantic analysis to determine what the program should do; optimization to improve the speed and performance of the compiler and code generation in some target language.

Various mathematical structures including partial orders, total orders, lattices and complete partial orders were considered. These are useful in the definition of the denotational semantics of a language, and in giving a mathematical interpretation of recursion.

References

Floyd R (1967) Assigning meanings to programs. Proc Symp Appl Math (19):19–32
Hoare CAR (1969) An axiomatic basis for computer programming. Commun ACM 12 (10):576–585

Meyer (1990) Introduction to the theory of programming languages. Prentice Hall

Naur P (ed) (1960) Report on the algorithmic language, ALGOL 60. Commun ACM 3(5):299–314

O'Regan G (2006) Mathematical approaches to software quality. Springer, Berlin

O'Regan G (2016) Introduction to the history of computing. Springer, Berlin

Plotkin G (1981) A structural approach to operational semantics. Technical Report DAIM FN-19. Computer Science Department. Aarhus University, Denmark

Stoy J (1977) Denotational semantics. the scott-strachey approach to programming language theory. MIT Press

Computability and Decidability

13

Key Topics

Computability
Completeness
Decidability
Formalism
Logicism

13.1 Introduction

It is impossible for a human or machine to write out all of the members of an infinite countable set, such as the set of natural numbers $\mathbb{N}$. However, humans can do something quite useful in the case of certain enumerable infinite sets: they can give explicit instructions (that may be followed by a machine or another human) to produce the nth member of the set for an arbitrary finite n. The problem remains that for all but a finite number of values of n it will be physically impossible for any human or machine to actually carry out the computation, due to the limitations on the time available for computation, the speed at which the individual steps in the computation may be carried out and due to finite materials.

The intuitive meaning of computability is in terms of an algorithm (or effective procedure) that specifies a set of instructions to be followed to complete the task. In other words, a function f is *computable* if there exists an algorithm that produces the value of f correctly for each possible argument of f. The computation of f for a particular argument x just involves following the instructions in the algorithm, and

© Springer Nature Switzerland AG 2020
G. O'Regan, *Mathematics in Computing*, Undergraduate Topics
in Computer Science, https://doi.org/10.1007/978-3-030-34209-8_13

it produces the result $f(x)$ in a finite number of steps if x is in the domain of f. If x is not in the domain of f then the algorithm may produce an answer saying so or it might run forever never halting. A computer program implements an algorithm.

The concept of computability may be made precise in several equivalent ways such as Church's *lambda calculus, recursive function theory* or by the theoretical *Turing machines*.[1] These are all equivalent and perhaps the most well known is the Turing machine (discussed in Chap. 23). This is a mathematical machine with a potentially infinite tape divided into frames (or cells) in which very basic operations can be carried out. The set of functions that are computable are those that are computable by a Turing machine.

Decidability is an important topic in contemporary mathematics. Church and Turing independently showed in 1936 that mathematics is not decidable. In other words, there is no mechanical procedure (i.e. algorithm) to determine whether an arbitrary mathematical proposition is true or false, and so the only way to determine the truth or falsity of a statement is trying to solve the problem. The fact that there is no general method to solve all instances of a specific problem, as well as the impossibility of proving or disproving certain statements within a formal system may indicate limitations to human and machine knowledge.

13.2   Logicism and Formalism

Gottlob Frege (Fig. 15.2) was a nineteenth-century German mathematician and logician who invented a formal system which is the basis of modern predicate logic. It included axioms, definitions, universal and existential quantification, and formalization of proof. His objective was to show that mathematics was reducible to logic (logicism) but his project failed as one of the axioms that he had added to his system led to inconsistency.

The inconsistency was pointed out by Bertrand Russell, and it is known as *Russell's paradox*[2] (see Chap. 3). Russell later introduced his theory of types to deal with the paradox, and he jointly published *Principia Mathematica* with Whitehead as an attempt to derive the truths of arithmetic from a set of logical axioms and rules of inference.

The sentences of Frege's logical system denote the truth-values of true or false. The sentences may include expressions such as equality $(x = y)$, and this returns true if x is the same as y, and false otherwise. Similarly, a more complex expression such as $f(x, y, z) = w$ is true if $f(x, y, z)$ is identical with w, and false otherwise. Frege represented statements such as '5 is a prime' by '$P(5)$' where $P()$ is termed a concept. The statement $P(x)$ returns true if x is prime and false otherwise. His

---

[1]The Church–Turing thesis states that anything that is computable is computable by a Turing machine.

[2]Russell's paradox considers the question as to whether the set of all sets that contain themselves as members is a set. In either case, there is a contradiction.

approach was to represent a predicate as a function of one variable which returns a Boolean value of true or false.

Formalism was proposed by Hilbert (Fig. 13.1) as a foundation for mathematics in the early twentieth century. The motivation for the program was to provide a secure foundation for mathematics, and to resolve the contradictions in the formalization of set theory identified by Russell's paradox. The presence of a contradiction in a theory means the collapse of the whole theory, and so it was seen as essential that there be a proof of the consistency of the formal system. The methods of proof in mathematics are formalized with axioms and rules of inference.

Formalism is a formal system that contains meaningless symbols together with rules for manipulating them. The individual formulas are certain finite sequences of symbols obeying the syntactic rules of the formal language. A formal system consists of

- A formal language,
- A set of axioms and
- Rules of inference.

The expressions in a formal system are terms, and a term may be simple or complex. A simple term may be an object such as a number and a complex term may be an arithmetic expression such as $4^3 + 1$. A complex term is formed via functions, and the expression above uses two functions, namely, the cube function with argument 4 and the plus function with two arguments.

Fig. 13.1 David Hilbert

A formal system is generally intended to represent some aspect of the real world. A rule of inference relates a set of formulas $(P_1, P_2, ..., P_k)$ called the premises to the consequence formula P called the conclusion. For each rule of inference, there is a finite procedure for determining whether a given formula Q is an immediate consequence of the rule from the given formulas $(P_1, P_2, ..., P_k)$. A *proof* in a formal system consists of a finite sequence of formulae, where each formula is either an axiom or derived from one or more preceding formulae in the sequence by one of the rules of inference.

Hilbert's program was concerned with the formalization of mathematics (i.e. the axiomatization of mathematics) together with a proof that the axiomatization is consistent (i.e. there is no formula A such that both A and $\neg A$ are deducible in the calculus). Its specific objectives were to

- Provide a formalism of mathematics.
- Show that the formalization of mathematics is *complete*, i.e. all mathematical truths can be proved in the formal system.
- Provide a proof that the formal system is *consistent* (i.e. that no contradictions may be derived).
- Show that mathematics is *decidable,* i.e. there is an algorithm to determine the truth or falsity of any mathematical statement.

The formalist movement in mathematics led to the formalization of large parts of mathematics, where theorems could be proved using just a few mechanical rules. The two most comprehensive formal systems developed were *Principia Mathematica* by Russell and Whitehead, and the axiomatization of set theory by Zermelo–Fraenkel (subsequently developed further by von Neumann).

Principia Mathematica is a comprehensive three volume work on the logical foundations of mathematics written by Bertrand Russel and Alfred Whitehead between 1910 and 1913. Its goal was to show that all of the concepts of mathematics can be expressed in logic, and that all of the theorems of mathematics can be proved using only the logical axioms and rules of inference of logic. It covered set theory, ordinal numbers and real numbers, and it showed that in principle that large parts of mathematics could be developed using *logicism.*

It avoided the problems with contradictions that arose with Frege's system by introducing the theory of types in the system. The theory of types meant that one could no longer speak of the set of all sets, as a set of elements is of a different type from that of each of its elements, and so Russell's paradox was avoided. It remained an open question at the time as to whether the *Principia* was consistent and complete. That is, is it possible to derive all the truths of arithmetic in the system and is it possible to derive a contradiction from the Principia's axioms? However, it was clear from the three volume work that the development of mathematics using the approach of the Principia was extremely lengthy and time-consuming.

13.3 Decidability

The question remained whether these axioms and rules of inference are sufficient to decide any mathematical question that can be expressed in these systems. Hilbert believed that every mathematical problem could be solved, and that the truth or falsity of any mathematical proposition could be determined in a finite number of steps. He outlined 23 key problems in 1900 that needed to be solved by mathematicians in the twentieth century.

He believed that the formalism of mathematics would allow a mechanical procedure (or algorithm) to determine whether a particular statement was true or false. The problem of the decidability of mathematics is known as the decision problem (*Entscheidungsproblem*).

The question of the decidability of mathematics had been considered by Leibnitz in the seventeenth century. He had constructed a mechanical calculating machine, and wondered if a machine could be built that could determine whether particular mathematical statements are true or false.

Definition 13.1 (Decidability) *Mathematics is decidable if the truth or falsity of any mathematical proposition may be determined by an algorithm.*

Church and Turing independently showed this to be impossible in 1936. Church developed the lambda calculus in the 1930s as a tool to study computability,[3] and he showed that anything that is computable is computable by the lambda calculus. Turing showed that decidability was related to the halting problem for Turing machines, and that, therefore, if first-order logic were decidable then the halting problem for Turing machines could be solved. However, he had already proved that there was no general algorithm to determine whether a given Turing machine halts. Therefore, first-order logic is undecidable.

The question as to whether a given Turing machine halts or not can be formulated as a first-order statement. If a general decision procedure exists for first-order logic, then the statement of whether a given Turing machine halts or not is within the scope of the decision algorithm. However, Turing had already proved that the halting problem for Turing machines is not computable, i.e. it is not possible algorithmically to decide whether or not any given Turing machine will halt or not. Therefore, since there is no general algorithm that can decide whether any given Turing machine halts, there is no general decision procedure for first-order logic. The only way to determine whether a statement is true or false is to try to solve it. However, if one tries but does not succeed this does not prove that an answer does not exist.

There are first-order theories that are decidable. However, first-order logic that includes Peano's axioms of arithmetic (or any formal system that includes addition and multiplication) cannot be decided by an algorithm. That is, there is no algorithm

---

[3]The Church–Turing thesis states that anytime that is computable is computable by lambda calculus or equivalently by a Turing machine.

Fig. 13.2 Kurt Gödel

to determine whether an arbitrary mathematical proposition is true or false. Propositional logic is decidable as there is a procedure (e.g. using a truth table) to determine whether an arbitrary formula is valid[4] in the calculus.

Gödel (Fig. 13.2) proved that first-order predicate calculus is *complete*, i.e. all truths in the predicate calculus can be proved in the language of the calculus.

Definition 13.2 (Completeness) *A formal system is complete if all the truths in the system can be derived from the axioms and rules of inference.*

Gödel's *first incompleteness theorem* showed that first-order arithmetic is incomplete, i.e. there are truths in first-order arithmetic that cannot be proved in the language of the axiomatization of first-order arithmetic. Gödel's *second incompleteness theorem* showed that any formal system extending basic arithmetic cannot prove its own consistency within the formal system.

Definition 13.3 (Consistency) *A formal system is consistent if there is no formula A such that A and ¬A are provable in the system (i.e. there are no contradictions in the system).*

---

[4]A well-formed formula is valid if it follows from the axioms of first-order logic. A formula is valid if and only if it is true in every interpretation of the formula in the model.

13.4   Computability

Alonzo Church (Fig. 13.3) developed the lambda calculus in the mid-1930s, as part of his work into the foundations of mathematics. Turing published a key paper on computability in 1936, which introduced the theoretical machine known as the Turing machine. This machine is computationally equivalent to the lambda calculus, and is capable of performing any conceivable mathematical problem that has an algorithm.

Definition 13.4 (Algorithm) *An algorithm (or effective procedure) is a finite set of unambiguous instructions to perform a specific task.*

A function is *computable* if there is an effective procedure or algorithm to compute f for each value of its domain. The algorithm is finite in length and sufficiently detailed so that a person can execute the instructions in the algorithm. The execution of the algorithm will halt in a finite number of steps to produce the value of $f(x)$ for all x in the domain of f. However, if x is not in the domain of f then the algorithm may produce an answer saying so, or it may get stuck, or it may run forever never halting.

The *Church–Turing thesis* states that *any computable function may be computed by a Turing machine*. There is overwhelming evidence in support of this thesis, including the fact that alternative formalizations of computability in terms of lambda calculus, recursive function theory and Post systems have all been shown to be equivalent to Turing machines.

Fig. 13.3 Alonzo Church

A Turing machine (discussed in Chap. 23) consists of a head and a potentially infinite tape that is divided into cells. Each cell on the tape may be either blank or printed with a symbol from a finite alphabet of symbols. The input tape may initially be blank or have a finite number of cells containing symbols.

At any step, the head can read the contents of a frame. The head may erase a symbol on the tape, leave it unchanged or replace it with another symbol. It may then move one position to the right, one position to the left or not at all. If the frame is blank, the head can either leave the frame blank or print one of the symbols.

Turing believed that a human with finite equipment and with an unlimited supply of paper could do every calculation. The unlimited supply of paper is formalized in the Turing machine by a tape marked off in cells.

The Turing machine is a simple theoretical machine, but it is equivalent to an actual physical computer in the sense that they both compute exactly the same set of functions. A Turing machine is easier to analyse and prove things about than a real computer. The formal definition of a Turing machine as a 7-tuple $M = (Q, \Gamma, b, \Sigma, \delta, q_0, F)$ is given in Chap. 23.

A Turing machine is essentially a Finite-State Machine (FSM) with an unbounded tape. The machine may read from and write to the tape and the tape provides memory and acts as the store. The finite-state machine is essentially the control unit of the machine, whereas the tape is a potentially infinite and unbounded store. A real computer has a large but finite store, whereas the store in a Turing machine is potentially infinite. However, the store in a real computer may be extended with backing tapes and discs, and in a sense may be regarded as unbounded. The maximum amount of tape that may be read or written within n steps is n.

A Turing machine has an associated set of rules that defines its behaviour. These rules are defined by the transition function that specifies the actions that a machine will perform with respect to a particular input. The behaviour will depend on the current state of the machine and the contents of the tape.

A Turing machine may be programmed to solve any problem for which there is an algorithm. However, if the problem is unsolvable then the machine will either stop in a non-accepting state or compute forever. The solvability of a problem may not be determined beforehand, but there is, of course, some answer (i.e. either the machine either halts or it computes forever).

Turing showed that there was no solution to the decision problem (*Entscheidungsproblem*) posed by Hilbert. Hilbert believed that the truth or falsity of a mathematical problem may always be determined by a mechanical procedure, and he believed that first-order logic is decidable, i.e. there is a decision procedure to determine if an arbitrary formula is a theorem of the logical system.

Turing also introduced the concept of a Universal Turing Machine and this machine is able to simulate any other Turing machine. Turing's results on computability were proved independently of Church's lambda calculus equivalent results in computability. Turing studied at Princeton University in 1937 and 1938 and was awarded a PhD from the university in 1938. His research supervisor was Alonzo Church.[5]

Question 13.1 (Halting Problem) *Given an arbitrary program is there an algorithm to decide whether the program will finish running or will it continue running forever? In other words, given a program and an input will the program eventually halt and produce an output or will it run forever?*

Note (Halting Problem) The halting problem was one of the first problems that was shown to be undecidable, i.e. there is no general decision procedure or algorithm that may be applied to an arbitrary program and input to decide whether the program halts or not when running with that input.

Proof We assume that there is an algorithm (i.e. a computable function $H(i, j)$) that takes any program i (program i refers to the i-th program in the enumeration of all the programs) and arbitrary input j to the program such that

$$H(i,j) = \begin{cases} 1 & \text{If program } i \text{ halts on input } j. \\ 0 & \text{otherwise} \end{cases}$$

We then employ a diagonalization argument[6] to show that every computable total function f with two arguments differs from the desired function H. First, we construct a partial function g from any computable function f with two arguments such that g is computable by some program e.

$$g(i) = \begin{cases} 0 & \text{if } f(i,i) = 0 \\ \text{undefined} & \text{otherwise} \end{cases}$$

There is a program e that computes g and this program is one of the programs in which the halting problem is defined. One of the following two cases must hold:

---

[5]Alonzo Church was a famous American mathematician and logician who developed the lambda calculus. He also showed that Peano arithmetic and first-order logic were undecidable. Lambda calculus is equivalent to Turing machines and whatever may be computed is computable by Lambda calculus or a Turing machine.

[6]This is similar to Cantor's diagonalization argument that shows that the real numbers are uncountable. This argument assumes that it is possible to enumerate all real numbers between 0 and 1, and it then constructs a number whose nth decimal differs from the nth decimal position in the nth number in the enumeration. If this holds for all n, then the newly defined number is not among the enumerated numbers.

$$g(e) = f(e,e) = 0 \tag{13.1}$$

In this case, $H(e, e) = 1$ because e halts on input e.

$$g(e) \text{ is undefined and } f(e,e) \neq 0 \tag{13.2}$$

In this case, $H(e, e) = 0$ because the program e does not halt on input e.

In either case, f is not the same function as H. Further, since f was an arbitrary total computable function all such functions must differ from H. Hence, the function H is not computable and there is no such algorithm to determine whether an arbitrary Turing machine halts for an input x. Therefore, the halting problem is not decidable.

13.5 Computational Complexity

An algorithm is of little practical use if it takes millions of years to compute particular instances. There is a need to consider the efficiency of the algorithm due to practical considerations. Chapter 10 discussed cryptography and the RSA algorithm, and the security of the RSA encryption algorithm is due to the fact that there is no known efficient algorithm to determine the prime factors of a large number.

There are often slow and fast algorithms for the same problem, and a measure of complexity is the number of steps in a computation. An algorithm is of *time complexity* $f(n)$ if for all n and all inputs of length n the execution of the algorithm takes at most $f(n)$ steps.

An algorithm is said to be *polynomially bounded* if there is a polynomial $p(n)$ such that for all n and all inputs of length n the execution of the algorithm takes at most $p(n)$ steps. The notation P is used for all problems that can be solved in polynomial time.

A problem is said to be *computationally intractable* if it may not be solved in polynomial time, that is, there is no known algorithm to solve the problem in polynomial time.

A problem L is said to be in the set *NP* (non-deterministic polynomial time problems) if any given solution to L can be verified quickly in polynomial time. A non-deterministic Turing machine may have several possibilities for its behaviour, and an input may give rise to several computations.

A problem is *NP-complete* if it is in the set NP of non-deterministic polynomial time problems and it is also in the class of *NP-hard* problems. A key characteristic to NP-complete problems is that there is no known fast solution to them, and the time required to solve the problem using known algorithms increases quickly as the size of the problem grows. Often, the time required to solve the problem is in billions or trillions of years. Although any given solution can be verified quickly there is no known efficient way to find a solution.

13.6   Review Questions

1. Explain computability and decidability.
2. What were the goals of logicism and formalism and how successful were these movement in mathematics?
3. What is a formal system?
4. Explain the difference between consistency, completeness and decidability.
5. Describe a Turing machine and explain its significance in computability.
6. Describe the halting problem and show that it is undecidable.
7. Discuss the complexity of an algorithm and explain terms such as 'polynomial bounded', 'computationally intractable' and 'NP-complete'.

13.7   Summary

This chapter provided an introduction to computability and decidability. The intuitive meaning of computability is that in terms of an algorithm (or effective procedure) that specifies a set of instructions to be followed to solve the problem. In other words, a function f is computable if there exists an algorithm that produces the value of f correctly for each possible argument of f. The computation of f for a particular argument x just involves following the instructions in the algorithm, and it produces the result $f(x)$ in a finite number of steps if x is in the domain of f.

The concept of computability may be made precise in several equivalent ways such as Church's lambda calculus, recursive function theory or by the theoretical Turing machines. The Turing machine is a mathematical machine with a potentially infinite tape divided into frames (or cells) in which very basic operations can be carried out. The set of functions that are computable are those that are computable by a Turing machine.

A formal system contains meaningless symbols together with rules for manipulating them, and is generally intended to represent some aspect of the real world. The individual formulas are certain finite sequences of symbols obeying the syntactic rules of the formal language. A formal system consists of a formal language, a set of axioms and rules of inference.

Church and Turing independently showed in 1936 that mathematics is not decidable. In other words, it is not possible to determine the truth or falsity of any mathematical proposition by an algorithm.

Turing had already proved that the halting problem for Turing machines is not computable, i.e. it is not possible algorithmically to decide whether a given Turing machine will halt or not. He then applied this result to first-order logic to show that it is undecidable. That is, the only way to determine whether a statement is true or false is to try to solve it.

The complexity of an algorithm was discussed, and it was noted that an algorithm is of little practical use if it takes millions of years to compute the solution. There is a need to consider the efficiency of the algorithm due to practical considerations. The class of polynomial time-bound problems and non-deterministic polynomial time problems was considered, and it was noted that the security of various cryptographic algorithms is due to the fact that there are no time-efficient algorithms to determine the prime factors of large integers.

The reader is referred to (Rozenberg and Salomaa 1994) for a more detailed account of decidability and computability.

Reference

Rozenberg G, Salomaa A (1994) Cornerstones of undecidability. Prentice Hall

Matrix Theory

14

14.1 Introduction

A *matrix* is a rectangular array of numbers that consists of horizontal rows and vertical columns. A matrix with m rows and n columns is termed an $m \times n$ matrix, where m and n are its dimensions. A matrix with an equal number of rows and columns (e.g. n rows and n columns) is termed a *square* matrix. Figure 14.1 is an example of a square matrix with four rows and four columns.

The entry in the i-th row and the j-th column of a matrix A is denoted by $A[i, j]$, $A_{i,j}$, or a_{ij}, and the matrix A may be denoted by the formula for its (i, j)-th entry, i.e. (a_{ij}) where i ranges from 1 to m and j ranges from 1 to n.

An $m \times 1$ matrix is termed a *column vector* and a $1 \times n$ matrix is termed a *row vector*. Any row or column of an $m \times n$ matrix determines a row or column vector which is obtained by removing the other rows (respectively, columns) from the matrix. For example, the row vector $(11, -5, 5, 3)$ is obtained from the matrix example by removing rows 1, 2 and 4 of the matrix.

© Springer Nature Switzerland AG 2020   221
G. O'Regan, *Mathematics in Computing*, Undergraduate Topics
in Computer Science, https://doi.org/10.1007/978-3-030-34209-8_14

$$\begin{pmatrix} 6 & 0 & -2 & 3 \\ 4 & 2 & 3 & 7 \\ 11 & -5 & 5 & 3 \\ 3 & -5 & -8 & 1 \end{pmatrix}$$

Fig. 14.1 Example of a 4×4 square matrix

Two matrices A and B are equal if they are both of the same dimensions, and if $a_{ij} = b_{ij}$ for each $i = 1, 2,..., m$ and each $j = 1, 2, ..., n$.

Matrices be added and multiplied (provided certain conditions are satisfied). There are identity matrices under the addition and multiplication binary operations such that the addition of the (additive) identity matrix to any matrix A yields A and similarly for the multiplicative identity. Square matrices have inverses (provided that their determinant is non-zero), and every square matrix satisfies its characteristic polynomial.

It is possible to consider matrices with infinite rows and columns, and although it is not possible to write down such matrices explicitly it is still possible to add, subtract and multiply by a scalar provided there is a well-defined entry in each (i, j)-th element of the matrix.

Matrices are an example of an algebraic structure known as an *algebra*. Chapter 6 discussed several algebraic structures such as groups, rings, fields and vector spaces. The matrix algebra for $m \times n$ matrices A, B, C and scalars λ, μ satisfies the following properties (there are additional multiplicative properties for square matrices):

1. $A + B = B + A$,
2. $A + (B + C) = (A + B) + C$,
3. $A + 0 = 0 + A = A$,
4. $A + (-A) = (-A) + A = 0$,
5. $\lambda(A + B) = \lambda A + \lambda B$,
6. $(\lambda + \mu)A = \lambda A + \mu B$,
7. $\lambda(\mu A) = (\lambda \mu)A$ and
8. $1A = A$.

Matrices have many applications including their use in graph theory to keep track of the distance between pairs of vertices in the graph; a rotation matrix may be employed to represent the rotation of a vector in three-dimensional space. The product of two matrices represents the composition of two linear transformations, and matrices may be employed to determine the solution to a set of linear equations. They also arise in computer graphics and may be employed to project a three-dimensional image onto a two-dimensional screen. It is essential to employ efficient algorithms for matrix computation, and this is an active area of research in the field of numerical analysis.

14.2   Two × Two Matrices

Matrices arose in practice as a means of solving a set of linear equations. One of the earliest examples of their use is in a Chinese text dating from between 300 BC and 200 AD. The Chinese text showed how matrices could be employed to solve simultaneous equations. Consider the set of equations:

$$ax + by = r$$
$$cx + dy = s$$

Then the coefficients of the linear equations in x and y above may be represented by the matrix A, where A is given by

$$A = \begin{bmatrix} a & b \\ c & d \end{bmatrix}$$

The linear equations may be represented as the multiplication of the matrix A and a vector $\underline{x}$ resulting in a vector $\underline{v}$:

$$A\underline{x} = \underline{v}$$

The matrix representation of the linear equations and its solution is as follows:

$$\begin{bmatrix} a & b \\ c & d \end{bmatrix} \begin{bmatrix} x \\ y \end{bmatrix} = \begin{bmatrix} r \\ s \end{bmatrix}$$

The vector $\underline{x}$ may be calculated by determining the inverse of the matrix A (provided that its inverse exists). The vector $\underline{x}$ is then given by

$$\underline{x} = A^{-1}\underline{v}$$

The solution to the set of linear equations is then given by

$$\begin{bmatrix} x \\ y \end{bmatrix} = \begin{bmatrix} a & b \\ c & d \end{bmatrix}^{-1} \begin{bmatrix} r \\ s \end{bmatrix}$$

The inverse of a matrix A exists if and only if its *determinant* is non-zero, and if this is the case, the vector $\underline{x}$ is given by

$$\begin{bmatrix} x \\ y \end{bmatrix} = \frac{1}{\det} \begin{bmatrix} d & -b \\ -c & a \end{bmatrix} \begin{bmatrix} r \\ s \end{bmatrix}$$

The determinant of a 2 × 2 matrix A is given by

$$\det A = ad - cb.$$

The determinant of a 2 × 2 matrix is denoted by

$$\begin{vmatrix} a & b \\ c & d \end{vmatrix}$$

A key property of determinants is that

$$\det(AB) = \det(A) \cdot \det(B)$$

The transpose of a 2 × 2 matrix A (denoted by A^T) involves exchanging rows and columns, and is given by

$$A^T = \begin{bmatrix} a & c \\ b & d \end{bmatrix}$$

The inverse of the matrix A (denoted by A^{-1}) is given by

$$A^{-1} = \frac{1}{\det} A \begin{bmatrix} d & -b \\ -c & a \end{bmatrix}$$

Further, $A \cdot A^{-1} = A^{-1} \cdot A = I$ where I is the identity matrix of the algebra of 2 × 2 matrices under multiplication. That is,

$$AA^{-1} = A^{-1}A = \begin{bmatrix} 1 & 0 \\ 0 & 1 \end{bmatrix}$$

The addition of two 2 × 2 matrices A and B is given by a matrix whose entries are the addition of the individual components of A and B. The addition of two matrices is commutative and we have

$$A + B = B + A = \begin{bmatrix} a+p & b+q \\ c+r & d+s \end{bmatrix}$$

where A, B are given by

$$A = \begin{bmatrix} a & b \\ c & d \end{bmatrix} \quad B = \begin{bmatrix} p & q \\ r & s \end{bmatrix}$$

The identity matrix under addition is given by the matrix whose entries are all 0, and it has the property that A + 0 = 0 + A = A.

$$\begin{bmatrix} 0 & 0 \\ 0 & 0 \end{bmatrix}$$

The multiplication of two 2×2 matrices is given by

$$AB = \begin{bmatrix} ap+br & aq+bs \\ cp+dr & cq+ds \end{bmatrix}$$

The multiplication of matrices is not commutative, i.e. AB $\neq$ BA. The multiplicative identity matrix I has the property that AI = IA = A, and it is given by

$$I = \begin{bmatrix} 1 & 0 \\ 0 & 1 \end{bmatrix}$$

A matrix A may be multiplied by a scalar λ, and this yields the matrix λA where each entry in A is multiplied by the scalar λ. That is, the entries in the matrix λA are λa_{ij}.

14.3 Matrix Operations

More general sets of linear equations may be solved with $m \times n$ matrices (i.e. a matrix with m rows and n columns) or square $n \times n$ matrices. In this section, we consider several matrix operations including addition, subtraction, multiplication of matrices, scalar multiplication and the transpose of a matrix.

The addition and subtraction of two matrices A, B are meaningful if and only if A and B have the same dimensions, i.e. they are both $m \times n$ matrices. In this case, A + B is defined by adding the corresponding entries:

$$(A+B)_{ij} = A_{ij} + B_{ij}$$

The additive identity matrix for the square $n \times n$ matrices is denoted by 0, where 0 is an $n \times n$ matrix whose entries are zero, i.e. $r_{ij} = 0$ for all i, j where $1 \leq i \leq n$ and $1 \leq j \leq n$.

The scalar multiplication of a matrix A by a scalar k is meaningful and the resulting matrix kA is given by

$$(kA)_{ij} = kA_{ij}$$

$$
\begin{pmatrix}
a_{11} & a_{12} & a_{13} & \cdots & a_{1n} \\
a_{21} & a_{22} & a_{23} & \cdots & a_{2n} \\
a_{31} & a_{32} & a_{33} & \cdots & a_{3n} \\
\cdots & \cdots & \cdots & \cdots & \cdots \\
\cdots & \cdots & \cdots & \cdots & \cdots \\
a_{m1} & a_{m2} & a_{m3} & \cdots & a_{mn}
\end{pmatrix}
\begin{pmatrix}
b_{11} & b_{12} & \cdots & b_{1p} \\
b_{21} & b_{22} & \cdots & b_{2p} \\
b_{31} & b_{32} & \cdots & b_{3p} \\
\cdots & \cdots & \cdots & \cdots \\
\cdots & \cdots & \cdots & \cdots \\
\cdots & \cdots & \cdots & \cdots \\
b_{n1} & b_{n2} & \cdots & b_{np}
\end{pmatrix}
=
\begin{pmatrix}
c_{11} & c_{12} & \cdots & c_{1p} \\
c_{21} & c_{22} & \cdots & c_{2p} \\
c_{31} & c_{32} & \cdots & c_{3p} \\
\cdots & \cdots & \cdots & \cdots \\
\cdots & \cdots & \cdots & \cdots \\
c_{m1} & c_{m2} & \cdots & c_{mp}
\end{pmatrix}
$$

      m rows, n columns   *n rows, p columns*   *m rows, p columns*

Fig. 14.2 Multiplication of two matrices

The multiplication of two matrices A and B is meaningful if and only if the number of columns of A is equal to the number of rows of B (Fig. 14.2), i.e. A is an $m \times n$ matrix and B is an $n \times p$ matrix and the resulting matrix AB is an $m \times p$ matrix.

Let $A = (a_{ij})$ where i ranges from 1 to m and j ranges from 1 to n, and let $B = (b_{jl})$ where j ranges from 1 to n and l ranges from 1 to p. Then AB is given by (c_{il}) where i ranges from 1 to m and l ranges from 1 to p with c_{il} given by

$$
c_{il} = \sum_{k=1}^{n} a_{ik} b_{kl}
$$

That is, the entry (c_{il}) is given by multiplying the i-th row in A by the lth column in B followed by a summation. Matrix multiplication is not commutative, i.e. $AB \neq BA$.

The identity matrix I is an $n \times n$ matrix and the entries are given by r_{ij} where $r_{ii} = 1$ and $r_{ij} = 0$ where $i \neq j$ (Fig. 14.3). A matrix that has non-zero entries only on the diagonal is termed as *diagonal matrix*. A triangular matrix is a square matrix in which all the entries above or below the main diagonal are zero. A matrix is an *upper triangular* matrix if all entries below the main diagonal are zero, and *lower triangular* if all of the entries above the main diagonal are zero. Upper triangular and lower triangular matrices form a sub-algebra of the algebra of square matrices.

A key property of the identity matrix is that for all $n \times n$ matrices A, we have

$$
AI = IA = A
$$

The inverse of an $n \times n$ matrix A is a matrix A^{-1} such that

$$
AA^{-1} = A^{-1}A = I
$$

The inverse A^{-1} exists if and only if the determinant of A is non-zero.

$$\begin{pmatrix} 1 & 0 & 0 & \cdots & 0 \\ 0 & 1 & 0 & \cdots & 0 \\ 0 & 0 & 1 & \cdots & 0 \\ \cdots & \cdots & \cdots & \cdots & \cdots \\ \cdots & \cdots & \cdots & \cdots & \cdots \\ \cdots & \cdots & \cdots & \cdots & \cdots \\ 0 & 0 & 0 & \cdots & 1 \end{pmatrix}$$

Fig. 14.3 Identity matrix I_n

$$\begin{pmatrix} a_{11} & a_{12} & a_{13} & \cdots & a_{1n} \\ a_{21} & a_{22} & a_{23} & \cdots & a_{2n} \\ a_{31} & a_{32} & a_{33} & \cdots & a_{3n} \\ \cdots & \cdots & \cdots & \cdots & \cdots \\ \cdots & \cdots & \cdots & \cdots & \cdots \\ a_{m1} & a_{m2} & a_{m3} & \cdots & a_{mn} \end{pmatrix}^{\mathrm{T}} = \begin{pmatrix} a_{11} & a_{21} & a_{31} & \cdots & a_{m1} \\ a_{12} & a_{22} & a_{32} & \cdots & a_{m2} \\ a_{13} & a_{23} & a_{33} & \cdots & a_{m3} \\ \cdots & \cdots & \cdots & \cdots & \cdots \\ \cdots & \cdots & \cdots & \cdots & \cdots \\ \cdots & \cdots & \cdots & \cdots & \cdots \\ a_{1n} & a_{2n} & a_{3n} & \cdots & a_{mn} \end{pmatrix}$$

m rows, n columns   *n rows, m columns*

Fig. 14.4 Transpose of a matrix

The *transpose* of a matrix $A = (a_{ij})$ involves changing the rows to columns and vice versa to form the transpose matrix A^{T}. The result of the operation is that the $m \times n$ matrix A is converted to the $n \times m$ matrix A^{T} (Fig. 14.4). It is defined by

$$\left(A^{\mathrm{T}}\right)_{ij} = \left(A_{ji}\right) \quad 1 \le j \le n \text{ and } 1 \le i \le m$$

A matrix is *symmetric* if it is equal to its transpose, i.e. $A = A^{\mathrm{T}}$.

14.4 Determinants

The determinant is a function defined on square matrices and its value is a scalar.
A key property of determinants is that a matrix is invertible if and only if its
determinant is non-zero. The determinant of a 2 × 2 matrix is given by

$$\begin{vmatrix} a & b \\ c & d \end{vmatrix} = ad - bc$$

The determinant of a 3 × 3 matrix is given by

$$\begin{vmatrix} a & b & c \\ d & e & f \\ g & h & i \end{vmatrix} = aei + bfg + cdh - afh - bdi - ceg$$

Cofactors

Let A be an $n \times n$ matrix. For $1 \le i, j \le n$, the (i, j) *minor* of A is defined to be the
$(n - 1) \times (n - 1)$ matrix obtained by deleting the i-th row and j-th column of
A (Fig. 14.5).

The shaded row is the i-th row and the shaded column is the j-th column. These
are both deleted from A to form the (i, j) minor of A, and this is a $(n - 1) \times (n - 1)$
matrix.

The (i, j) *cofactor* of A is defined to be $(-1)^{i+j}$ times the determinant of the
(i, j) minor. The (i, j) cofactor of A is denoted by $K_{ij}(A)$.

The cofactor matrix C of A is formed in this way where the (i, j)-th element in
the cofactor matrix is the (i, j) cofactor of A.

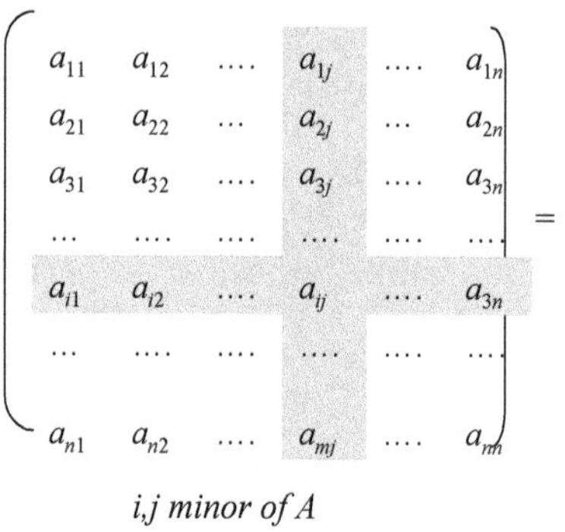

Fig. 14.5 Determining the
(i, j) minor of A

i,j minor of A

Definition of Determinant

The determinant of a matrix is defined as

$$\det A = \sum_{j=1}^{n} A_{ij} K_{ij}$$

In other words, the determinant of A is determined by taking any row of A and multiplying each element by the corresponding cofactor and adding the results. The determinant of the product of two matrices is the product of their determinants.

$$\det(AB) = \det A \times \det B$$

Definition The *adjugate* of A is the $n \times n$ matrix $Adj(A)$ whose (i, j) entry is the (j, i) cofactor K_{ji} (A) of A. That is, the adjugate of A is the transpose of the cofactor matrix of A.

Inverse of A

The inverse of A is determined from the determinant of A and the adjugate of A. That is,

$$A^{-1} = \frac{1}{\det A} Adj A = \frac{1}{\det A} (CofA)^{T}$$

A matrix is invertible if and only if its determinant is non-zero, i.e. A is invertible if and only if $\det(A) \neq 0$.

Cramer's Rule

Cramer's rule is a theorem that expresses the solution to a system of linear equations with several unknowns using the determinant of a matrix. There is a unique solution if the determinant of the matrix is non-zero.

For a system of linear equations of the $A\underline{x} = \underline{v}$ where $\underline{x}$ and $\underline{v}$ are n-dimensional column vectors, then if $\det A \neq 0$ then the unique solution for each x_i is

$$x_i = \frac{\det U_i}{\det A}$$

where U_i is the matrix obtained from A by replacing the i-th column in A by the v-column.

Characteristic Equation

For every $n \times n$ matrix A, there is a polynomial equation of degree n satisfied by A. The *characteristic polynomial* of A is a polynomial in x of degree n. It is given by

$$cA(x) = \det(xI - A)$$

Cayley–Hamilton Theorem

Every matrix A satisfies its characteristic polynomial, i.e. $p(A) = 0$ where $p(x)$ is the characteristic polynomial of A.

14.5   Eigenvectors and Values

A number λ is an eigenvalue of an $n \times n$ matrix A if there is a non-zero vector v such that the following equation holds:

$$Av = \lambda v$$

The vector v is termed an eigenvector and the equation is equivalent to

$$(A - \lambda I)v = 0$$

This means that $(A - \lambda I)$ is a zero divisor, and hence it is not an invertible matrix. Therefore,

$$\det(A - \lambda I) = 0$$

The polynomial function $p(\lambda) = \det (A - \lambda I)$ is called the characteristic polynomial of A, and it is of degree n. The characteristic equation is $p(\lambda) = 0$ and as the polynomial is of degree n there are at most n roots of the characteristic equation, and so there are at most n eigenvalues.

The *Cayley–Hamilton theorem* states that every matrix satisfies its characteristic equation, i.e. the application of the characteristic polynomial to the matrix A yields the zero matrix.

$$p(A) = 0$$

14.6   Gaussian Elimination

Gaussian elimination with backward substitution is an important method used in solving a set of linear equations. A matrix is used to represent the set of linear equations, and Gaussian elimination reduces the matrix to a *triangular* or *reduced form*, which may then be solved by backward substitution.

This allows the set of n linear equations (E_1 to E_n) defined below to be solved by applying operations to the equations to reduce the matrix to triangular form. This reduced form is easier to solve and it provides exactly the same solution as the original set of equations. The set of equations is defined as

$$E_1 : \quad a_{11}x_1 + a_{12}x_2 + \ldots + a_{1n}x_n = b_1$$
$$E_2 : \quad a_{21}x_1 + a_{22}x_2 + \ldots + a_{2n}x_n = b_2$$
$$\vdots \quad \vdots \quad \vdots \qquad \qquad \vdots \quad \vdots$$
$$E_n : \quad a_{n1}x_1 + a_{n2}x_2 + \ldots + a_{nn}x_n = b_n$$

Three operations are permitted on the equations and these operations transform the linear system into a reduced form. They are

(a) Any equation may be multiplied by a non-zero constant.
(b) An equation E_i may be multiplied by a constant and added to another equation E_j, with the resulting equation replacing E_j.
(c) Equations E_i and E_j may be transposed with E_j replacing E_i and vice versa.

This method for solving a set of linear equations is best illustrated by an example, and we consider an example taken from [BuF:89 (1959)]. Then the solution to a set of linear equations with four unknowns may be determined as follows:

$$E_1 : \quad x_1 + x_2 + 3x_4 = 4$$
$$E_2 : \quad 2x_1 + x_2 - x_3 + x_4 = 1$$
$$E_3 : \quad 3x_1 - x_2 - x_3 + 2x_4 = -3$$
$$E_4 : \quad -x_1 + 2x_2 + 3x_3 - x_4 = 4$$

First, the unknown x_1 is eliminated from E_2, E_3 and E_4 and this is done by replacing E_2 with $E_2 - 2E_1$; replacing E_3 with $E_3 - 3E_1$ and replacing E_4 with $E_4 + E_1$. The resulting system is

$$E_1 : \quad x_1 + x_2 + 3x_4 = 4$$
$$E_2 : \quad -x_2 - x_3 - 5x_4 = -7$$
$$E_3 : \quad -4x_2 - x_3 - 7x_4 = -15$$
$$E_4 : \quad 3x_2 + 3x_3 + 2x_4 = 8$$

The next step is then to eliminate x_2 from E_3 and E_4. This is done by replacing E_3 with $E_3 - 4E_2$ and replacing E_4 with $E_4 + 3E_2$. The resulting system is now in triangular form and the unknown variable may be solved easily by backward substitution. That is, we first use equation E_4 to find the solution to x_4 and then we use equation E_3 to find the solution to x_3. We then use equations E_2 and E_1 to find the solutions to x_2 and x_1.

$$E_1 : \quad x_1 + x_2 \qquad \quad + 3x_4 = 4$$
$$E_2 : \quad -x_2 - x_3 \qquad -5x_4 = -7$$
$$E_3 : \qquad \qquad \quad 3x_3 + 13x_4 = 13$$
$$E_4 : \qquad \qquad \qquad \quad -13x_4 = -13$$

The usual approach to Gaussian elimination is to do it with an augmented matrix. That is, the set of equations is an $n \times n$ matrix and it is augmented by the column vector to form the augmented $n \times n + 1$ matrix. Gaussian elimination is then applied to the matrix to put it into triangular form, and it is then easy to solve the unknowns.

The other common approach to solving a set of linear equation is to employ Cramer's rule, which was discussed in Sect. 14.4. Finally, another possible (but computationally expensive) approach to solving the set of linear equations $A\underline{x} = \underline{v}$ is to compute the determinant and inverse of A, and to then compute $\underline{x} = A^{-1}\underline{v}$.

14.7   Review Questions

1. Show how 2×2 matrices may be added and multiplied.
2. What is the additive identity for 2×2 matrices? The multiplicative identity?
3. What is the determinant of a 2×2 matrix?
4. Show that a 2×2 matrix is invertible if its determinant is non-zero.
5. Describe general matrix algebra including addition and multiplication, determining the determinant and inverse of a matrix.
6. What is Cramer's rule?
7. Show how Gaussian elimination may be used to solve a set of linear equations.
8. Write a program to find the inverse of a 3×3 and then a $(n \times n)$ matrix.

14.8   Summary

A matrix is a rectangular array of numbers that consists of horizontal rows and vertical columns. A matrix with m rows and n columns is termed an $m \times n$ matrix, where m and n are its dimensions. A matrix with an equal number of rows and columns (e.g. n rows and n columns) is termed as square matrix.

Matrices arose in practice as a means of solving a set of linear equations, and one of the earliest examples of their use is from a Chinese text dating from between 300 BC and 200 AD.

Matrices of the same dimensions may be added, subtracted and multiplied by a scalar. Two matrices A and B may be multiplied provided that the number of columns of A equals the number of rows in B.

Matrices have an identity matrix under addition and multiplication, and a square matrix has an inverse provided that its determinant is non-zero. The inverse of a matrix involves determining its determinant, constructing the cofactor matrix and transposing the cofactor matrix.

The solution to a set of linear equations may be determined by Gaussian elimination to convert the matrix to upper triangular form, and then employing backward substitution. Another approach is to use Cramer's rule.

Eigenvalues and eigenvectors lead to the characteristic polynomial and every matrix satisfies its characteristic polynomial. The characteristic polynomial is of degree n, and a square $n \times n$ matrix has at most n eigenvalues.

Reference

BuF:89 (1959) Numerical analysis. In: 4th Edn. Burden RL, Faires JD PWS Kent

A Short History of Logic

15

Key Topics

Syllogistic logic
Fallacies
Paradoxes
Stoic logic
Boole's symbolic logic
Digital computing
Propositional logic
Predicate logic
Universal and existential quantifiers

15.1 Introduction

Logic is concerned with reasoning and with establishing the validity of arguments. It allows conclusions to be deduced from premises according to logical rules, and the logical argument establishes the truth of the conclusion provided that the premises are true.

The origins of logic are with the Greeks who were interested in the nature of truth. The sophists (e.g. Protagoras and Gorgias) were teachers of rhetoric, who taught their pupils techniques in winning an argument and convincing an audience. Plato explores the nature of truth in some of his dialogues, and he is critical of the position of the sophists who argue that there is no absolute truth, and that truth instead is always relative to some frame of reference. The classic sophist position is stated by Protagoras '*Man is the measure of all things: of things which are, that*

© Springer Nature Switzerland AG 2020
G. O'Regan, *Mathematics in Computing*, Undergraduate Topics
in Computer Science, https://doi.org/10.1007/978-3-030-34209-8_15

they are, and of things which are not, that they are not'. In other words, what is true for you is true for you, and what is true for me is true for me.

Socrates had a reputation for demolishing an opponent's position, and the Socratean enquiry consisted of questions and answers in which the opponent would be led to a conclusion incompatible with his original position. The approach was similar to a *reductio ad absurdum* argument, although Socrates was a moral philosopher who did no theoretical work on logic.

Aristotle did important work on logic, and he developed a system of logic, called *syllogistic logic*, that remained in use up to the nineteenth century. Syllogistic logic is a 'term-logic', with letters used to stand for the individual terms. A syllogism consists of two premises and a conclusion, where the conclusion is a valid deduction from the two premises. Aristotle also did some early work on modal logic.

The Stoics developed an early form of propositional logic, where the assertibles (propositions) have a truth-value such that at any time they are either true or false. The assertibles may be simple or non-simple, and various connectives such as conjunctions, disjunctions and implication are used in forming more complex assertibles.

George Boole developed his symbolic logic in the mid-1800s, and it later formed the foundation for digital computing. Boole argued that logic should be considered as a separate branch of mathematics, rather than a part of philosophy. He argued that there are mathematical laws to express the operation of reasoning in the human mind, and he showed how Aristotle's syllogistic logic could be reduced to a set of algebraic equations.

Frege is considered (along with Boole) to be one of the founders of modern logic. He also made important contributions to the foundations of mathematics, and he attempted to show that all of the basic truths of mathematics (or at least of arithmetic) could be derived from a limited set of logical axioms.

Logic plays a key role in reasoning and deduction in mathematics, but it is considered a separate discipline to mathematics. There were attempts in the early twentieth century to show that all mathematics can be derived from formal logic, and that the formal system of mathematics would be complete, with all the truths of mathematics provable in the system (see Chap. 13). However, this program failed when the Austrian logician, Kurt Gödel, showed that first-order arithmetic is incomplete.

15.2   Syllogistic Logic

Early work on logic was done by Aristotle in the fourth century B.C. in the *Organon* (Ackrill 1994). Aristotle regarded logic as a useful tool of enquiry into any subject, and he developed *syllogistic logic*. This is a form of reasoning in which a conclusion is drawn from two premises, where each premise is in a subject–predicate form. A common or middle term is present in each of the two premises but not in the conclusion. For example,

Table 15.1 Types of syllogistic premises

Type	Symbol	Example
Universal affirmative	G A M	All Greeks are mortal
Universal negative	G E M	No Greek is mortal
Particular affirmative	G I M	Some Greeks are mortal
Particular negative	G O M	Some Greeks are not mortal

Table 15.2 Forms of syllogistic premises

	Form (i)	Form (ii)	Form (iii)	Form (iv)
Premise 1	M P	P M	P M	M P
Premise 2	M S	S M	M S	S M
Conclusion	S P	S P	S P	S P

All Greeks are mortal.
Socrates is a Greek

— — — — — — — — — — — —

Therefore Socrates is mortal

The common (or middle) term in this example is 'Greek'. It occurs in both premises but not in the conclusion. The above argument is valid, and Aristotle studied and classified the various types of syllogistic arguments to determine those that were valid or invalid. Each premise contains a subject and a predicate, and the middle term may act as subject or a predicate. Each premise is a positive or negative affirmation, and an affirmation may be universal or particular. The universal and particular affirmations and negatives are described in Table 15.1.

This leads to four basic forms of syllogistic arguments (Table 15.2) where the middle is the subject of both premises; the predicate of both premises and the subject of one premise and the predicate of the other premise.

There are four types of premises (A, E, I, O), and therefore 16 sets of premise pairs for each of the forms above. However, only some of these premise pairs will yield a valid conclusion. Aristotle went through every possible premise pair to determine if a valid argument may be derived. The syllogistic argument above is of form (iv) and is valid:

$$G A M$$
$$S I G$$
— — —
$$S I M$$

Syllogistic logic is a 'term-logic' with letters used to stand for the individual terms. Syllogistic logic was the first attempt at a science of logic and it remained in use up to the nineteenth century. There are many limitations to what it may express, and on its suitability as a representation of how the mind works.

15.3 Paradoxes and Fallacies

A paradox is a statement that apparently contradicts itself, and it presents a situation that appears to defy logic. Some logical paradoxes have a solution, whereas others are contradictions or invalid arguments. There are many examples of paradoxes, and they often arise due to self-reference in which one or more statements refer to each other. We discuss several paradoxes such as the *liar paradox* and the *sorites paradox*, which were invented by Eubulides of Miletus and the *barber paradox*, which was introduced by Russell to explain the contradictions in naïve set theory.

An example of the *liar paradox* is the statement 'Everything that I say is false', which is made by the liar. This looks like a normal sentence but it is also saying something about itself as a sentence. If the statement is true, then the statement must be false, since the meaning of the sentence is that every statement (including the current statement) made by the liar is false. If the current statement is false, then the statement that everything that I say is false, and so this must be a true statement.

The *Epimenides paradox* is a variant of the liar paradox. Epimenides was a Cretan who allegedly stated '*All Cretans are liars*'. If the statement is true, then since Epimenides is Cretan, he must be a liar, and so the statement is false and we have a contradiction. However, if we assume that the statement is false and that Epimenides is lying about all Cretan being liars, then we may deduce (without contradiction) that there is at least one Cretan who is truthful. So, in this case, the paradox can be avoided.

The *sorites paradox* (paradox of the heap) involves a heap of sand in which grains are individually removed. It is assumed that removing a single grain of sand does not turn a heap into a non-heap, and the paradox is to consider what happens after when the process is repeated often enough. Is a single remaining grain a heap? When does it change from being a heap to a non-heap? This paradox may be avoided by specifying a fixed boundary of the number of grains of sand required to form a heap, or to define a heap as a collection of multiple grains (≥ 2 grains). Then any collection of grains of sand less than this boundary is not a heap.

The *barber paradox* is a variant of Russell's paradox (a contradiction in naïve set theory), which was discussed in Chap. 3. In a village, there is a barber who shaves everyone who does not shave himself, and no one else. Who shaves the barber? The answer to this question results in a contradiction, as the barber cannot shave himself, since he shaves only those who do not shave themselves. Further, as the barber does not shave himself then he falls into the group of people who would be shaved by the barber (himself). Therefore, we conclude that there is no such barber (or that the barber has a beard).

The purpose of a debate is to convince an audience of the correctness of your position, and to challenge and undermine your opponent's position. Often, the arguments made are factual, but occasionally individuals skilled in rhetoric and persuasion introduce bad arguments as a way to persuade the audience. Aristotle studied and classified bad arguments (known as *fallacies*), and these include fallacies such as the *ad hominem* argument, the *appeal to authority* argument and the *straw man* argument. The fallacies are described in more detail in Table 15.3.

Table 15.3 Fallacies in arguments

Fallacy	Description/Example
Hasty/accident generalization	This is a bad argument that involves a generalization that disregards exceptions
Slippery Slope	This argument outlines a chain reaction leading to a highly undesirable situation that will occur if a certain situation is allowed. The claim is that even if one step is taken onto the slippery slope then we will fall all the way down to the bottom
Against the Person Ad Hominem	The focus of this argument is to attack the person rather than the argument that the person has made
Appeal to People *Ad Populum*	This argument involves an appeal to popular belief to support an argument, with a claim that the majority of the population supports this argument. However, popular opinion is not always correct
Appeal to Authority (*Ad Verecundiam*)	This argument is when an appeal is made to an authoritative figure to support an argument, and where the authority is not an expert in this area
Appeal to Pity (*Ad Misericordiam*)	This is where the arguer tries to get people to accept a conclusion by making them feel sorry for someone
Appeal to Ignorance	The arguer makes the case that there is no conclusive evidence on the issue at hand and that, therefore, his conclusion should be accepted
Straw Man Argument	The arguer sets up a version of an opponent's position of his argument and defeats this watered down version of his opponent's position
Begging the Question (Petitio Principii)	This is a circular argument where the arguer relies on a premise that says the same thing as the conclusion and without providing any real evidence for the conclusion
Red Herring	The arguer goes off on a tangent that has nothing to do with the argument in question
False Dichotomy	The arguer presents the case that there are only two possible outcomes (often there are more). One of the possible outcomes is then eliminated leading to the desired outcome. The argument suggests that there is only one outcome

15.4 Stoic Logic

The Stoic school[1] was found in the Hellenistic period by Zeno of Citium (in Cyprus) in the late fourth/ early third century B.C. The school presented its philosophy as a way of life, and it emphasized ethics as the main focus of human knowledge. The Stoics stressed the importance of living a good life in harmony with nature.

The Stoics recognized the importance of reason and logic, and Chrysippus, the head of the Stoics in the third century B.C., developed an early version of propositional logic. This was a system of deduction in which the smallest unanalyzed expressions are assertibles (Stoic equivalent of propositions). The assertibles have a truth-value such that at any moment of time they are either true or false. True assertibles are viewed as facts in the Stoic system of logic and false assertibles are defined as the contradictories of true ones (Fig. 15.1).

Truth is temporal and assertions may change their truth-value over time. The assertibles may be simple or non-simple (more than one assertible), and there may be present tense, past tense and future tense assertibles. Chrysippus distinguished between simple and compound propositions, and he introduced a set of logical connectives for conjunction, disjunction and implication that are used to form non-simple assertibles from existing assertibles.

The conjunction connective is of the form '*both .. and ..*', and it has two conjuncts. The disjunction connective is of the form '*either .. or .. or ..*', and it consists of two or more disjuncts. Conditionals are formed from the connective '*if .., ..*' and they consist of an antecedent and a consequence.

His deductive system included various logical argument forms such as *modus ponens* and *modus tollens*. His propositional logic differed from syllogistic logic, in that the Stoic logic was based on propositions (or statements) as distinct from Aristotle's term logic. However, he could express the universal affirmation in syllogistic logic (e.g. all As are B) by rephrasing it as a conditional statement that if something is A then it is B.

Chrysippus's propositional logic did not replace Aristotle's syllogistic logic, and syllogistic logic remained in use up to the mid-nineteenth century. George Boole developed his symbolic logic in the mid-1800s, and this logic is discussed in the next section.

---

[1]The origin of the word Stoic is from the *Stoa Poikile* (Στοα Ποιλικη), which was a covered walkway in the Agora of Athens. Zeno taught his philosophy in a public space at this location, and his followers became known as Stoics.

Fig. 15.1 Zeno of Citium

15.5   Boole's Symbolic Logic

We discussed Boole's symbolic logic in Chap. 2. Boole inherited his father's interest in knowledge, and he was self-taught in mathematics and Greek. He taught at various schools near Lincoln, and he developed his mathematical knowledge by working his way through Newton's Principia, as well as applying himself to the work of mathematicians such as Laplace and Lagrange.

He developed his symbolic algebra, which is the foundation for modern computing, and he is considered (along with Babbage) to be one of the grandfathers of computing. His work was theoretical, and he never actually built a computer or calculating machine. *However, Boole's symbolic logic was the perfect mathematical model for switching theory, and for the design of digital circuits.*

Boole published a pamphlet titled 'Mathematical Analysis of Logic' in 1847 (Boole 1848). This short book developed novel ideas on a logical method, and he argued that logic should be considered as a separate branch of mathematics, rather than a part of philosophy. He argued that there are mathematical laws to express the operation of reasoning in the human mind, and he showed how Aristotle's syllogistic logic could be reduced to a set of algebraic equations. He corresponded regularly on logic with Augustus De Morgan.[2]

He introduced two quantities '0' and '1' with the quantity 1 used to represent the universe of thinkable objects (i.e. the universal set), and the quantity 0 represents the absence of any objects (i.e. the empty set). He then employed symbols such as x, y, z, etc. to represent collections or classes of objects given by the meaning attached to adjectives and nouns. Next, he introduced three operators ($+$, $-$ and $\times$) that combined classes of objects.

He showed that these symbols obeyed a rich collection of algebraic laws and could be added, multiplied, etc. in a manner that is similar to real numbers. These symbols may be used to reduce propositions to equations, and algebraic rules may be employed to solve the equations.

Boole applied the symbols to encode Aristotle's Syllogistic Logic, and he showed how the syllogisms could be reduced to equations. This allowed conclusions to be derived from premises by eliminating the middle term in the syllogism. He refined his ideas on logic further in his book '*An Investigation of the Laws of Thought*' (Boole 1958). This book aimed to identify the fundamental laws underlying reasoning in the human mind, and to give expression to these laws in the symbolic language of calculus.

He considered the equation $x^2 = x$ to be fundamental laws of thought. It allows the principle of contradiction to be expressed (i.e. for an entity to possess an attribute and at the same time not to possess it).

Boole's logic appeared to have no practical use, but this changed with Claude Shannon's 1937 Master's Thesis, which showed its applicability to switching theory and to the design of digital circuits.

15.5.1   Switching Circuits and Boolean Algebra

Claude Shannon (Fig. 2.11) showed in his famous Master's Thesis ('*A Symbolic Analysis of Relay and Switching Circuits*) Claude (1937) that Boole's symbolic algebra provided the perfect mathematical model for switching theory and for the design of digital circuits. He realized that you could combine switches in circuits in such a manner as to carry out symbolic logic operations. This allowed binary arithmetic and more complex mathematical operations to be performed by relay circuits. He designed a circuit, which could add binary numbers, and he later designed circuits that could make comparisons, and thus is capable of performing a conditional statement. *This was the birth of digital logic and the digital computing age.*

---

[2]De Morgan was a nineteenth-century British mathematician based at University College London. De Morgan's laws in Set Theory and Logic state that $(A \cup B)^c = A^c \cap B^c$ and $\neg (A \vee B) \equiv \neg A \wedge \neg B$.

He showed that the binary digits (i.e. 0 and 1) can be represented by electrical switches. The implication of this was enormous, as it allowed binary arithmetic and more complex mathematical operations to be performed by relay circuits. This provided electronics engineers with the mathematical tool they needed to design digital electronic circuits, and provided the foundation of digital electronic design.

His *Master's Thesis is a key milestone in computing* and it shows how to lay out circuits according to Boolean principles. It provides the theoretical foundation of switching circuits, and *his insight of using the properties of electrical switches to do Boolean logic is the basic concept that underlies all electronic digital computers.*

The use of the properties of electrical switches to process logic is the basic concept that underlies all modern electronic digital computers. Digital computers use the binary digits 0 and 1, and Boolean logical operations may be implemented by electronic AND, OR and NOT gates. More complex circuits (e.g. arithmetic) may be designed from these fundamental building blocks.

15.6 Frege

Gottlob Frege (Fig. 15.2) was a German mathematician and logician who is considered (along with Boole) to be one of the founders of modern logic. He also made important contributions to the foundations of mathematics, and he attempted to show that all of the basic truths of mathematics (or at least of arithmetic) could be derived from a limited set of logical axioms (this approach is known as *logicism*).

He invented predicate logic and the universal and existential quantifiers, and predicate logic was a significant advance on Aristotle's syllogistic logic. Predicate logic is described in more detail in Chap. 16.

Frege's first logical system contained nine axioms and one rule of inference. It was the first axiomatization of logic, and it was complete in its treatment of propositional logic and first-order predicate logic. He published several important books on logic, including *Begriffsschrift*, in 1879; *Die Grundlagen der Arithmetik* (The Foundations of Arithmetic) in 1884 and the two-volume work *Grundgesetze der Arithmetik* (Basic Laws of Arithmetic), which were published in 1893 and 1903. These books described his invention of axiomatic predicate logic, the use of quantified variables and the application of his logic to the foundations of arithmetic.

Frege presented his predicate logic in his books, and he began to use it to define the natural numbers and their properties. He had intended producing three volumes of the Basic Laws of Arithmetic, with the later volumes dealing with the real numbers and their properties. However, Bertrand Russell discovered a contradiction in Frege's system (see Russell's paradox in Chap. 3), which he communicated to Frege shortly before the publication of the second volume. Frege was astounded by the contradiction and he struggled to find a satisfactory solution, and Russell later introduced the theory of types in the *Principia Mathematica* as a solution.

Fig. 15.2 Gottlob Frege

15.7   Review Questions

1. What is logic?
2. What is a fallacy?
3. Give examples of fallacies in arguments in natural language (e.g. in politics, marketing, debates).
4. Investigate some of the early paradoxes (for example, the Tortoise and Achilles paradox or the arrow in flight paradox) and give your interpretation of the paradox.
5. What is syllogistic logic and explain its relevance.
6. What is stoic logic and explain its relevance.
7. Explain the significance of the equation $x^2 = x$ in Boole's symbolic logic.
8. Describe how Boole's symbolic logic provided the foundation for digital computing.
9. Describe Frege's contributions to logic.

15.8   Summary

This chapter gave a short introduction to logic, and logic is concerned with reasoning and with establishing the validity of arguments. It allows conclusions to be deduced from premises according to logical rules, and the logical argument establishes the truth of the conclusion provided that the premises are true.

The origins of logic are with the Greeks who were interested in the nature of truth. Aristotle did important work on logic, and he developed a system of logic, *syllogistic logic*, that remained in use up to the nineteenth century. Syllogistic logic is a 'term-logic', with letters used to stand for the individual terms. A syllogism consists of two premises and a conclusion, where the conclusion is a valid deduction from the two premises. He also did some early work on modal logic.

The Stoics developed an early form of propositional logic, where the assertibles (propositions) have a truth-value such that at any time they are either true or false.

George Boole developed his symbolic logic in the mid-1800s, and it later formed the foundation for digital computing. Boole argued that logic should be considered as a separate branch of mathematics, rather than a part of philosophy. He argued that there are mathematical laws to express the operation of reasoning in the human mind, and he showed how Aristotle's syllogistic logic could be reduced to a set of algebraic equations.

Gottlob Frege made important contributions to logic and to the foundations of mathematics. He attempted to show that all of the basic truths of mathematics (or at least of arithmetic) could be derived from a limited set of logical axioms (this approach is known as *logicism*). He invented predicate logic and the universal and existential quantifiers, and predicate logic was a significant advance on Aristotle's syllogistic logic.

References

Ackrill JL (1994) Aristotle the philosopher. Clarendon Press Oxford

Boole G (1848) The calculus of logic. Cambridge Dublin Math J III:183–198

Boole G (1958) An investigation into the laws of thought. Dover Publications (First published in 1854)

Claude S (1937) A symbolic analysis of relay and switching circuits. Masters Thesis. Massachusetts Institute of Technology

Propositional and Predicate Logic

16

Key Topics

Propositions
Truth tables
Semantic tableaux
Natural deduction
Proof
Predicates
Universal quantifiers
Existential quantifiers

16.1 Introduction

Logic is the study of reasoning and the validity of arguments, and it is concerned with the truth of statements (propositions) and the nature of truth. Formal logic is concerned with the form of arguments and the principles of valid inference. Valid arguments are truth preserving, and for a valid deductive argument the conclusion will always be true if the premises are true.

Propositional logic is the study of propositions, where a proposition is a statement that is either true or false. Propositions may be combined with other propositions (with a logical connective) to form compound propositions. Truth tables are used to give operational definitions of the most important logical connectives, and they provide a mechanism to determine the truth-values of more complicated logical expressions.

© Springer Nature Switzerland AG 2020
G. O'Regan, *Mathematics in Computing*, Undergraduate Topics
in Computer Science, https://doi.org/10.1007/978-3-030-34209-8_16

Propositional logic may be used to encode simple arguments that are expressed in natural language, and to determine their validity. The validity of an argument may be determined from truth tables, or using the inference rules such as modus ponens to establish the conclusion via deductive steps.

Predicate logic allows complex facts about the world to be represented, and new facts may be determined via deductive reasoning. Predicate calculus includes predicates, variables and quantifiers, and a *predicate* is a characteristic or property that the subject of a statement can have. A predicate may include variables, and statements with variables become propositions once the variables are assigned values.

The universal quantifier is used to express a statement such as that all members of the domain of discourse have property P. This is written as $(\forall x)\ P(x)$, and it expresses the statement that the property $P(x)$ is true for all x.

The existential quantifier states that there is at least one member of the domain of discourse that has property P. This is written as $(\exists x)\ P(x)$.

16.2 Propositional Logic

Propositional logic is the study of propositions where a proposition is a statement that is either true or false. There are many examples of propositions such as '$1 + 1 = 2$' which is a true proposition, and the statement that 'Today is Wednesday' which is true if today is Wednesday and false otherwise. The statement $x > 0$ is not a proposition as it contains a variable x, and it is only meaningful to consider its truth or falsity only when a value is assigned to x. Once the variable x is assigned a value it becomes a proposition. The statement 'This sentence is false' is not a proposition as it contains a self-reference that contradicts itself. Clearly, if the statement is true it is false, and if it is false it is true.

A propositional variable may be used to stand for a proposition (e.g. let the variable P stand for the proposition '$2 + 2 = 4$' which is a true proposition). A propositional variable takes the value or false. The negation of a proposition P (denoted $\neg P$) is the proposition that is true if and only if P is false, and is false if and only if P is true.

A well-formed formula (*wff*) in propositional logic is a syntactically correct formula created according to the syntactic rules of the underlying calculus. A well-formed formula is built up from variables, constants, terms and logical connectives such as conjunction (and), disjunction (or), implication (if .. then ..), equivalence (if and only if) and negation. A distinguished subset of these well-formed formulae is the *axioms* of the calculus, and there are *rules of inference* that allow the truth of new formulae to be derived from the axioms and from formulae that have already demonstrated to be true in the calculus.

A formula in propositional calculus may contain several propositional variables, and the truth or falsity of the individual variables needs to be known prior to determining the truth or falsity of the logical formula.

Table 16.1 Truth table for formula W

A	B	W (A, B)
T	T	T
T	F	F
F	T	F
F	F	T

Each propositional variable has two possible values, and a formula with n-propositional variables has 2^n values associated with the n-propositional variables. The set of values associated with the n variables may be used to derive a truth table with 2^n rows and $n + 1$ columns. Each row gives each of the 2^n truth-values that the n variables may take, and column $n + 1$ gives the result of the logical expression for that set of values of the propositional variables. For example, the propositional formula W defined in Table 16.1 has two propositional variables, A and B, with $2^2 = 4$ rows for each of the values that the two propositional variables may take. There are $2 + 1 = 3$ columns with W defined in the third column.

A rich set of connectives is employed in the calculus to combine propositions and to build up the well-formed formulae. This includes the conjunction of two propositions $(A \wedge B)$, the disjunction of two propositions $(A \vee B)$ and the implication of two propositions $(A \rightarrow B)$. These connectives allow compound propositions to be formed, and the truth of the compound propositions is determined from the truth-values of its constituent propositions and the rules associated with the logical connective. The meaning of the logical connectives is given by truth tables.[1]

Mathematical logic is concerned with inference, and it involves proceeding in a methodical way from the axioms and using the rules of inference to derive further truths.

The rules of inference allow new propositions to be deduced from a set of existing propositions. A valid argument (or deduction) is truth preserving, i.e. for a valid logical argument if the set of premises is true then the conclusion (i.e. the deduced proposition) will also be true. The rules of inference include rules such as *modus ponens*, and this rule states that given the truths of the proposition A, and the proposition $A \rightarrow B$, then the truth of proposition B may be deduced.

The propositional calculus is employed in reasoning about propositions, and it may be applied to formalize arguments in natural language. *Boolean algebra* is used in computer science, and it is named after George Boole, who was the first Professor of Mathematics at Queens College, Cork.[2]

---

[1] Basic truth tables were first used by Frege, and developed further by Post and Wittgenstein.
[2] This institution is now known as University College Cork and has approximately 18,000 students.

16.2.1 Truth Tables

Truth tables give operational definitions of the most important logical connectives, and they provide a mechanism to determine the truth-values of more complicated compound expressions. Compound expressions are formed from propositions and connectives, and the truth-values of a compound expression containing several propositional variables are determined from the underlying propositional variables and the logical connectives.

The conjunction of A and B (denoted $A \wedge B$) is true if and only if both A and B are true, and is false in all other cases (Table 16.2). The disjunction of two propositions A and B (denoted $A \vee B$) is true if at least one of A and B is true, and false in all other cases (Table 16.3). The disjunction operator is known as the '*inclusive or*' operator as it is also true when both A and B are true; there is also an *exclusive or* operator that is true exactly when one of A or B is true, and is false otherwise.

Example 16.1 Consider proposition A given by 'An orange is a fruit' and the proposition B given by '$2 + 2 = 5$', then A is true and B is false. Therefore,

(i) $A \wedge B$ (i.e. An orange is a fruit and $2 + 2 = 5$) is false.
(ii) $A \vee B$ (i.e. An orange is a fruit or $2 + 2 = 5$) is true.

The implication operation ($A \rightarrow B$) is true if whenever A is true means that B is also true, and also whenever A is false (Table 16.4). It is equivalent (as shown by a truth table) to $\neg A \vee B$. The equivalence operation ($A \leftrightarrow B$) is true whenever both A and B are true, or whenever both A and B are false (Table 16.5).

The not operator ($\neg$) is a unary operator (i.e. it has one argument) and is such that $\neg A$ is true when A is false, and is false when A is true (Table 16.6).

Example 16.2 Consider proposition A given by 'Jaffa cakes are biscuits' and the proposition B given by '$2 + 2 = 5$', then A is true and B is false. Therefore,

Table 16.2 Conjunction

A	B	$A \wedge B$
T	T	T
T	F	F
F	T	F
F	F	F

Table 16.3 Disjunction

A	B	$A \vee B$
T	T	T
T	F	T
F	T	T
F	F	F

Table 16.4 Implication

A	B	A → B
T	T	T
T	F	F
F	T	T
F	F	T

Table 16.5 Equivalence

A	B	A ↔ B
T	T	T
T	F	F
F	T	F
F	F	T

Table 16.6 Not operation

A	¬A
T	F
F	T

(i) $A \rightarrow B$ (i.e. Jaffa cakes are biscuits implies $2 + 2 = 5$) is false.
(ii) $A \leftrightarrow B$ (i.e. Jaffa cakes are biscuits is equivalent to $2 + 2 = 5$) is false.
(iii) $\neg B$ (i.e. $2 + 2 \neq 5$) is true.

Creating a Truth Table

The truth table for a well-formed formula $W(P_1, P_2, \ldots, P_n)$ is a table with 2^n rows and $n + 1$ columns. Each row lists a different combination of truth-values of the propositions $P_1, P_2, \ldots, P_n$ followed by the corresponding truth-value of W.

The example in Table 16.7 gives the truth table for a formula W with three propositional variables (meaning that there are $2^3 = 8$ rows in the truth table).

Table 16.7 Truth table for W(P, Q, R)

P	Q	R	W(P, Q, R)
T	T	T	F
T	T	F	F
T	F	T	F
T	F	F	T
F	T	T	T
F	T	F	F
F	F	T	F
F	F	F	F

16.2.2 Properties of Propositional Calculus

There are many well-known properties of the propositional calculus such as the commutative, associative and distributive properties. These ease the evaluation of complex expressions, and allow logical expressions to be simplified.

The *commutative property* holds for the conjunction and disjunction operators, and it states that the order of evaluation of the two propositions may be reversed without affecting the resulting truth-value, i.e.

$$A \wedge B = B \wedge A$$
$$A \vee B = B \vee A$$

The *associative property* holds for the conjunction and disjunction operators. This means that order of evaluation of a sub-expression does not affect the resulting truth-value, i.e.

$$(A \wedge B) \wedge C = A \wedge (B \wedge C)$$
$$(A \vee B) \vee C = A \vee (B \vee C)$$

The conjunction operator *distributes* over the disjunction operator and vice versa.

$$A \wedge (B \vee C) = (A \wedge B) \vee (A \wedge C)$$
$$A \vee (B \wedge C) = (A \vee B) \wedge (A \vee C)$$

The result of the logical conjunction of two propositions is false if one of the propositions is false (irrespective of the value of the other proposition).

$$A \wedge F = F \wedge A = F$$

The result of the logical disjunction of two propositions is true if one of the propositions is true (irrespective of the value of the other proposition).

$$A \vee T = T \vee A = T$$

The result of the logical disjunction of two propositions, where one of the propositions is known to be false is given by the truth-value of the other proposition. That is, the Boolean value 'F' acts as the identity for the disjunction operation.

$$A \vee F = A = F \vee A$$

The result of the logical conjunction of two propositions, where one of the propositions is known to be true, is given by the truth-value of the other proposition. That is, the Boolean value 'T' acts as the identity for the conjunction operation.

$$A \wedge T = A = T \wedge A$$

The $\wedge$ and $\vee$ operators are *idempotent*. That is, when the arguments of the conjunction or disjunction operator are the same proposition A the result is A. The idempotent property allows expressions to be simplified.

$$A \wedge A = A$$
$$A \vee A = A$$

The *law of the excluded middle* is a fundamental property of the propositional calculus. It states that a proposition A is either true or false, i.e. there is no third logical value.

$$A \vee \neg A$$

We mentioned earlier that $A \rightarrow B$ is logically equivalent to $\neg A \vee B$ (same truth table), and clearly $\neg A \vee B$ is the same as $\neg A \vee \neg\neg B = \neg\neg B \vee \neg A$ which is logically equivalent to $\neg B \rightarrow \neg A$. In other words, $A \rightarrow B$ is logically equivalent to $\neg B \rightarrow \neg A$ (this is known as the *contrapositive*).

De Morgan was a contemporary of Boole in the nineteenth century, and the following law is known as De Morgan's law:

$$\neg(A \wedge B) \equiv \neg A \vee \neg B$$
$$\neg(A \vee B) \equiv \neg A \wedge \neg B$$

Certain well-formed formulae are true for all values of their constituent variables. This can be seen from the truth table when the last column of the truth table consists entirely of true values.

A proposition that is true for all values of its constituent propositional variables is known as a *tautology*. An example of a tautology is the proposition $A \vee \neg A$ (Table 16.8).

A proposition that is false for all values of its constituent propositional variables is known as a *contradiction*. An example of a contradiction is the proposition $A \wedge \neg A$.

16.2.3 Proof in Propositional Calculus

Logic enables further truths to be derived from existing truths by rules of inference that are truth preserving. Propositional calculus is both *complete* and *consistent*.

Table 16.8 Tautology $B \vee \neg B$

B	$\neg B$	$B \vee \neg B$
T	F	T
F	T	T

The completeness property means that all true propositions are deducible in the calculus, and the consistency property means that there is no formula A such that both A and $\neg A$ are deducible in the calculus.

An argument in propositional logic consists of a sequence of formulae that are the premises of the argument and a further formula that is the conclusion of the argument. One elementary way to see if the argument is valid is to produce a truth table to determine if the conclusion is true whenever all of the premises are true.

Consider a set of premises $P_1, P_2, \ldots, P_n$ and conclusion Q. Then to determine if the argument is valid using a truth table involves adding a column in the truth table for each premise $P_1, P_2, \ldots, P_n$, and then to identify the rows in the truth table for which these premises are all true. The truth-value of the conclusion Q is examined in each of these rows, and if Q is true for each case for which $P_1, P_2, \ldots, P_n$ are all true then the argument is valid. This is equivalent to $P_1 \wedge P_2 \wedge \cdots \wedge P_n \rightarrow Q$ is a tautology.

An alternate approach to proof with truth tables is to assume the negation of the desired conclusion (i.e. $\neg Q$) and to show that the premises and the negation of the conclusion result in a contradiction (i.e. $P_1 \wedge P_2 \wedge \cdots \wedge P_n \wedge \neg Q$) is a contradiction.

The use of truth tables becomes cumbersome when there are a large number of variables involved, as there are 2^n truth table entries for n-propositional variables.

Procedure for Proof by Truth Table

(i)   Consider argument $P_1, P_2, \ldots, P_n$ with conclusion Q.
(ii)   Draw truth table with column in truth table for each premise $P_1, P_2, \ldots, P_n$.
(iii) Identify rows in truth table for when these premises are all true.
(iv)   Examine truth-value of Q for these rows.
(v)   If Q is true for each case that $P_1, P_2, \ldots, P_n$ are true then the argument is valid.
(vi)   That is, $P_1 \wedge P_2 \wedge \cdots \wedge P_n \rightarrow Q$ is a tautology.

Example 16.3 (*Truth Tables*) Consider the argument adapted from Kelly (1997) and determine if it is valid.

If the pianist plays the concerto then crowds will come if the prices are not too high.
If the pianist plays the concerto then the prices will not be too high.
Therefore, if the pianist plays the concerto then crowds will come.

Solution We will adopt a common proof technique that involves showing that the negation of the conclusion is incompatible (inconsistent) with the premises, and from this we deduce the conclusion must be true. First, we encode the argument in propositional logic:

Table 16.9 Proof of argument with a truth table

P	C	H	¬H	¬H → C	P → (¬H → C)	P → ¬H	P → C	¬(P → C)	*
T	T	T	F	T	T	F	T	F	F
T	T	F	T	T	T	T	T	F	F
T	F	T	F	T	T	F	F	T	F
T	F	F	T	F	F	T	F	T	F
F	T	T	F	T	T	T	T	F	F
F	T	F	T	T	T	T	T	F	F
F	F	T	F	T	T	T	T	F	F
F	F	F	T	F	T	T	T	F	F

Let P stand for 'The pianist plays the concerto'; C stands for 'Crowds will come' and H stands for 'Prices are too high'. Then the argument may be expressed in propositional logic as follows:

$$P \rightarrow (\neg H \rightarrow C)$$
$$P \rightarrow \neg H$$
$$P \rightarrow C$$

Then we negate the conclusion $P \rightarrow C$ and check the consistency of $P \rightarrow (\neg H \rightarrow C) \wedge (P \rightarrow \neg H) \wedge \neg (P \rightarrow C)*$ using a truth table (Table 16.9).

It can be seen from the last column in the truth table that the negation of the conclusion is incompatible with the premises, and therefore it cannot be the case that the premises are true and the conclusion false. Therefore, the conclusion must be true whenever the premises are true, and we conclude that the argument is valid.

Logical Equivalence and Logical Implication

The laws of mathematical reasoning are truth preserving, and are concerned with deriving further truths from existing truths. Logical reasoning is concerned with moving from one line in mathematical argument to another, and involves deducing the truth of another statement Q from the truth of P.

The statement Q may be in some sense be logically equivalent to P and this allows the truth of Q to be immediately deduced. In other cases, the truth of P is sufficiently strong to deduce the truth of Q; in other words, P logically implies Q. This leads naturally to a discussion of the concepts of logical equivalence ($W_1 \equiv W_2$) and logical implication ($W_1 \vdash W_2$).

Logical Equivalence

Two well-formed formulae $W1$ and W_2 with the same propositional variables (P, Q, R ...) are logically equivalent ($W1 \equiv W_2$) if they are always simultaneously true or false for any given truth-values of the propositional variables.

Table 16.10 Logical
equivalence of two WFFs

P	Q	$P \wedge Q$	$\neg P$	$\neg Q$	$\neg P \vee \neg Q$	$\neg(\neg P \vee \neg Q)$
T	T	T	F	F	F	T
T	F	F	F	T	T	F
F	T	F	T	F	T	F
F	F	F	T	T	T	F

Table 16.11 Logical
implication of two WFFs

P Q R	$(P \wedge Q) \vee (Q \wedge \neg R)$	$Q \vee R$
T T T	T	T
T T F	T	T
T F T	F	T
T F F	F	F
F T T	F	T
F T F	T	T
F F T	F	T
F F F	F	F

If two well-formed formulae are logically equivalent then it does not matter which of W_1 and W_2 is used, and $W_1 \leftrightarrow W_2$ is a tautology. In Table 16.10, we see that $P \wedge Q$ is logically equivalent to $\neg(\neg P \vee \neg Q)$.

Logical Implication

For two well-formed formulae W_1 and W_2 with the same propositional variables (P, Q, R ...) W_1 logically implies W_2 ($W_1 \vdash W_2$) if any assignment to the propositional variables which makes W_1 true also makes W_2 true. That is, $W_1 \rightarrow W_2$ is a tautology.

Example 16.4 Show by truth tables that $(P \wedge Q) \vee (Q \wedge \neg R) \vdash (Q \vee R)$.

The formula $(P \wedge Q) \vee (Q \wedge \neg R)$ is true on rows 1, 2 and 6 and formula $(Q \vee R)$ is also true on these rows (Table 16.11). Therefore, $(P \wedge Q) \vee (Q \wedge \neg R) \vdash (Q \vee R)$.

16.2.4 Semantic Tableaux in Propositional Logic

We showed in Example 16.3 how truth tables may be used to demonstrate the validity of a logical argument. However, the problem with truth tables is that they can get extremely large very quickly (as the size of the table is 2^n where n is the number of propositional variables), and so in this section, we will consider an alternate approach known as semantic tableaux.

The basic idea of semantic tableaux is to determine if it is possible for a conclusion to be false when all of the premises are true. If this is not possible, then the conclusion must be true when the premises are true, and so the conclusion is *semantically entailed* by the premises. The method of semantic tableaux is a technique to expose inconsistencies in a set of logical formulae, by identifying conflicting logical expressions.

We present a short summary of the rules of semantic tableaux in Table 16.12, and we then proceed to provide a proof for Example 16.3 using semantic tableaux instead of a truth table.

Whenever a logical expression A and its negation $\neg A$ appear in a branch of the tableau, then an inconsistency has been identified in that branch, and the branch is said to be *closed*. If all of the branches of the semantic tableaux are closed, then the logical propositions from which the tableau was formed are mutually inconsistent and cannot be true together.

The method of proof with semantic tableaux is to negate the conclusion, and to show that all branches in the semantic tableau are closed, and thus it is not possible

Table 16.12 Rules of semantic tableaux

Rule no.	Definition	Description
1	$A \wedge B$ A B	If $A \wedge B$ is true then both A and B are true, and may be added to the branch containing $A \wedge B$
2.	$A \vee B$ $A \qquad B$	If $A \vee B$ is true then either A or B is true, and we add two new branches to the tableaux, one containing A and one containing B
3.	$A \to B$ $\neg A \qquad B$	If $A \to B$ is true then either $\neg A$ or B is true, and we add two new branches to the tableaux, one containing $\neg A$ and one containing B
4.	$A \leftrightarrow B$ $A \wedge B \qquad \neg A \wedge \neg B$	If $A \leftrightarrow B$ is true then either $A \wedge B$ or $\neg A \wedge \neg B$ is true, and we add two new branches, one containing $A \wedge B$ and one containing $\neg A \wedge \neg B$
5.	$\neg\neg A$ A	If $\neg\neg A$ is true then A may be added to the branch containing $\neg\neg A$
6.	$\neg(A \wedge B)$ $\neg A \qquad \neg B$	If $\neg(A \wedge B)$ is true then either $\neg A$ or $\neg B$ is true, and we add two new branches to the tableaux, one containing $\neg A$ and one containing $\neg B$
7.	$\neg(A \vee B)$ $\neg A$ $\neg B$	If $\neg(A \vee B)$ is true then both $\neg A$ and $\neg B$ are true, and may be added to the branch containing $\neg(A \vee B)$
8.	$\neg(A \to B)$ A $\neg B$	If $\neg(A \to B)$ is true then both A and $\neg B$ are true, and may be added to the branch containing $\neg(A \to B)$

for the premises of the argument to be true and for the conclusion to be false. Therefore, the argument is valid and the conclusion follows from the premises.

Example 16.5 (*Semantic Tableaux*) Perform the proof for Example 16.3 using semantic tableaux.

Solution We formalized the argument previously as

$$
\begin{aligned}
&\text{(Premise 1)} && P \rightarrow (\neg H \rightarrow C) \\
&\text{(Premise 2)} && P \rightarrow \neg H \\
&\text{(Conclusion)} && P \rightarrow C
\end{aligned}
$$

We negate the conclusion to get $\neg(P \rightarrow C)$ and we show that all branches in the semantic tableau are closed, and that, therefore, it is not possible for the premises of the argument to be true and for the conclusion false. Therefore, the argument is valid, and the truth of the conclusion follows from the truth of the premises.

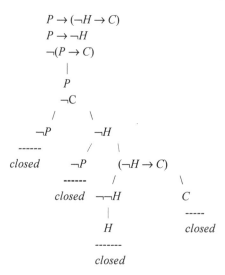

We have showed that all branches in the semantic tableau are closed, and that, therefore, it is not possible for the premises of the argument to be true and for the conclusion to be false. Therefore, the argument is valid as required.

16.2.5 Natural Deduction

The German mathematician, Gerhard Gentzen (Fig. 16.1), developed a method for logical deduction known as '*Natural Deduction*', and this formal approach aims to be as close as possible to natural reasoning. Gentzen worked as an assistant to David Hilbert (Hilbert's program was discussed in Chap. 13) at the University of

Fig. 16.1 Gerhard Gentzen

Göttingen, and he died of malnutrition in Prague at the end of the Second World War.

Natural deduction includes rules for $\wedge$, $\vee$, $\rightarrow$ introduction and elimination and also for *reductio ad absurdum*. There are ten inference rules in the system, and they include two inference rules for each of the five logical operators $\wedge$, $\vee$, $\neg$, $\rightarrow$ and $\leftrightarrow$. There are two inference rules per operator (an introduction rule and an elimination rule), and the rules are defined in Table 16.13.

Natural deduction may be employed in logical reasoning and it is described in detail in Gries (1981), Kelly (1997).

16.2.6 Sketch of Formalization of Propositional Calculus

Truth tables provide an informal approach to proof and the proof is provided in terms of the meanings of the propositions and logical connectives. The formalization of propositional logic includes the definition of an alphabet of symbols and well-formed formulae of the calculus, the axioms of the calculus and rules of inference for logical deduction.

The deduction of a new formula Q is via a sequence of well-formed formulae P_1, P_2, ... P_n (where $P_n = Q$) such that each P_i is either an axiom, a hypothesis or deducible from an earlier pair of formula P_j, P_k, (where P_k is of the form $P_j \Rightarrow P_i$) and modus ponens. *Modus ponens* is a rule of inference that states that given propositions A, and $A \Rightarrow B$ then proposition B may be deduced. The deduction of a formula Q from a set of hypothesis H is denoted by $H \vdash Q$, and where Q is deducible from the axioms alone this is denoted by $\vdash Q$.

The *deduction theorem* of propositional logic states that if $H \cup \{P\} \vdash Q$, then H $\vdash P \rightarrow Q$, and the converse of the theorem, is also true, i.e. if $H \vdash P \rightarrow Q$ then $H \cup \{P\} \vdash Q$. Formalism (this approach was developed by the German mathematician, David Hilbert) allows reasoning about symbols according to rules, and to

Table 16.13 Natural deduction rules

Rule	Definition	Description
$\wedge$ I	$\dfrac{P_1, P_2, \ldots P_n}{P_1 \wedge P_2 \wedge \cdots \wedge P_n}$	Given the truth of propositions P_1, P_2, $\cdots$ P_n then the truth of the conjunction $P_1 \wedge P_2 \wedge \cdots \wedge P_n$ follows. This rule shows how conjunction can be introduced
$\wedge$ E	$\dfrac{P_1 \wedge P_2 \wedge \cdots \wedge P_n}{P_i}$ where $i \in \{1, \ldots, n\}$	Given the truth the conjunction $P_1 \wedge P_2 \wedge \cdots \wedge P_n$ then the truth of proposition P_i ($1 \leq i \leq n$) follows. This rule shows how a conjunction can be eliminated
$\vee$ I	$\dfrac{P_i}{P_1 \vee P_2 \vee \cdots \vee P_n}$	Given the truth of propositions P_i then the truth of the disjunction $P_1 \vee P_2 \vee \cdots \vee P_n$ follows. This rule shows how a disjunction can be introduced
$\vee$ E	$\dfrac{P_1 \vee \cdots \vee P_n, P_1 \to E, \ldots P_n \to E}{E}$	Given the truth of the disjunction $P_1 \vee P_2 \vee \cdots \vee P_n$, and that each disjunct implies E, then the truth of E follows. This rule shows how a disjunction can be eliminated
$\to$ I	$\dfrac{\text{From } P_1, P_2, \ldots P_n \text{ infer } P}{(P_1 \wedge P_2 \wedge \cdots \wedge P_n) \to P}$	This rule states that if we have a theorem that allows P to be inferred from the truth of premises P_1, P_2, $\ldots$ P_n (or previously proved) then we can deduce $(P_1 \wedge P_2 \wedge \cdots \wedge P_n) \to P$. This is known as the *Deduction Theorem*
$\to$ E	$\dfrac{P_i \to P_j P_i}{P_j}$	This rule is known as *modus ponens*. The consequence of an implication follows if the antecedent is true (or has been previously proved)
$\equiv$ I	$\dfrac{P_i \to P_j, P_j \to P_i}{P_i \leftrightarrow P_j}$	If proposition P_i implies proposition P_j and vice versa then they are equivalent (i.e., $P_i \leftrightarrow P_j$)
$\equiv$ E	$\dfrac{P_i \leftrightarrow P_j}{P_i \to P_j, P_j \to P_i}$	If proposition P_i is equivalent to proposition P_j then proposition P_i implies proposition P_j and vice versa
$\neg$ I	$\dfrac{\text{From } P \text{ infer } P_1 \wedge \neg P_1}{\neg P}$	If the proposition P allows a contradiction to be derived, then $\neg P$ is deduced. This is an example of a *proof by contradiction*
$\neg$ E	$\dfrac{\text{From } \neg P \text{ infer } P_1 \wedge \neg P_1}{P}$	If the proposition $\neg P$ allows a contradiction to be derived, then P is deduced. This is an example of a proof by contradiction

derive theorems from formulae irrespective of the meanings of the symbols and formulae.

Propositional calculus is *sound*, i.e. any theorem derived using the Hilbert approach is true. Further, the calculus is also *complete*, and every tautology has a proof (i.e. is a theorem in the formal system). The propositional calculus is *consistent* (i.e. it is not possible that both the well-formed formulae A and $\neg A$ are deducible in the calculus).

Propositional calculus is *decidable*, i.e. there is an algorithm (truth table) to determine for any well-formed formula A whether A is a theorem of the formal system. The Hilbert style system is slightly cumbersome in conducting proof and is quite different from the normal use of logic in mathematical deduction.

16.2.7 Applications of Propositional Calculus

Propositional calculus may be employed in reasoning with arguments in natural language. First, the premises and conclusion of the argument are identified and formalized into propositions. Propositional logic is then employed to determine if the conclusion is a valid deduction from the premises.

Consider, for example, the following argument that aims to prove that Superman does not exist.

'*If Superman were able and willing to prevent evil, he would do so. If Superman were unable to prevent evil he would be impotent; if he were unwilling to prevent evil he would be malevolent; Superman does not prevent evil. If superman exists he is neither malevolent nor impotent; therefore Superman does not exist*'.

First, letters are employed to represent the propositions as follows:

a: Superman is able to prevent evil,
w: Superman is willing to prevent evil,
i: Superman is impotent,
m: Superman is malevolent,
p: Superman prevents evil and
e: Superman exists.

Then, the argument above is formalized in propositional logic as follows:

Premises		
	P_1	$(a \wedge w) \to p$
	P_2	$(\neg a \to i) \wedge (\neg w \to m)$
	P_3	$\neg p$
	P_4	$e \to \neg i \wedge \neg m$
Conclusion		$P_1 \wedge P_2 \wedge P_3 \wedge P_4 \Rightarrow \neg e$

Proof that Superman does not exist

1.	$a \wedge w \to p$	Premise 1
2.	$(\neg a \to i) \wedge (\neg w \to m)$	Premise 2
3.	$\neg p$	Premise 3
4.	$e \to (\neg i \wedge \neg m)$	Premise 4
5.	$\neg p \to \neg(a \wedge w)$	1, Contrapositive
6.	$\neg(a \wedge w)$	3, 5 Modus Ponens
7.	$\neg a \vee \neg w$	6, De Morgan's Law
8.	$\neg (\neg i \wedge \neg m) \to \neg e$	4, Contrapositive
9.	$i \vee m \to \neg e$	8, De Morgan's Law
10.	$(\neg a \to i)$	2, $\wedge$ Elimination

(continued)

11.	$(\neg w \rightarrow m)$	2, $\wedge$ *Elimination*
12.	$\neg\neg a \vee i$	10, A$\rightarrow$ B equivalent to $\neg$A$\vee$ B
13.	$\neg\neg a \vee i \vee m$	11, $\vee$ Introduction
14.	$\neg\neg a \vee (i \vee m)$	
15.	$\neg a \rightarrow (i \vee m)$	14, A $\rightarrow$ B equivalent to $\neg$A$\vee$ B
16.	$\neg\neg w \vee m$	11, A $\rightarrow$ B equivalent to $\neg$A$\vee$ B
17.	$\neg\neg w \vee (i \vee m)$	
18.	$\neg w \rightarrow (i \vee m)$	17, A $\rightarrow$ B equivalent to $\neg$A$\vee$ B
19.	$(i \vee m)$	7, 15, 18 $\vee$*Elimination*
20.	$\neg e$	9, 19 *Modus Ponens*

Second Proof

1.	$\neg p$	P_3
2.	$\neg(a \wedge w) \vee p$	P_1 (A $\rightarrow$ B $\equiv$ $\neg$A $\vee$ B)
3.	$\neg(a \wedge w)$	1, 2 A $\vee$ B, $\neg$B $\vdash$ A
4.	$\neg a \vee \neg w$	3, De Morgan's Law
5.	$(\neg a \rightarrow i)$	P_2 ($\wedge$-Elimination)
6.	$\neg a \rightarrow i \vee m$	5, $x \rightarrow y \vdash x \rightarrow y \vee z$
7.	$(\neg w \rightarrow m)$	P_2 ($\wedge$-Elimination)
8.	$\neg w \rightarrow i \vee m$	7, $x \rightarrow y \vdash x \rightarrow y \vee z$
9.	$(\neg a \vee \neg w) \rightarrow (i \vee m)$	8, $x \rightarrow z, y \rightarrow z \vdash x \vee y \rightarrow z$
10.	$(i \vee m)$	4, 9 Modus Ponens
11.	$e \rightarrow \neg(i \vee m)$	P_4 (De Morgan's Law)
12.	$\neg e \vee \neg (i \vee m)$	11, (A $\rightarrow$ B $\equiv$ $\neg$A $\vee$ B)
13.	$\neg e$	10, 12 A $\vee$ B, $\neg$B $\vdash$ A

Therefore, the conclusion that Superman does not exist is a valid deduction from the given premises.

16.2.8 Limitations of Propositional Calculus

The propositional calculus deals with propositions only. It is incapable of dealing with the syllogism 'All Greeks are mortal; Socrates is a Greek; therefore Socrates is mortal'. This would be expressed in propositional calculus as three propositions A, B therefore C, where A stands for 'All Greeks are mortal', B stands for 'Socrates is a Greek' and C stands for 'Socrates is mortal'. Propositional logic does not allow the conclusion that all Greeks are mortal to be derived from the two premises.

Predicate calculus deals with these limitations by employing variables and terms, and using universal and existential quantification to express that a particular property is true of all (or at least one) values of a variable.

16.3 Predicate Calculus

Predicate logic is a richer system than propositional logic, and it allows complex facts about the world to be represented. It allows new facts about the world to be derived in a way that guarantees that if the initial premises are true then the conclusions are true. Predicate calculus includes predicates, variables, constants and quantifiers.

A *predicate* is a characteristic or property that an object can have, and we are predicating some property of the object. For example, '*Socrates is a Greek*' could be expressed as $G(s)$, with capital letters standing for predicates and small letters standing for objects. A predicate may include variables, and a statement with a variable becomes a proposition once the variables are assigned values. For example, $G(x)$ states that the variable x is a Greek, whereas $G(s)$ is an assignment of values to x. The set of values that the variables may take is termed the universe of discourse (the variables take values from this set).

Predicate calculus employs quantifiers to express properties such as all members of the domain have a particular property, e.g. $(\forall x)P(x)$, or that there is at least one member that has a particular property, e.g. $(\exists x)P(x)$. These are referred to as the *universal and existential quantifiers.*

The syllogism 'All Greeks are mortal; Socrates is a Greek; therefore Socrates is mortal' may be easily expressed in predicate calculus by

$$(\forall x)(G(x) \rightarrow M(x))$$
$$G(s)$$

$$M(s)$$

In this example, the predicate $G(x)$ stands for x is a Greek and the predicate $M(x)$ stands for x is mortal. The formula $G(x) \rightarrow M(x)$ states that if x is a Greek then x is mortal, and the formula $(\forall x)(G(x) \rightarrow M(x))$ states for any x that if x is a Greek then x is mortal. The formula $G(s)$ states that Socrates is a Greek and the formula $M(s)$ states that Socrates is mortal.

Example 16.6 (*Predicates*) A predicate may have one or more variables. A predicate that has only one variable (i.e. a unary or 1-place predicate) is often related to sets; a predicate with two variables (a 2-place predicate) is a relation and a predicate with n variables (a n-place predicate) is a n-ary relation. Propositions do not contain variables and so they are 0-place predicates. The following are examples of predicates:

(i)   The predicate *Prime(x)* states that x is a prime number (with the natural numbers being the universe of discourse).

(ii)   *Lawyer(a)* may stand for a is a lawyer.

(iii) Mean(m, x, y) states that m is the mean of x and y, i.e. $m = \frac{1}{2}(x + y)$.

(iv)   LT(x, 6) states that x is less than 6.

(v)   G(x, π) states that x is greater than π (where π is the constant 3.14159).

(vi)   G(x, y) states that x is greater than y.

(vii) EQ(x, y) states that x is equal to y.

(viii) LE(x, y) states that x is less than or equal to y.

(ix)   Real(x) states that x is a real number.

(x)   Father(x, y) states that x is the father of y.

(xi) $\neg(\exists x)(Prime(x) \wedge B(x, 32, 36))$ states that there is no prime number between 32 and 36.

Universal and Existential Quantification

The universal quantifier is used to express a statement such as that all members of the domain have property P. This is written as $(\forall x)P(x)$ and expresses the statement that the property $P(x)$ is true for all x. Similarly, $(\forall x_1, x_2, \ldots, x_n) P(x_1, x_2, \ldots, x_n)$ states that property $P(x_1, x_2, \ldots, x_n)$ is true for all $x_1, x_2, \ldots, x_n$. Clearly, the predicate $(\forall x) P(a, b)$ is identical to $P(a, b)$ since it contains no variables, and the predicate $(\forall y \in \mathbb{N}) (x \leq y)$ is true if $x = 1$ and false otherwise.

The existential quantifier states that there is at least one member in the domain of discourse that has property P. This is written as $(\exists x)P(x)$ and the predicate $(\exists x_1, x_2, \ldots, x_n) P(x_1, x_2, \ldots, x_n)$ states that there is at least one value of $(x_1, x_2, \ldots, x_n)$ such that $P(x_1, x_2, \ldots, x_n)$ is true.

Example 16.7 (*Quantifiers*)

(i) $(\exists p)$ (Prime(p) $\wedge$ $p > 1{,}000{,}000$) is trueIt expresses the fact that there is at least one prime number greater than a million, which is true as there are infinite number of primes.

(ii) $(\forall x) (\exists y) x < y$ is true
       This predicate expresses the fact that given any number x we can always find a larger number, e.g. take $y = x + 1$.

(iii) $(\exists y) (\forall x) x < y$ is false
       This predicate expresses the statement that there is a natural number y such that all natural numbers are less than y. Clearly, this statement is false since there is no largest natural number, and so the predicate $(\exists y) (\forall x) x < y$ is false.

Comment 16.1 It is important to be careful with the order in which quantifiers are written, as the meaning of a statement may be completely changed by the simple transposition of two quantifiers.

The well-formed formulae in the predicate calculus are built from terms and predicates, and the rules for building the formulae are described briefly in Sect. 16.3.1. Examples of well-formed formulae include the following:

$$(\forall x)(x > 2)$$
$$(\exists x)x^2 = 2$$
$$(\forall x)(x > 2 \wedge x < 10)$$
$$(\exists y)x^2 = y$$
$$(\forall x)(\exists y)\, Love(y,x) \qquad \text{(everyone is loved by someone)}$$
$$(\exists y)(\forall x)\, Love(y,x) \qquad \text{(someone loves everyone)}$$

The formula $(\forall x)(x > 2)$ states that every x is greater than the constant 2; $(\exists x)\, x^2 = 2$ states that there is an x that is the square root of 2; $(\forall x)(\exists y)\, x^2 = y$ states that for every x there is a y such that the square of x is y.

16.3.1 Sketch of Formalization of Predicate Calculus

The formalization of predicate calculus includes the definition of an alphabet of symbols (including constants and variables), the definition of function and predicate letters, logical connectives and quantifiers. This leads to the definitions of the terms and well-formed formulae of the calculus.

The predicate calculus is built from an alphabet of constants, variables, function letters, predicate letters and logical connectives (including the logical connectives discussed in propositional logic, and universal and existential quantifiers).

The definition of terms and well-formed formulae specifies the syntax of the predicate calculus, and the set of well-formed formulae gives the language of the predicate calculus. The terms and well-formed formulae are built from the symbols, and these symbols are not given meaning in the formal definition of the syntax.

The language defined by the calculus needs to be given an *interpretation* in order to give a meaning to the terms and formulae of the calculus. The interpretation needs to define the domain of values of the constants and variables, provide meaning to the function letters, the predicate letters and the logical connectives.

Terms are built from constants, variables and function letters. A constant or variable is a term, and if $t_1, t_2, \ldots, t_k$ are terms, then $f_i^k(t_1, t_2, \ldots, t_k)$ is a term (where f_i^k is a k-ary function letter). Examples of terms include

x^2 where x is a variable and square is a 1-ary function letter

$x^2 + y^2$ where $x^2 + y^2$ is shorthand for the function add(square(x), square(y))
               where add is a 2-ary function letter and square is a 1-ary function letter

The well-formed formulae are built from terms as follows. If P_i^k is a k-ary predicate letter, $t_1, t_2, \ldots, t_k$ are terms, then $P_i^k(t_1, t_2, \ldots, t_k)$ is a well-formed formula. If A and B are well-formed formulae then so are $\neg A$, $A \wedge B$, $A \vee B$, $A \rightarrow B$, $A \leftrightarrow B$, $(\forall x)A$ and $(\exists x)A$.

There is a set of axioms for predicate calculus and two rules of inference used for the deduction of new formulae from the existing axioms and previously deduced formulae. The deduction of a new formula Q is via a sequence of well-formed formulae $P_1, P_2, \ldots P_n$ (where $P_n = Q$) such that each P_i is either an axiom, a hypothesis, or deducible from one or more of the earlier formulae in the sequence.

The two rules of inference are *modus ponens* and *generalization*. Modus ponens is a rule of inference that states that given predicate formulae A, and $A \Rightarrow B$ then the predicate formula B may be deduced. Generalization is a rule of inference that states that given predicate formula A, then the formula $(\forall x)A$ may be deduced where x is any variable.

The deduction of a formula Q from a set of hypothesis H is denoted by $H \vdash Q$, and where Q is deducible from the axioms alone this is denoted by $\vdash Q$. The *deduction theorem* states that if $H \cup \{P\} \vdash Q$ then $H \vdash P \rightarrow Q$[3] and the converse of the theorem is also true, i.e. if $H \vdash P \rightarrow Q$ then $H \cup \{P\} \vdash Q$.

The approach allows reasoning about symbols according to rules, and to derive theorems from formulae irrespective of the meanings of the symbols and formulae. Predicate calculus is *sound*, i.e. any theorem derived using the approach is true and the calculus is also *complete*.

Scope of Quantifiers

The scope of the quantifier $(\forall x)$ in the well-formed formula $(\forall x)A$ is A. Similarly, the scope of the quantifier $(\exists x)$ in the well-formed formula $(\exists x)B$ is B. The variable x that occurs within the scope of the quantifier is said to be a *bound variable*. If a variable is not within the scope of a quantifier it is *free*.

Example 16.8 (*Scope of Quantifiers*)

(i) x is free in the well-formed formula $\forall y \, (x^2 + y > 5)$.
(ii) x is bound in the well-formed formula $\forall x \, (x^2 > 2)$.

A well-formed formula is *closed* if it has no free variables. The substitution of a term t for x in A can only take place only when no free variable in t will become bound by a quantifier in A through the substitution. Otherwise, the interpretation of A would be altered by the substitution.

---

[3]This is stated more formally that if $H \cup \{P\} \vdash Q$ by a deduction containing no application of generalization to a variable that occurs free in P then $H \vdash P \rightarrow Q$.

A term t is free for x in A if no free occurrence of x occurs within the scope of a quantifier ($\forall y$) or ($\exists y$) where y is free in t. This means that the term t may be substituted for x without altering the interpretation of the well-formed formula A.

For example, suppose A is $\forall y\,(x^2 + y^2 > 2)$ and the term t is y, then t is not free for x in A as the substitution of t for x in A will cause the free variable y in t to become bound by the quantifier $\forall y$ in A, thereby altering the meaning of the formula to$\forall y\,(y^2 + y^2 > 2)$.

16.3.2 Interpretation and Valuation Functions

An *interpretation* gives meaning to a formula and it consists of a *domain of discourse* and a *valuation function*. If the formula is a sentence (i.e. does not contain any free variables) then the given interpretation of the formula is either true or false. If a formula has free variables, then the truth or falsity of the formula depends on the values given to the free variables. A formula with free variables essentially describes a relation say, $R(x_1, x_2, \ldots x_n)$ such that $R(x_1, x_2, \ldots x_n)$ is true if $(x_1, x_2, \ldots x_n)$ is in relation R. If the formula is true irrespective of the values given to the free variables, then the formula is true in the interpretation.

A *valuation* (meaning) *function gives meaning to the logical symbols and connectives*. Thus, associated with each constant c is a constant c_Σ in some universe of values Σ, and with each function symbol f of parity k, we have a function symbol f_Σ in Σ and $f_\Sigma : \Sigma^k \to \Sigma$ and for each predicate symbol P of parity k a relation $P_\Sigma \subseteq \Sigma^k$. *The valuation function, in effect, gives the semantics of the language of the predicate calculus L.*

The truth of a predicate P is then defined in terms of the meanings of the terms, the meanings of the functions, predicate symbols and the normal meanings of the connectives.

Mendelson (1987) provides a technical definition of truth in terms of *satisfaction* (with respect to an interpretation M). Intuitively a formula F is *satisfiable* if it is *true* (in the intuitive sense) for some assignment of the free variables in the formula F. If a formula F is satisfied for every possible assignment to the free variables in F, then it is *true* (in the technical sense) for the interpretation M. An analogous definition is provided for *false* in the interpretation M.

A formula is *valid* if it is true in every interpretation; however, as there may be an uncountable number of interpretations, it may not be possible to check this requirement in practice. M is said to be a model for a set of formulae if and only if every formula is true in M.

There is a distinction between proof-theoretic and model-theoretic approaches in predicate calculus. *Proof theoretic* is essentially syntactic, and there is a list of axioms with rules of inference. The theorems of the calculus are logically derived (i.e. $\vdash A$) and the logical truths are as a result of the syntax or form of the formulae, rather than the *meaning* of the formulae. *Model theoretical*, in contrast, is essentially semantic. The truth derives from the meaning of the symbols and connectives, rather than the logical structure of the formulae. This is written as $\vdash_M A$.

A calculus is *sound* if all of the logically valid theorems are true in the interpretation, i.e. proof theoretic $\Rightarrow$ model theoretic. A calculus is *complete* if all the truths in an interpretation are provable in the calculus, i.e. model theoretic $\Rightarrow$ proof theoretic. A calculus is *consistent* if there is no formula A such that $\vdash A$ and $\vdash \neg A$.

The predicate calculus is sound, complete and consistent. *Predicate calculus is not decidable*, i.e. there is no algorithm to determine for any well-formed formula A whether A is a theorem of the formal system. The undecidability of the predicate calculus may be demonstrated by showing that if the predicate calculus is decidable then the halting problem (of Turing machines) is solvable. The halting problem was discussed in Chap. 13.

16.3.3 Properties of Predicate Calculus

The following are properties of the predicate calculus:

(i) $(\forall x)P(x) \equiv (\forall y)P(y)$,
(ii) $(\forall x)P(x) \equiv \neg(\exists x)\neg P(x)$,
(iii) $(\exists x)P(x) \equiv \neg(\forall x)\neg P(x)$,
(iv) $(\exists x)P(x) \equiv (\exists y)P(y)$,
(v) $(\forall x)(\forall y)P(x, y) \equiv (\forall y)(\forall x)P(x, y)$,
(vi) $(\exists x)(P(x) \vee Q(x)) \equiv (\exists x)P(x) \vee (\exists y)Q(y)$ and
(vii) $(\forall x)(P(x) \wedge Q(x)) \equiv (\forall x)P(x) \wedge (\forall y)Q(y)$.

16.3.4 Applications of Predicate Calculus

The predicate calculus may be employed to formally state the system requirements of a proposed system. It may be used to conduct formal proof to verify the presence or absence of certain properties in a specification.

It may also be employed to define piecewise-defined functions such as $f(x, y)$ where $f(x, y)$ is defined by

$$
\begin{array}{ll}
f(x, y) = x^2 - y^2 & \text{where } x \leq 0 \wedge y < 0 \\
f(x, y) = x^2 + y^2 & \text{where } x > 0 \wedge y < 0 \\
f(x, y) = x + y & \text{where } x \geq 0 \wedge y = 0 \\
f(x, y) = x - y & \text{where } x < 0 \wedge y = 0 \\
f(x, y) = x + y & \text{where } x \leq 0 \wedge y > 0 \\
f(x, y) = x^2 + y^2 & \text{where } x > 0 \wedge y > 0
\end{array}
$$

The predicate calculus may be employed for program verification, and to show that a code fragment satisfies its specification. The statement that a program F is correct with respect to its precondition P and postcondition Q is written as $P\{F\}Q$. The objective of program verification is to show that if the precondition is true before execution of the code fragment, then this implies that the postcondition is true after execution of the code fragment.

A program fragment a is *partially correct* for precondition P and postcondition Q if and only if whenever a is executed in any state in which P is satisfied and execution terminates, then the resulting state satisfies Q. Partial correctness is denoted by $P\{F\}Q$, and Hoare's Axiomatic Semantics is based on partial correctness. It requires proof that the postcondition is satisfied if the program terminates.

A program fragment a is *totally correct* for precondition P and postcondition Q, if and only if whenever a is executed in any state in which P is satisfied then the execution terminates and the resulting state satisfies Q. It is denoted by $\{P\}F\{Q\}$, and Dijkstra's calculus of weakest preconditions is based on total correctness (Dijkstra 1976; Gries 1981). It is required to prove that if the precondition is satisfied then the program terminates and the postcondition is satisfied.

16.3.5   Semantic Tableaux in Predicate Calculus

We discussed the use of semantic tableaux for determining the validity of arguments in propositional logic earlier in this chapter, and its approach is to negate the conclusion of an argument and to show that this results in inconsistency with the premises of the argument.

The use of semantic tableaux is similar with predicate logic, except that there are some additional rules to consider. As before, if all branches of a semantic tableau are closed, then the premises and the negation of the conclusion are mutually inconsistent. From this, we deduce that the conclusion must be true.

The rules of semantic tableaux for propositional logic were presented in Table 16.12, and the additional rules specific to predicate logic are detailed in Table 16.14.

Example 16.9 (*Semantic Tableaux*) Show that the syllogism 'All Greeks are mortal; Socrates is a Greek; therefore Socrates is mortal' is a valid argument in predicate calculus.

Table 16.14 Extra rules of semantic tableaux (for predicate calculus)

Rule no.	Definition	Description
1	$(\forall x)\ A(x)$ $A(t)$ where t is a term	Universal instantiation
2.	$(\exists x)\ A(x)$ $A(t)$ where t is a term that has not been used in the derivation so far	Rule of existential instantiation. The term 't' is often a constant 'a'
3.	$\neg(\forall x)\ A(x)$ $(\exists x)\ \neg A(x)$	
4.	$\neg(\exists x)\ A(x)$ $(\forall x)\neg A(x)$	

Solution We expressed this argument previously as $(\forall x)(G(x) \rightarrow M(x))$; $G(s); M(s)$. Therefore, we negate the conclusion (i.e. $\neg M(s)$) and try to construct a closed tableau.

$$(\forall x)(G(x) \rightarrow M(x))$$
$$G(s)$$
$$\neg M(s).$$
$$G(s) \rightarrow M(s) \qquad\qquad\qquad \text{Universal Instantiation}$$
$$\wedge$$
$$\neg G(s) \quad M(s)$$
$$\text{-----} \qquad \text{--------}$$
$$closed \qquad closed$$

Therefore, as the tableau is closed we deduce that the negation of the conclusion is inconsistent with the premises, and that, therefore, the conclusion follows from the premises.

Example 16.10 (*Semantic Tableaux*) Determine whether the following argument is valid:

All lecturers are motivated,
Anyone who is motivated and clever will teach well,
Joanne is a clever lecturer and
Therefore, Joanne will teach well.

Solution We encode the argument as follows:

$L(x)$ stands for 'x is a lecturer',
$M(x)$ stands for 'x is motivated',
$C(x)$ stands for 'x is clever' and
$W(x)$ stands for 'x will teach well'.

We therefore wish to show that

$$(\forall x)(L(x) \rightarrow M(x)) \wedge (\forall x)((M(x) \wedge C(x)) \rightarrow W(x)) \wedge L(joanne) \wedge C(joanne) \mid =$$
$$W(joanne)$$

Therefore, we negate the conclusion (i.e. $\neg W(joanne)$) and try to construct a closed tableau.

1. $(\forall x)(L(x) \rightarrow M(x))$
2. $(\forall x)((M(x) \wedge C(x)) \rightarrow W(x))$
3. $L(joanne)$
4. $C(joanne)$
5. $\neg W(joanne)$
6. $L(joanne) \rightarrow M(joanne)$ Universal Instantiation (line 1)
7. $(M(joanne) \wedge C(joanne)) \rightarrow W(joanne)$ Universal Instantiation (line 2)

                          /\
8. $\neg L(joanne)$ $M(joanne)$ From line 6

          Closed

                              /\
9. $\neg (M(joanne) \wedge C(joanne))$ $W(joanne)$ From line 7

                              Closed

                  / \
10. $\neg M(joanne)$ $\neg C(joanne)$
    ---------------   -------------

        Closed   Closed

Therefore, since the tableau is closed we deduce that the argument is valid.

16.4   Review Questions

1. Draw a truth table to show that $\neg(P \rightarrow Q) \equiv P \wedge \neg Q$.
2. Translate the sentence 'Execution of program P begun with $x < 0$ will not terminate' into propositional form.
3. Prove the following theorems using the inference rules of natural deduction:

    (a) From b infer $b \vee \neg c$,
    (b) From $b \Rightarrow (c \wedge d)$, b infer d.

4. Explain the difference between the universal and the existential quantifier.
5. Express the following statements in the predicate calculus:

    a. All natural numbers are greater than 10.
    b. There is at least one natural number between 5 and 10.
    c. There is a prime number between 100 and 200.

6. Which of the following predicates are true?

  a. $\forall i \in \{10,\ldots,50\}.\, i^2 < 2000 \wedge i < 100$,
  b. $\exists i \in \mathbb{N}.\, i > 5 \wedge i^2 = 25$ and
  c. $\exists i \in \mathbb{N}.\, i^2 = 25$.

7. Use semantic tableaux to show that $(A \rightarrow A) \vee (B \wedge \neg B)$ is true.
8. Determine if the following argument is valid:
  If Pilar lives in Cork, she lives in Ireland. Pilar lives in Cork. Therefore,
  Pilar lives in Ireland.

16.5 Summary

Propositional logic is the study of propositions, and a proposition is a statement that is either true or false. A formula in propositional calculus may contain several variables, and the truth or falsity of the individual variables, and the meanings of the logical connectives determine the truth or falsity of the logical formula.

  A rich set of connectives is employed to combine propositions and to build up the well-formed formulae of the calculus. This includes the conjunction of two propositions, the disjunction of two propositions and the implication of two propositions. These connectives allow compound propositions to be formed, and the truth of the compound propositions is determined from the truth-values of the constituent propositions and the rules associated with the logical connectives. The meaning of the logical connectives is given by truth tables.

  Propositional calculus is both complete and consistent with all true propositions deducible in the calculus, and there is no formula A such that both A and $\neg A$ are deducible in the calculus.

  An argument in propositional logic consists of a sequence of formulae that are the premises of the argument and a further formula that is the conclusion of the argument. One elementary way to see if the argument is valid is to produce a truth table to determine if the conclusion is true whenever all of the premises are true. Other ways are to use semantic tableaux and natural deduction.

  Predicates are statements involving variables and these statements become propositions once the variables are assigned values. Predicate calculus allows expressions such as all members of the domain have a particular property or that there is at least one member that has a particular property.

Predicate calculus may be employed to specify the requirements for a proposed system and to give the definition of a piecewise-defined function. Semantic tableaux may be used for determining the validity of arguments in propositional or predicate logic, and its approach is to negate the conclusion of an argument and to show that this results in inconsistency with the premises of the argument.

References

Dijkstra EW (1976) A disciple of programming. Prentice Hall

Gries D (1981) The science of programming. Springer, Berlin

Kelly J (1997) The essence of logic. Prentice Hall

Mendelson E (1987) Introduction to mathematical logic. Wadsworth and Cole/Brook, Advanced Books & Software

Advanced Topics in Logic

17

17.1 Introduction

In this chapter, we consider some advanced topics in logic including fuzzy logic, temporal logic, intuitionistic logic, approaches that deal with undefined values, and logic and AI. Fuzzy logic is an extension of classical logic that acts as a mathematical model for vagueness, and it handles the concept of partial truth where truth-values lie between completely true and completely false. Temporal logic is concerned with the expression of properties that have time dependencies, and it allows temporal properties about the past, present and future to be expressed.

Brouwer and others developed intuitionistic logic which was a controversial theory of the foundations of mathematics based on a rejection of the law of the excluded middle and an insistence on constructive existence. Martin Löf successfully applied intuitionistic logic to type theory in the 1970s.

© Springer Nature Switzerland AG 2020
G. O'Regan, *Mathematics in Computing*, Undergraduate Topics
in Computer Science, https://doi.org/10.1007/978-3-030-34209-8_17

Partial functions arise naturally in computer science, and such functions may fail to be defined for one or more values in their domain. One approach to dealing with partial functions is to employ a precondition, which restricts the application of the function to values where it is defined. We consider three approaches to deal with undefined values, including the logic of partial functions, Dijkstra's approach with his *cand* and *cor* operators and Parnas's approach which preserves a classical two-valued logic.

We examine the contribution of logic to the AI field, with a short discussion of the work of John McCarthy and PROLOG logic programming language.

17.2 Fuzzy Logic

Fuzzy logic is a branch of *many-valued logic* that allows inferences to be made when dealing with vagueness, and it can handle problems with imprecise or incomplete data. It differs from classical two-valued propositional logic, in that it is based on degrees of truth, rather than on the standard binary truth-values of 'true or false' (1 or 0) of propositional logic. That is, while statements made in propositional logic are either true or false (1 or 0), the truth-value of a statement made in fuzzy logic is a value between 0 and 1. Its value expresses the extent to which the statement is true, with a value of 1 expressing absolute truth and a value of 0 expressing absolute falsity.

Fuzzy logic uses *degrees of truth* as a mathematical model for vagueness, and this is useful since statements made in natural language are often vague and have a certain (rather than an absolute) degree of truth. It is an extension of classical logic to deal with the concept of partial truth, where the truth-value lies between completely true and completely false. Lofti Zadeh developed fuzzy logic at Berkley in the 1960s, and it has been successfully applied to Expert Systems and other areas of Artificial Intelligence.

For example, consider the statement 'John is tall'. If John is 6 foot, 4 in., then we would say that this is a true statement (with a truth-value of 1) since John is well above average height. However, if John is 5 feet, 9 in. tall (around average height) then this statement has a degree of truth, and this could be indicated by a fuzzy truth-value of 0.6. Finally, if John's height is 4 feet, 10 in., then we would say that this is a false statement with truth-value 0. Similarly, the statement that 'today is sunny' may be assigned a truth-value of 1 if there are no clouds, 0.8 if there are a small number of clouds and 0 if it is raining all day.

Propositions in fuzzy logic may be combined together to form compound propositions. Suppose X and Y are propositions in fuzzy logic, then compound propositions may be formed from the conjunction, disjunction and implication operators. The usual definition in fuzzy logic of the truth-values of the compound propositions formed from X and Y is given by

$$\text{Truth}(\neg X) = 1 - \text{Truth}(X)$$
$$\text{Truth}(X \text{ and } Y) = \min(\text{Truth}(X), \text{Truth}(Y))$$
$$\text{Truth}(X \text{ or } Y) = \max(\text{Truth}(X), \text{Truth}(Y))$$
$$\text{Truth}(X \rightarrow Y) = \text{Truth}(\neg X \text{ or } Y))$$

Another way in which the operators may be defined is in terms of multiplication:

$$\text{Truth}(X \text{ and } Y) = \text{Truth}(X) * \text{Truth}(Y)$$
$$\text{Truth}(X \text{ or } Y) = 1 - (1 - \text{Truth}(X)) * (1 - \text{Truth}(Y))$$
$$\text{Truth}(X \rightarrow Y) = \max\{z | \text{Truth}(X) * z \leq \text{Truth}(Y)\} \text{ where } 0 \leq z \leq 1$$

Under these definitions, fuzzy logic is an extension of classical two-valued logic, which preserves the usual meaning of the logical connectives of propositional logic when the fuzzy values are just {0, 1}.

Fuzzy logic has been very useful in expert system and artificial intelligence applications. The first fuzzy logic controller was developed in England in the mid-1970s. Fuzzy logic has also been applied to the aerospace and automotive sectors, and the medical, robotics and transport sectors.

17.3 Temporal Logic

Temporal logic is concerned with the expression of properties that have time dependencies, and the various temporal logics can express facts about the past, present and future. Temporal logic has been applied to specify temporal properties of natural language, artificial intelligence as well as the specification and verification of program and system behaviour. It provides a language to encode temporal properties in artificial intelligence applications, and it plays a useful role in the formal specification and verification of temporal properties (e.g. liveness and fairness) in safety-critical systems.

The statements made in temporal logic can have a truth-value that varies over time. In other words, sometimes the statement is true and sometimes it is false, but it is never true or false at the same time. The two main types of temporal logics are *linear time logics* (reason about a single timeline) and *branching time logics* (reason about multiple timelines).

The roots of temporal logic lie in work done by Aristotle in the fourth century B. C., when he considered whether a truth-value should be given to a statement about a future event that may or may not occur. For example, what truth-value (if any) should be given to the statement that *'There will be a sea battle tomorrow'*. Aristotle argued against assigning a truth-value to such statements in the present time.

Newtonian mechanics assumes an absolute concept of time independent of space, and this viewpoint remained dominant until the development of the theory of relativity in the early twentieth century (when spacetime became the dominant paradigm).

Arthur Prior began analysing and formalizing the truth-values of statements concerning future events in the 1950s, and he introduced Tense Logic (a temporal logic) in the early 1960s. Tense logic contains four modal operators (strong and weak) that express events in the future or in the past:

- P (it has at some time been the case that),
- F (it will be at some time be the case that),
- H (it has always been the case that) and
- G (it will always be the case that).

The P and F operators are known as weak tense operators, while the H and G operators are known as strong tense operators. The two pairs of operators are interdefinable via the equivalences:

$$P\phi \cong \neg H\neg\phi$$
$$H\phi \cong \neg P\neg\phi$$
$$F\phi \cong \neg G\neg\phi$$
$$G\phi \cong \neg F\neg\phi$$

The set of formulae in Prior's temporal logic may be defined recursively, and they include the connectives used in classical logic (e.g. $\neg$, $\wedge$, $\vee$, $\rightarrow$, $\leftrightarrow$). We can express a property ϕ that is always true as $A\phi \cong H\phi \wedge \phi \wedge G\phi$ and a property that is sometimes true as $E\phi \cong P\phi \vee \phi \vee F\phi$. Various extensions of Prior's tense logic have been proposed to enhance its expressiveness. These include the binary *since* temporal operator 'S', and the binary *until* temporal operator 'U'. For example, the meaning of $\phi S\psi$ is that ϕ has been true since a time when ψ was true.

Temporal logics are applicable in the specification of computer systems, as a specification may require *safety, fairness* and *liveness properties* to be expressed. For example, a fairness property may state that it will always be the case that a certain property will hold sometime in the future. The specification of temporal properties often involves the use of special temporal operators.

We discuss common temporal operators that are used, including an operator to express properties that will always be true, properties that will eventually be true and a property that will be true in the next time instance. For example,

□ P P is always true,
◊ P P will be true sometimes in the future,
○ P P is true in the next time instant (*discrete time*).

Linear Temporal Logic (LTL) was introduced by Pnueli in the late-1970s, and is useful in expressing safety and liveness properties. Branching time logics assume a non-deterministic branching future for time (with a deterministic, linear past).

Computation Tree Logic (CTL and CTL*) was introduced in the early 1980s by Emerson and others.

It is also possible to express temporal operations directly in classical mathematics, and the well-known computer scientist, Parnas, prefers this approach. He is critical of computer scientists for introducing unnecessary formalisms when classical mathematics already possesses the ability to do this. For example, the value of a function f at a time instance prior to the current time t is defined as

$$Prior(f,t) = \lim_{\varepsilon \to 0} f(t - \varepsilon)$$

For more detailed information on temporal logic, the reader is referred to the excellent article on temporal logic in http://plato.stanford.edu/entries/logic-temporal/.

17.4 Intuitionistic Logic

The controversial school of intuitionistic mathematics was found by the Dutch mathematician, L. E. J. Brouwer, who was a famous topologist, and well known for his fixpoint theorem in topology. This constructive approach to mathematics proved to be highly controversial, as its acceptance as a foundation of mathematics would have led to the rejection of many accepted theorems in classical mathematics (including his own fixed-point theorem).

Brouwer was deeply interested in the foundations of mathematics and the problems arising from the paradoxes of set theory. He was determined to provide a secure foundation for mathematics, and his view was that an existence theorem in mathematics that demonstrates the proof of a mathematical object has no validity, unless the proof is constructive and accompanied by a procedure to construct the object. He, therefore, rejected indirect proof and the law of the excluded middle $(P \vee \neg P)$ or equivalently $(\neg \neg P \to P)$, and he insisted on an explicit construction of the mathematical object.

The problem with the Law of the Excluded Middle (LEM) arises in dealing with properties of infinite sets. For finite sets, one can decide if all elements of the set possess a certain property P by testing each one. However, this procedure is no longer possible for infinite sets. We may know that a certain element of the infinite set does not possess the property, or it may be the actual method of construction of the set that allows us to prove that every element has the property. However, the application of the law of the excluded middle is invalid for infinite sets, as we cannot conclude from the situation where not all elements of an infinite set possesses a property P that there exists at least one element which does not have the property P. In other words, the law of the excluded middle may only be applied in cases where the conclusion can be reached in a finite number of steps.

Consequently, if the Brouwer view of the world was accepted, then many of the classical theorems of mathematics (including his own well-known results in topology) could no longer be said to be true. His approach to the foundations of mathematics hardly made him popular with other mathematicians (the differences were so fundamental that it was more like a war), and intuitionism never became mainstream in mathematics. It led to deep and bitter divisions between Hilbert and Brouwer, with Hilbert accusing Brouwer (and Weyl) of trying to overthrow everything that did not suit them in mathematics, and that intuitionism was treason · to science. Hilbert argued that a suitable foundation for mathematics should aim to preserve most of mathematics. Brouwer described Hilbert's formalist program as a false theory that would produce nothing of mathematical value. For Brouwer, 'to exist' is synonymous with 'constructive existence', and constructive mathematics is relevant to computer science, as a program may be viewed as the result obtained from a constructive proof of its specification.

Brouwer developed one of the more unusual logics that has been invented (intuitionistic logic), in which many of the results of classical mathematics were no longer true. Intuitionistic logic may be considered the logical basis of constructive mathematics, and formal systems for intuitionistic propositional and predicate logic were developed by Heyting and others (Heyting 1966).

Consider a hypothetical mathematical property $P(x)$ of which there is no known proof (i.e. it is unknown whether $P(x)$ is true or false for arbitrary x where x ranges over the natural numbers). Therefore, the statement $\forall x\ (P(x) \lor \neg P(x))$ cannot be asserted with the present state of knowledge, as neither $P(x)$ nor $\neg P(x)$ has been proved. That is, unproved statements in intuitionistic logic are not given an intermediate truth-value, and they remain of an unknown truth-value until they have been either proved or disproved.

The intuitionistic interpretation of the logical connectives is different from classical propositional logic. A sentence of the form $A \lor B$ asserts that either a proof of A or a proof of B has been constructed, and $A \lor B$ is not equivalent to $\neg\ (\neg A \land \neg B)$. Similarly, a proof of $A \land B$ is a pair whose first component is a proof of A, and whose second component is a proof of B. The statement $\neg\ \forall x\ \neg P(x)$ is not equivalent to $\exists x\ P(x)$ in intuitionistic logic.

Intuitionistic logic was applied to Type Theory by Martin Löf in the 1970s (Martin Löf 1984). Intuitionistic type theory is based on an analogy between propositions and types, where $A \land B$ is identified with $A \times B$, the Cartesian product of A and B. The elements in the set $A \times B$ are of the form (a, b) where $a \in A$ and $b \in B$. The expression $A \lor B$ is identified with $A + B$, the disjoint union of A and B. The elements in the set $A + B$ are got from tagging elements from A and B, and they are of the form inl(a) for $a \in A$, and inr(b) for $b \in B$. The left and right injections are denoted by inl and inr.

17.5   Undefined Values

Total functions $f : X \rightarrow Y$ are functions that are defined for every element in their domain, and total functions are widely used in mathematics. However, there are functions that are undefined for one or more elements in their domain, and one example is the function $y = 1/x$. This function is undefined at $x = 0$.

Partial functions arise naturally in computer science, and such functions may fail to be defined for one or more values in their domain. One approach to dealing with partial functions is to employ a precondition, which restricts the application of the function to where it is defined. This makes it possible to define a new set (a proper subset of the domain of the function) for which the function is total over the new set.

Undefined terms often arise[1] and need to be dealt with. Consider the example of the square root function $\sqrt{x}$ taken from Parnas (1993). The domain of this function is the positive real numbers, and the following expression is undefined:

$$((x > 0) \wedge (y = \sqrt{x})) \vee ((x \leq 0) \wedge (y = \sqrt{-x}))$$

The reason this is undefined is since the usual rules for evaluating such an expression involves evaluating each sub-expression, and then performing the Boolean operations. However, when $x < 0$ the sub-expression $y = \sqrt{x}$ is undefined, whereas when $x > 0$ the sub-expression $y = \sqrt{-x}$ is undefined. Clearly, it is desirable that such expressions be handled, and that for the example above, the expression would evaluate to true.

Classical two-valued logic does not handle this situation adequately, and there have been several proposals to deal with undefined values. Dijkstra's approach is to use the **cand** and **cor** operators in which the value of the left-hand operand determines whether the right-hand operand expression is evaluated or not. Jones' logic of partial functions (Jones 1986) uses a three-valued logic[2] and Parnas's[3] approach is an extension to the predicate calculus to deal with partial functions that preserve the two-valued logic.

17.5.1   Logic of Partial Functions

Jones (1986) has proposed the Logic of Partial Functions (LPFs) as an approach to deal with terms that may be undefined. This is a three-valued logic and a logical term may be true, false or undefined (denoted $\perp$). The definition of the truth functional operators used in classical two-valued logic is extended to three-valued logic. The truth tables for conjunction and disjunction are defined in Fig. 17.1.

---

[1] It is best to avoid undefinedness by taking care with the definitions of terms and expressions.
[2] The above expression would evaluate to true under Jones' three-valued logic of partial functions.
[3] The above expression evaluates to true for Parnas logic (a two-valued logic).

The conjunction of P and Q is true when both P and Q are true, false if one of P or Q is false and undefined otherwise. The operation is commutative. The disjunction of P and Q ($P \vee Q$) is true if one of P or Q is true, false if both P and Q are false and undefined otherwise. The implication operation ($P \rightarrow Q$) is true when P is false or when Q is true, false when P is true and Q is false and undefined otherwise. The equivalence operation ($P \leftrightarrow Q$) is true when both P and Q are true or false, it is false when P is true and Q is false (and vice versa) and it is undefined otherwise (Fig. 17.2).

The not operator ($\neg$) is a unary operator such that $\neg A$ is true when A is false, false when A is true and undefined when A is undefined (Fig. 17.3).

The result of an operation may be known immediately after knowing the value of one of the operands (e.g. disjunction is true if P is true irrespective of the value of Q). The law of the excluded middle, i.e. $A \vee \neg A$ does not hold in the three-valued logic, and Jones (1986) argues that this is reasonable as one would not expect the following to be true:

$$\left(\tfrac{1}{0} = 1\right) \vee \left(\tfrac{1}{0} \neq 1\right)$$

$\wedge$	Q = T	Q = F	Q = ⊥
P		P∧Q	
T	T	F	⊥
F	F	F	F
⊥	⊥	F	⊥

$\vee$	Q = T	Q = F	Q = ⊥
P		P∨Q	
T	T	T	T
F	T	F	⊥
⊥	T	⊥	⊥

Fig. 17.1 Conjunction and disjunction operators

$\rightarrow$	Q = T	Q = F	Q = ⊥
P		P→Q	
T	T	F	⊥
F	T	T	T
⊥	T	⊥	⊥

$\leftrightarrow$	Q = T	Q = F	Q = ⊥
P		P↔Q	
T	T	F	⊥
F	F	T	⊥
⊥	⊥	⊥	⊥

Fig. 17.2 Implication and equivalence operators

⊥	⊥

Fig. 17.3 Negation

There are other well-known laws that fail to hold such as

(i) $E \Rightarrow E$.
(ii) Deduction Theorem $E_1 \vdash E_2$ does not justify $\vdash E_1 \Rightarrow E_2$ unless it is known that E_1 is defined.

Many of the tautologies of standard logic also fail to hold.

17.5.2   Parnas Logic

Parnas's approach is based on classical two-valued logic, and his philosophy is that truth-values should be true or false only,[4] and that there is no third logical value. His system is an extension to predicate calculus to deal with partial functions. The evaluation of a logical expression yields the value 'true' or 'false' irrespective of the assignment of values to the variables in the expression. This allows the expression: $((x > 0) \land (y = \sqrt{x})) \lor ((x \leq 0) \land (y = \sqrt{-x}))$ that is undefined in classical logic to yield the value true.

The advantages of his approach are that no new symbols are introduced into the logic, and that the logical connectives retain their traditional meaning. This makes it easier for engineers and computer scientists to understand, as it is closer to their intuitive understanding of logic.

The meaning of predicate expressions is given by first defining the meaning of the primitive expressions. These are used as the building blocks for predicate expressions. The evaluation of a primitive expression $R_j(V)$ (where V is a comma separated set of terms with some elements of V involving the application of partial functions) is false if the value of an argument of a function used in one of the terms of V is not in the domain of that function.[5] The following examples (Tables 17.1 and 17.2) should make this clearer.

These primitive expressions are used to build the predicate expressions, and the standard logical connectives are used to yield truth-values for the predicate expression. Parnas logic is defined in detail in Parnas (1993).

The power of Parnas logic may be seen by considering a tabular expression example (Parnas 1993). Figure 17.4 specifies the behaviour of a program that searches the array B for the value x. It describes the properties of the values of 'j' and '$present$'. There are two cases to consider.

1. There is an element in the array with the value of x.
2. There is no such element in the array with the value of x.

Clearly, from the example above, the predicate expressions $\exists i, B[i] = x$ and $\neg (\exists i, B[i] = x)$ are defined. One disadvantage of the Parnas approach is that some common relational operators (e.g. $>$, $\geq$, $\leq$ and $<$) are not primitive in the logic.

---

[4]It seems strange to assign the value false to the primitive predicate calculus expression $y = 1/0$.
[5]The approach avoids the undefined logical value ($\bot$) and preserves the two-valued logic.

Table 17.1 Examples of Parnas evaluation of undefinedness

Expression	$x < 0$	$x \geq 0$
$y = \sqrt{x}$	False	True if $y = \sqrt{x}$, False otherwise
$y = {}^1\!/_0$	False	False
$y = x^2 + \sqrt{x}$	False	True if $y = x^2 + \sqrt{x}$, False otherwise

Table 17.2 Example of undefinedness in array

Expression	$i \in \{1 \ldots N\}$	$i \notin \{1 \ldots N\}$
$B[i] = x$	True if $B[i] = x$	False
$\exists i, B[i] = x$	True if $B[i] = x$ for some i, False otherwise	False

1. There is an element in the array with the value of x
2. There is no such element in the array with the value of x.

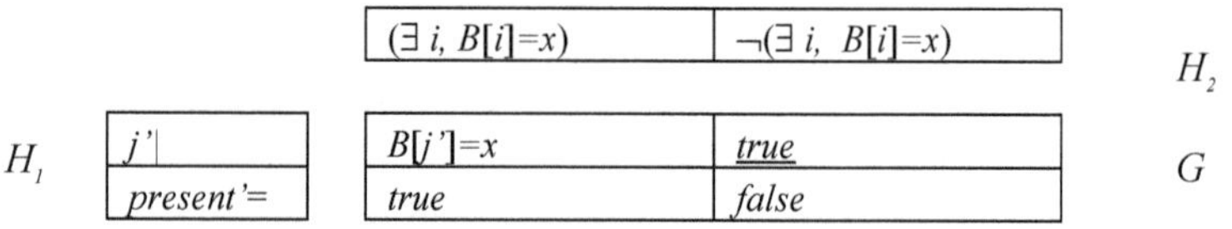

Fig. 17.4 Finding index in array

However, these relational operators are then constructed from primitive operators. Further, the axiom of reflection does not hold in the logic.

17.5.3 Dijkstra and Undefinedness

The **cand** and **cor** operators were introduced by Dijkstra (Fig. 17.5) to deal with undefined values. They are non-commutative operators and allow the evaluation of predicates that contain undefined values.

Consider the following expression:

$$y = 0 \vee (x/y = 2)$$

Then this expression is undefined when $y = 0$ as x/y is undefined, since the logical disjunction operation is not defined when one of its operands is undefined. However, there is a case for giving meaning to such an expression when $y = 0$, since in that case, the first operand of the logical or operation is true. Further, the logical *disjunction* operation is defined to be true if either of its operands is true. This motivates the introduction of the **cand** and **cor** operators. These operators are associative and their truth tables are defined in Tables 17.3 and 17.4.

Fig. 17.5 Edsger Dijkstra.
Courtesy of Brian Randell

Table 17.3 *a* **cand** *b*

a	*b*	*a* **cand** *b*
T	T	T
T	F	F
T	U	U
F	T	F
F	F	F
F	U	F
U	T	U
U	F	U
U	U	U

Table 17.4 *a* **cor** *b*

a	*b*	*a* **cor** *b*
T	T	T
T	F	T
T	U	T
F	T	T
F	F	F
F	U	U
U	T	U
U	F	U
U	U	U

The order of the evaluation of the operands for the ***cand*** operation is to *evaluate the first operand*; if the first operand is true then the result of the operation is the second operand; otherwise, the result is false. The expression *a* **cand** *b* is equivalent to

$$a \, cand \, b \cong \textbf{if} \ a \ \textbf{then} \ b \ \textbf{else} \ F$$

The order of the evaluation of the operands for the **cor** operation is to evaluate the first operand. If the first operand is true then the result of the operation is true; otherwise, the result of the operation is the second operand. The expression *a* **cor** *b* is equivalent to

$$a \, \textbf{cor} \, b \cong \textbf{if} \ a \ \textbf{then} \ T \ \textbf{else} \ b$$

The **cand** and **cor** operators satisfy the following laws:

• *Associativity*

The cand and cor operators are associative.

$$(A \, \textbf{cand} \, B) \, \textbf{cand} \, C = A \, \textbf{cand} \, (B \, \textbf{cand} \, C)$$
$$(A \, \textbf{cor} \, B) \, \textbf{cor} \, C = A \, \textbf{cor} (B \, \textbf{cor} \, C)$$

• *Distributivity*

The **cand** operator distributes over the **cor** operator and vice versa.

$$A \, \textbf{cand} \, (B \, \textbf{cor} \, C) = (A \, \textbf{cand} \, B) \, \textbf{cor} \, (A \, \textbf{cand} \, C)$$
$$A \, \textbf{cor} \, (B \wedge C) = (A \, \textbf{cor} \, B) \, \textbf{cand} \, (A \, \textbf{cor} \, C)$$

De Morgan's law enables logical expressions to be simplified.

$$\neg(A \, \textbf{cand} \, B) = \neg A \, \textbf{cor} \, \neg B$$
$$\neg(A \, \textbf{cor} \, B) = \neg A \, \textbf{cand} \, \neg B$$

17.6 Logic and AI

The long-term goal of Artificial Intelligence is to create a thinking machine that is intelligent, has consciousness, has the ability to learn, has free will and is ethical. Artificial Intelligence is a young field and John McCarthy and others coined the term in 1956. Alan Turing devised the Turing Test in the early 1950s as a way to determine whether a machine was conscious and intelligent. Turing believed that machines would eventually be developed that would stand a good chance of passing the 'Turing Test'.

Fig. 17.6 John McCarthy.
Courtesy of John McCarthy

There are deep philosophical problems in Artificial Intelligence, and some researchers believe that its goals are impossible or incoherent. Even if Artificial Intelligence is possible there are moral issues to consider such as the exploitation of artificial machines by humans and whether it is ethical to do this. Weizenbaum argues that AI is a threat to human dignity, and that AI should not replace humans in positions that require respect and care.

John McCarthy[6] has long advocated the use of logic in AI, and mathematical logic has been used in the AI field to formalize knowledge, and in guiding the design of mechanized reasoning systems. Logic has been used as an analytic tool, as a knowledge representation formalism and as a programming language (Fig. 17.6).

McCarthy's long-term goal was to formalize common-sense reasoning, i.e. the normal reasoning that is employed in problem-solving and dealing with normal events in the real world. McCarthy (1959) argues that it is reasonable for logic to play a key role in the formalization of common-sense knowledge, and this includes the formalization of basic facts about actions and their effects; facts about beliefs and desires and facts about knowledge and how it is obtained. His approach allows common-sense problems to be solved by logical reasoning.

Its formalization requires sufficient understanding of the common-sense world, and often the relevant facts to solve a particular problem are unknown. It may be that knowledge thought relevant may be irrelevant and vice versa. A computer may have millions of facts stored in its memory, and the problem is how to determine which of these should be chosen from its memory to serve as premises in logical deduction.

McCarthy's influential 1959 paper discusses various common-sense problems such as getting home from the airport. Mathematical logic is the standard approach to express premises, and it includes rules of inferences that are used to deduce valid

---

[6]John McCarthy received the Turing Award in 1971 for his contributions to Artificial Intelligence. He also developed the programming language LISP.

conclusions from a set of premises. Its rigorous deductive reasoning shows how new formulae may be logically deduced from a set or premises.

McCarthy's approach to programs with common sense has been criticized by Bar-Hillel and others on the grounds that common sense is fairly elusive, and the difficulty that a machine would have in determining which facts are relevant to a particular deduction from its known set of facts. However, McCarthy's approach has shown how logical techniques can contribute to the solution of specific AI problems.

Logic programming languages describe what is to be done, rather than how it should be done. These languages are concerned with the statement of the problem to be solved, rather than how the problem will be solved. These languages use mathematical logic as a tool in the statement of the problem definition. Logic is a useful tool in developing a body of knowledge (or theory), and it allows rigorous mathematical deduction to derive further truths from the existing set of truths. The theory is built up from a small set of axioms or postulates and rules of inference derive further truths logically.

The objective of logic programming is to employ mathematical logic to assist with computer programming. Many problems are naturally expressed as a theory, and the statement of a problem to be solved is often equivalent to determining if a new hypothesis is consistent with an existing theory. Logic provides a rigorous way to determine this, as it includes a rigorous process for conducting proof.

Computation in logic programming is essentially logical deduction, and logic programming languages use first-order[7] predicate calculus. They employ theorem proving to derive a desired truth from an initial set of axioms. These proofs are constructive[8] in that an actual object that satisfies the constraints is produced rather than a pure existence theorem. Logic programming specifies the objects, the relationships between them and the constraints that must be satisfied for the problem.

- The set of objects involved in the computation.
- The relationships that hold between the objects.
- The constraints of the particular problem.

The language interpreter decides how to satisfy the particular constraints. Artificial Intelligence influenced the development of logic programming, and John McCarthy demonstrated that mathematical logic could be used for expressing knowledge. The first logic programming language was Planner developed by Carl Hewitt at MIT in 1969. It uses a procedural approach for knowledge representation rather than McCarthy's declarative approach.

---

[7]First-order logic allows quantification over objects but not functions or relations. Higher order logics allow quantification of functions and relations.

[8]For example, the statement $\exists x$ such that $x = \sqrt{4}$ states that there is an x such that x is the square root of 4, and the constructive existence yields that the answer is that $x = 2$ or $x = -2$, i.e. constructive existence provides more the truth of the statement of existence, and an actual object satisfying the existence criteria is explicitly produced.

The best-known logic programming language is Prolog, which was developed in the early 1970s by Alain Colmerauer and Robert Kowalski. It stands for *pro*gramming in *log*ic. It is a goal-oriented language that is based on predicate logic. Prolog became an ISO standard in 1995. The language attempts to solve a goal by tackling the sub-goals that the goal consists of

$$\text{goal:- subgoal}_1, \ldots, \text{subgoal}_n.$$

That is, in order to prove a particular goal it is sufficient to prove subgoal_1 through subgoal_n. Each line of a Prolog program consists of a rule or a fact, and the language specifies what exists rather than how. The following program fragment has one rule and two facts:

> grandmother(G, S):- parent(P, S), mother(G, P).
> mother(sarah, isaac).
> parent(isaac, jacob).

The first line in the program fragment is a rule that states that G is the grand-mother of S if there is a parent P of S and G is the mother of P. The next two statements are facts stating that Isaac is a parent of Jacob, and that Sarah is the mother of Isaac. A particular goal clause is true if all of its subclauses are true:

$$\text{goalclause}(V_g):- \text{clause}_1(V_1), \ldots, \text{clause}_m(V_m)$$

A Horn clause consists of a goal clause and a set of clauses that must be proven separately. Prolog finds solutions by *unification,* i.e. by binding a variable to a value. For an implication to succeed, all goal variables Vg on the left side of :- must find a solution by binding variables from the clauses which are activated on the right side. When all clauses are examined and all variables in Vg are bound, the goal succeeds. But if a variable cannot be bound for a given clause, then that clause fails. Following the failure, Prolog *backtracks,* and this involves going back to the left to previous clauses to continue trying to unify with alternative bindings. Backtracking gives Prolog the ability to find multiple solutions to a given query or goal.

Logic programming languages generally use a simple searching strategy to consider alternatives:

–   If a goal succeeds and there are more goals to achieve, then remember any untried alternatives and go on to the next goal.
–   If a goal is achieved and there are no more goals to achieve then stop with success.
–   If a goal fails and there are alternative ways to solve it then try the next one.
–   If a goal fails and there are no alternate ways to solve it, and there is a previous goal, then go back to the previous goal.

– If a goal fails and there are no alternate ways to solve it, and no previous goal, then stop with failure.

Constraint programming is a programming paradigm where relations between variables can be stated in the form of constraints. Constraints specify the properties of the solution, and differ from the imperative programming languages in that they do not specify the sequence of steps to execute.

17.7 Review Questions

1. What is fuzzy logic?
2. What is intuitionistic logic and how is it different from classical logic?
3. Discuss the problem of undefinedness and the advantages and disadvantages of three-valued logics. Describe the approaches of Parnas, Dijkstra and Jones.
4. What is temporal logic?
5. Show how the temporal operators may be expressed in classical mathematics. Discuss the merits of temporal operators.
6. Discuss the applications of logic to AI.

17.8 Summary

We discussed some advanced topics in logic in this chapter, including fuzzy logic, temporal logic, intuitionistic logic, undefined values, logic and AI, and theorem provers. Fuzzy logic is an extension of classical logic that acts as a mathematical model for vagueness, whereas temporal logic is concerned with the expression of properties that have time dependencies

Intuitionism was a controversial school of mathematics that aimed to provide a solid foundation for mathematics. Its adherents rejected the law of the excluded middle, and insisted that for an entity to exist there must be a constructive proof of its existence. Martin Löf applied intuitionistic logic to type theory in the 1970s.

Partial functions arise naturally in computer science, and such functions may fail to be defined for one or more values in their domain. There are a number of approaches to deal with undefined values, including the logic of partial functions, Dijkstra's approach with his cand and cor operators and Parnas's approach which preserves a classical two-valued logic.

We discussed temporal logic and its applications to the specification of properties with time dependencies. We discussed the application of logic to the AI field, where logic has been used to formalize knowledge in AI systems.

References

Heyting A (1966) Intuitionist logic. An introduction. North-Holland Publishing
Jones C (1986) Systematic software development using VDM. Prentice Hall International
Martin Löf P (1984) Intuitionist type theory. Notes by Giovanni Savin of lectures given in Padua, June, 1980. Bibliopolis. Napoli
McCarthy J (1959) Programs with common sense. In: Proceedings of the Teddington conference on the mechanization of thought processes
Parnas DL (1993) Predicate calculus for software engineering. IEEE Trans Softw Eng 19(9)
Stanford Encyclopedia of Philosophy. Temporal logic. http://plato.stanford.edu/entries/logic-temporal/

The Nature of Theorem Proving

18

Key Topics

Mathematical proof
Formal proof
Automated theorem prover
Interactive theorem prover
Logic theorist
Resolution
Proof checker

18.1 Introduction

The word '*proof*' is generally interpreted as facts or evidence that support a particular proposition or belief, and such proofs are conducted in natural language. Several premises (which are self-evident or already established) are presented, and from these premises (via deductive or inductive reasoning) further propositions are established until finally the conclusion is established.

The proof of a theorem in mathematics requires additional rigour, and such proofs consist of a mixture of natural language and mathematical argument. It is common to skip over the trivial steps in the proof, and independent mathematicians conduct peer reviews to provide additional confidence in the correctness of the proof, and to ensure that no unwarranted assumptions or errors in reasoning have

© Springer Nature Switzerland AG 2020
G. O'Regan, *Mathematics in Computing*, Undergraduate Topics
in Computer Science, https://doi.org/10.1007/978-3-030-34209-8_18

been made. Proofs conducted in logic are extremely rigorous with every step in the proof explicit.[1]

Mathematical proof dates back to the Greeks, and most students are familiar with Euclid's work (*The Elements*) in geometry, where from a small set of axioms and postulates and definitions he derived many of the well-known theorems of geometry. Euclid was a Hellenistic mathematician based in Alexandria around 300 BC, and his style of proof was mainly constructive, i.e. in addition to the proof of the existence of an object, he actually constructed the object in the proof. Euclidean geometry remained unchallenged for over 2000 years, until the development of the non-Euclidean geometries in the nineteenth century, and these geometries were based on a rejection of Euclid's controversial fifth postulate (the parallels postulate).

Mathematical proof may employ a '*divide-and-conquer*' technique, i.e. breaking the conjecture down into sub-goals and then attempting to prove each of the sub-goals. Another common proof technique is *indirect proof* where we assume the opposite of what we wish to prove, and we show that this results in a contradiction (e.g. see Chap. 4 where we consider the proof that there are infinite number of primes or Chap. 5 for the proof that there is no rational number whose square is 2). Other proof techniques used are the method *of mathematical induction*, where it involves a proof of the base case and inductive step (see Chap. 8).

Aristotle developed *syllogistic logic* in the fourth century BC, and the rules of reasoning with valid syllogisms remained dominant in logic up to the nineteenth century. Boole developed his mathematical logic in the mid-nineteenth century, and he aimed to develop a calculus of reasoning to verify the correctness of arguments using logical connectives. Predicate logic (including universal and existential quantifiers) was introduced by Frege in the late nineteenth century as part of his efforts to derive mathematics from purely logical principles. Russell and Whitehead continued this attempt in *Principia Mathematica*, and Russell introduced the theory of types to deal with the paradoxes in set theory, which he identified in Frege's system.

The *formalists* introduced extensive axioms in addition to logical principles, and Hilbert's program led to the definition of a *formal mathematical proof* as a sequence of formulae, where each element is either an axiom or derived from a previous element in the series by applying a fixed set of mechanical rules (e.g. *modus ponens*). The last line in the proof is the theorem to be proved, and the formal proof is essentially syntactic following rules with the formulae simply a string of symbols and the meaning of the symbols is unimportant.

The formalists later ran into problems in trying to prove that a formal system powerful enough to include arithmetic was both complete and consistent, and the results of Gödel showed that such a system would be *incomplete* (and one of the propositions without a proof is that of its own *consistency*). Turing later showed (with his Turing machine) that mathematics is *undecidable*, i.e. there is no

---

[1]Perhaps a good analogy might be that a mathematical proof is like a program written in a high-level language such as C, whereas a formal mathematical proof in logic is like a program written in assembly language.

Fig. 18.1 Idea of automated
theorem proving

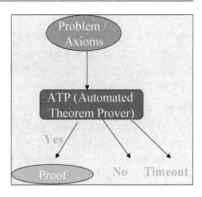

algorithm or mechanical procedure that may be applied in a finite number of steps
to determine if an arbitrary mathematical proposition is true or false.

The proofs employed in mathematics are rarely formal (in the sense of Hilbert's
program), and whereas they involve deductions from a set of axioms, these
deductions are rarely expressed as the application of individual rules of logical
inference.

The application of formal methods (see Chap. 20) in an industrial environment
requires the use of machine-assisted proof, since thousands of proof obligations
arise from a formal specification, and theorem provers are essential in resolving
these efficiently. Many proofs in formal methods are concerned with cross-checking
the details of the specification, or checking the validity of the refinement steps, or
checking that certain properties are satisfied by the specification. There are often
many tedious lemmas to be proved, and theorem provers[2] are essential in dealing
with these. Machine proof is explicit and reliance on some brilliant insight is
avoided. Proofs by hand in formal methods are notorious for containing errors or
jumps in reasoning, whereas machine proofs are explicit but are often extremely
lengthy and essentially unreadable.

Automated Theorem Proving (ATP) is difficult, as often mathematicians prove a
theorem with an initial intuitive feeling that the theorem is true (Fig. 18.1). Human
intervention to provide guidance or intuition improves the effectiveness of the
theorem prover. There are several tools available to support theorem proving, and
these include the Boyer–Moore theorem prover (known as NQTHM), the Isabelle
theorem prover and the HOL system.

The proof of various properties about a program increases confidence in its
correctness. However, an absolute proof of correctness[3] is unlikely except for the
most trivial of programs. A program may consist of legacy software that is assumed
to work; a compiler that is assumed to work correctly creates it. Theorem provers

---

[2]Most existing theorem provers are difficult to use and are for specialist use only. There is a need to
improve the usability of theorem provers.
[3]This position is controversial with others arguing that if correctness is defined mathematically
then the mathematical definition (i.e. formal specification) is a theorem, and the task is to prove
that the program satisfies the theorem. They argue that the proofs for non-trivial programs exist,

are programs that are assumed to function correctly. The best that mathematical proof in formal methods can claim is increased confidence in the correctness of the software, rather than an absolute proof of correctness.

18.2   Early Automation of Proof

Early work on the automation of proof began in the 1950s with the beginning of work in the Artificial Intelligence field, where the early AI practitioners were trying to develop a '*thinking machine*'. One of the earliest programs developed was the *Logic Theorist*, which was presented at the Dartmouth Conference on Artificial Intelligence in 1956 (MacKensie 1995).

It was developed by Allen Newell and Herbert Simon, and it could prove 38 of the first 52 theorems from Russell and Whitehead's *Principia Mathematica* (Russell and Whitehead 1910).[4] Russell and Whitehead had attempted to derive all mathematics from axioms and the inference rules of logic, and the LT program conducted proof from a small set of propositional axioms and deduction rules. Its approach was to start with the theorem to be proved, and to then search for relevant axioms and operators to prove the theorem. The Logic Theorist proved theorems in the propositional calculus, but it did not support predicate calculus. It used the five basic axioms of propositional logic and three rules of inference from the *Principii* to prove theorems.[5]

LT demonstrated that computers had the ability to encode knowledge and information, and to perform intelligent operations such as solving theorems in mathematics. The heuristic approach of the LT program tried to emulate human mathematicians, but could not guarantee that a proof could be found for every valid theorem.

If no immediate one-step proof could be found, then a set of sub-goals was generated (these are formulae from which the theorem may be proved in one step) and proofs of these were then searched for, and so on. The program could use previously proved theorems in the course of developing a proof of a new theorem. Newell and Simon were hoping that the Logic Theorist would do more than just prove theorems in logic, and their goal was that it would attempt to prove theorems in a human-like way and especially with a selective search.

---

and that the reason why there are not many examples of such proofs is due to the lack of mathematical specifications.

[4]Russell is said to have remarked that he was delighted to see that the Principia Mathematica could be done by machine, and that if he and Whitehead had known this in advance that they would not have wasted 10 years doing this work by hand in the early twentieth century.

[5]Another possibility (though an inefficient and poor simulation of human intelligence) would be to start with the five axioms of the *Principia*, and to apply the three rules of inference to logically derive all possible sequences of valid deductions. This is known as the British Museum algorithm (as sensible as putting monkeys in front of typewriters to reproduce all of the books of the British Museum).

However, in practice, the Logic Theorist search was not very selective in its approach, and the subproblems were considered in the order in which they were generated, and so there was no actual heuristic procedure (as in human problem-solving) to guess at which subproblem was most likely to yield an actual proof. This meant that the Logic Theorist could, in practice, find only very short proofs, since as the number of steps in the proof increased, the amount of search required to find the proof exploded.

The Geometry Machine was developed by Herbert Gelernter at the IBM Research Center in New York in the late 1950s, with the goal of developing intelligent behaviour in machines. It differed from the Logic Theorist in that it selected only the valid sub-goals (i.e. it ignored the invalid ones), and attempted to find a proof of these. The Geometry Machine was successful in finding the solution to a large number of geometry problems taken from high-school textbooks in plane geometry.

The logicians Hao Wang and Evert Beth (the inventor of semantic tableaux which was discussed in Chap. 16) were critical of the approaches of the AI pioneers, and believed that mathematical logic could do a lot more. Wang and others developed a theorem prover for first-order predicate calculus in 1960, but it had serious limitations due to the combinatorial explosion.

Alan Robinson's work on theorem provers in the early 1960s led to a proof procedure termed '*resolution*', which appeared to provide a breakthrough in the automation of predicate calculus theorem provers. A resolution theorem prover is essentially provided with the axioms of the field of mathematics in question, and the negation of the conjecture whose proof is sought. It then proceeds until a contradiction is reached, where there is no possible way for the axioms to be true and for the conjecture to be false.

The initial success of resolution led to excitement in the AI field where pioneers such as John McCarthy (see Chap. 17) believed that human knowledge could be expressed in predicate calculus,[6] and that, therefore, if resolution was indeed successful for efficient automated theorem provers, then the general problem of Artificial Intelligence was well on the way to a solution. However, while resolution led to improvements with the state explosion problem, it did not eliminate the problem.

This led to a fall-off in research into resolution-based approaches to theorem proving, and other heuristic-based techniques were investigated by Bledsoe in the late 1970s. The field of logic programming began in the early 1970s with the development of the Prolog programming language (see Chap. 17). Prolog is in a sense an application of automated theorem proving, where problems are stated in the form of goals (or theorems) in which the system tries to prove using a resolution theorem prover. The theorem prover generally does not need to be very powerful as

---

[6]McCarthy's viewpoint that predicate logic was the solution for the AI field was disputed by Minsksy and others (resulting in a civil war between the logicists and the proceduralists). The proceduralists argued that formal logic was an inadequate representation of knowledge for AI, and that predicate calculus was an overly rigid and inadequate framework. They argued that an alternative approach such as the procedural representation of knowledge was required.

many Prolog programs require only a very limited search, and a *depth-first* search from the goal backwards to the hypotheses is conducted.

The Argonne Laboratory developed the Aura System in the early 1980s (it was later replaced by Otter), as an improved resolution-based automated theorem prover, and this led to renewed interest in resolution-based approaches to theorem proving. There is a more detailed account of the nature of proof and theorem proving in (MacKensie 1995).

18.3   Interactive Theorem Provers

The challenges in developing efficient automated theorem provers led researchers to question whether an effective fully automated theorem prover was possible, and if it made more sense to develop a theorem prover that could be guided by a human in its search for a proof. This led to the concept of *Interactive Theorem Proving* (ITP) which involves developing formal proofs by man–machine collaboration, and is (in a sense) a new way of doing mathematics in front of a computer.

Such a system is potentially useful in mathematical research in formalizing and checking proofs, and it allows the user to concentrate on the creative parts of the proof and relieves the user of the need of carrying out the trivial steps in the proof. It is also a useful way of verifying the correctness of published mathematical proofs by acting as a *proof checker*, where the ITP is provided with a formal proof constructed by a human, which may then be checked for correctness.[7] Such a system is important in *program verification* in showing that the program satisfies its specification, especially in the safety/security-critical fields.

A group at Princeton developed a series of systems called Semi-automated Mathematics (SAM) in the late 1960s, which combined logic routines with human guidance and control. Their approach placed the mathematician at the heart of the theorem proving, and it was a departure from the existing theorem-proving approaches where the computer attempted to find proofs unaided. SAM provided a proof of an unproven conjecture in lattice theory (SAM's lemma), and this is regarded as the first contribution of automated reasoning systems to mathematics (MacKensie 1995).

De Bruijn and others at the Technische Hogeschool in Eindhoven in the Netherlands commenced development of the Automath system in the late 1960s. This was a large-scale project for the automated verification of mathematics, and it was tested by treating a full textbook. Automath systematically checked the proofs from Landau's text *Grundlagen der Analysis* (this foundation of analysis text was first published in 1930).

---

[7]A formal mathematical proof (of a normal proof) is difficult to write down and can be lengthy. Mathematicians were not really interested in these proof checkers.

The typical components of an Interactive Theorem Prover include an interactive proof editor to allow editing of proofs, formulae and terms in a formal theory of mathematics, and a large library of results which is essential for achieving complex results.

The Gypsy verification environment and its associated theorem prover were developed at the University of Texas in the 1980s, and it achieved early success in program verification with its verification of the encrypted packet interface program (a 4200 line program). It supports the development of software systems and formal mathematical proof of their behaviour.

The Boyer–Moore Theorem prover (NQTHM) was developed in the 1970s/1980s at the University of Texas. B.S. Boyer and J.S. Moore developed it in the early 1970s (Boyer and Moore 1979). It has been improved since then and it is currently known as NQTHM (it has been superseded by ACL2 available from the University of Texas). It supports mathematical induction as a rule of inference, and induction is a useful technique in proving properties of programs. The axioms of Peano arithmetic are built into the theorem prover, and new axioms added to the system need to pass a '*correctness test*' to prevent the introduction of inconsistencies.

It is far more automated than many other interactive theorem provers, but it requires detailed human guidance (with suggested lemmas) for difficult proofs. The user, therefore, needs to understand the proof being sought and the internals of the theorem prover.

It has been effective in proving well-known theorems such as Gödel's Incompleteness Theorem, the insolvability of the Halting problem, a formalization of the Motorola MC 68020 Microprocessor and many more.

Computational Logic Inc. was a company found by Boyer and Moore in 1983 to share the benefits of a formal approach to software development with the wider computing community. It was based in Austin, Texas, and provided services in the mathematical modelling of hardware and software systems. This involved the use of mathematics and logic to formally specify microprocessors and other systems. The use of its theorem prover was to formally verify that the implementation meets its specification, i.e. to prove that the microprocessor or other system satisfies its specification.

The HOL system was developed at Cambridge University in the U.K., and it is an environment for interactive theorem proving in a higher order logic. It has been applied to the formalization of mathematics and to the verification of hardware (including the verification of microprocessor design). It requires skilled human guidance and is one of the most widely used theorem provers. It was originally developed in the early 1980s and HOL 4 is the latest version. It is an open-source project and is used by academia and industry.

Isabelle is a theorem-proving environment developed at Cambridge University by Larry Paulson and Tobias Nipkow of the Technical University of Munich. It allows mathematical formulas to be expressed in a formal language and provides tools for proving those formulas. The main application is the formalization of mathematical proofs, proving the correctness of computer hardware or software with respect to its specification and proving properties of computer languages and protocols.

Isabelle is a generic theorem prover in the sense that it has the capacity to accept a variety of formal calculi, whereas most other theorem provers are specific to a specific formal calculus. Isabelle is available free of charge under an open-source license.

There is a steep learning curve with the theorem provers above and it generally takes a couple of months for users to become familiar with them. However, automated theorem proving has become a useful tool in the verification of integrated circuit design. Several semiconductor companies use automated theorem proving to demonstrate the correctness of division and other operators on their processors. We present a selection of theorem provers in the next section.

18.4   A Selection of Theorem Provers

Table 18.1 presents a small selection of the available automated and interactive theorem provers.

18.5   Review Questions

1. What is a mathematical proof?
2. What is a formal mathematical proof?
3. What approaches are used to prove a theorem?
4. What is a theorem prover?
5. What role can theorem provers play in software development?
6. What is the difference between an automated theorem prover and an interactive theorem prover?
7. Investigate and give a detailed description of one of the theorem provers in Table 15.1.

18.6   Summary

A mathematical proof includes natural language and mathematical symbols, and often many of the tedious details of the proof are omitted. The proofs in mathematics are rarely formal as such, and many proofs in formal methods are concerned

Table 18.1 Selection of theorem provers

Theorem prover	Description
ACL2	A Computational Logic for Applicative Common Lisp (ACL2) is part of the Boyer–Moore family of theorem provers. It is a software system consisting of a programming language (LISP) and an interactive theorem prover. It was developed in the mid-1990s as an industrial strength successor to the Boyer–Moore theorem prover (NQTHM). It is used in the verification of safety-critical hardware and software, and in industrial applications such as the verification of the floating-point module of a microprocessor
OTTER	OTTER is a resolution-style theorem prover for first-order logic developed at the Argonne Laboratory at the University of Chicago (it was the successor to Aura). It has been mainly applied to abstract algebra and formal logic
PVS	The Prototype Verification System (PVS) is a mechanized environment for formal specification and verification. It includes a specification language integrated with support tools and an interactive theorem prover. It was developed by SRI in California. The specification language is based on higher order logic, and the theorem prover is guided by the user in conducting proof. It has been applied to the verification of hardware and software
Theorem Proving System (TPS)	TPS is an automated theorem prover for first-order and higher order logic (it can also prove theorems interactively). It was developed at Carnegie Mellon University, and is used for hardware and software verification
HOL and Isabelle	HOL and Isabelle were developed by the Automated Reasoning Group at the University of Cambridge. The HOL system is an environment for interactive theorem proving in a higher order logic, and it has been applied to hardware verification. Isabelle is a generic proof assistant which allows mathematical formulae to be expressed in a formal language, and it provides tools for proving those formulae in a logical calculus
Boyer–Moore	The Boyer–Moore theorem prover (NQTHM) was developed at the University of Texas in the 1970s with the goal of checking the correctness of computer systems. It has been used to verify the correctness of microprocessors, and it has been superseded by ACL2

with cross-checking the details of the specification, or checking the validity of the refinement steps, or checking that certain properties are satisfied by the specification.

Machine proof is explicit and reliance on some brilliant insight is avoided. Proofs by hand often contain errors or jumps in reasoning, while machine proofs are often extremely lengthy and unreadable. The application of formal methods in an industrial environment requires the use of machine-assisted proof, since thousands of proof obligations arise from a formal specification, and theorem provers are essential in resolving these efficiently. The proof of various properties about a program increases confidence in its correctness. However, an absolute proof of correctness is unlikely except for the most trivial of programs.

Automated theorem proving is difficult, as often mathematicians prove a theorem with an initial intuitive feeling that the theorem is true. Human intervention to provide guidance or intuition improves the effectiveness of the theorem prover. Early work on the automation of proof began in the 1950s with the beginning of work in the Artificial Intelligence field, and one of the earliest programs developed was the Logic Theorist, which was presented at the Dartmouth Conference on Artificial Intelligence in 1956.

The challenges in developing effective automated theorem provers led researchers to investigate whether it made more sense to develop a theorem prover that could be guided by a human in its search for a proof. This led to the development of interactive theorem proving which involved developing formal proofs by man–machine collaboration.

The typical components of an Interactive Theorem Prover include an interactive proof editor to allow editing of proofs, formulae and terms in a formal theory of mathematics, and a large library of results which is essential for achieving complex results.

An interactive theorem prover allows the user to concentrate on the creative parts of the proof, and relieves the user of the need to carry out and verify the trivial steps in the proof. It is also a useful way of verifying the correctness of published mathematical proofs by acting as a proof checker, and is also useful in program verification in showing that the program satisfies its specification, especially in the safety/security-critical fields.

References

Boyer R, Moore JS (1979) A computational logic. The Boyer Moore Theorem Prover. Academic Press

MacKensie D (1995) The automation of proof. A historical and sociological exploration. IEEE Ann History Comput 17(3)

Russell B, Whitehead AN (1910) Principia mathematica. Cambridge University Press, Cambridge

Software Engineering Mathematics

19

Key Topics

Birth of software engineering
Software engineering mathematics
Floyd
Hoare
Formal methods
Software inspections and testing
Project management
Software process maturity models

19.1 Introduction

The computer sector in the 1960s was dominated by several large mainframe computer manufacturers. Computers were large, expensive and difficult to use for a non-specialist. The software used on the mainframes of the 1960s was proprietary and the hardware of manufacturers was generally incompatible with one another. It was usually necessary to rewrite all existing software application programs for a new computer if a business decided to change to a new manufacturer or upgrade to a more powerful machine from its existing manufacturer.

Software projects tended to be written once-off for specific customers, and large projects were often characterized by underestimation and overexpectations. There was no independent software sector in the 1960s with software and training included as part of the computer hardware delivered to the customers. IBM's dominant position in the market led to antitrust inquiries by the US Justice

© Springer Nature Switzerland AG 2020
G. O'Regan, *Mathematics in Computing*, Undergraduate Topics
in Computer Science, https://doi.org/10.1007/978-3-030-34209-8_19

Department, and this led IBM to 'unbundle' its software and services from its hardware sales. It then began charging separately for software, training and hardware, and this led to the creation of a multi-billion-dollar software industry, and to a major growth of software suppliers.

The NATO Science Committee organized two famous conferences on software engineering in the late 1960s. The first conference was held in Garmisch, Germany, in 1968, and it was followed by a second conference in Rome in 1969. The Garmisch conference was attended by over fifty people from eleven countries.

The conferences highlighted the problems that existed in the software sector in the late 1960s, and the term '*software crisis*' was coined to refer to these problems. These included budget and schedule overruns of projects, and problems with the quality and reliability of the delivered software. This conference led to the birth of *software engineering* as a separate discipline, and the realization that programming is quite distinct from science and mathematics.

Programmers are like engineers in the sense that they design and build products. Therefore, they need an appropriate software engineering education (not just on the latest technologies but on the fundamentals of engineering) in order to properly design and develop software. This led to the birth of software engineering as a discipline in its own right, and to the subsequent development of a plethora of techniques to elicit requirements, design, develop and test software to meet customer needs.

The construction of bridges was problematic in the nineteenth century, and many people who presented themselves as qualified to design and construct bridges did not have the required knowledge and expertise. Consequently, many bridges collapsed, endangering the lives of the public. This led to legislation requiring an engineer to be licensed by the Professional Engineering prior to practicing as an engineer. These engineering associations identify a core body of knowledge that the engineer is required to possess, and the licensing body verifies that the engineer has the required qualifications and experience. The licensing of engineers by most branches of engineering ensures that only personnel competent to design and build products actually do so. This in turn leads to products that the public can safely use. In other words, the engineer has a responsibility to ensure that the products are properly built, and are safe for the public to use.

Parnas argues that traditional engineering be contrasted with the software engineering discipline where there is no licensing mechanism, and where individuals with no qualifications can participate in the design and building of software products.[1] However, best practice in modern HR places a strong emphasis on the qualification of staff.

---

[1]Modern HR recruitment specifies the requirements for a particular role, and the interviews establish whether the candidate is suitably qualified, and has the appropriate experience for the role. Parnas is arguing against the content of courses that emphasize the latest technologies rather than the fundamentals of engineering.

The Standish group has conducted research since the late 1990s (Standish Group Research Note 1999) on the extent of problems with schedule and budget overruns of IT projects. The results indicate serious problems with on-time delivery, cost overruns and quality.[2] Fred Brooks has argued that software is inherently complex, and that there is no silver bullet that will resolve all of the problems associated with software projects such as schedule overruns and software quality problems (Brooks 1975, 1986).

Poor quality software can at best cause minor irritation to clients, and in some circumstances, it may seriously disrupt the work of the client organization leading to injury or even the death of individuals (e.g. as in the case of the Therac-25[3] radiotherapy machine). The Y2K problem occurred due to poor design, as the representation of the date used two digits to record the year rather than four. Its correction required major rework, as it was necessary to examine all existing software code to determine how the date was represented and to make appropriate corrections. Clearly, well-designed programs would have hidden the representation of the date thereby minimizing the changes required for year 2000 compliance.

Mathematics plays a key role in engineering, and it may potentially assist software engineers in delivering high-quality software products that are safe to use. Several mathematical approaches that may assist in delivering high-quality software are described in O'Regan (2006). However, it is important to recognize that while the use of mathematics is suitable for some areas of software engineering (especially in the safety and security-critical fields), less rigorous techniques (such as software inspections and testing) are sufficient in most other areas of software engineering.

There is a lot of industrial interest in approaches to mature software engineering practices in software organizations (e.g. the use of software process maturity models such as the CMMI). These include approaches to assess and mature the software engineering processes in software companies, and they are described in O'Regan (2010, 2014).[4] Software process improvement focuses mainly on improving the effectiveness of the management, engineering and organization practices related to software engineering.

---

[2]It should be noted that these are IT projects covering diverse sectors including banking, telecommunications, etc., rather than pure software companies. Mature software companies using the CMM tend to be more consistent in project delivery with high quality.

[3]Therac-25 was a radiotherapy machine produced by the Atomic Energy of Canada Limited (AECL). It was involved in at least six accidents between 1985 and 1987 in which patients were given massive overdoses of radiation. The dose given was over 100 times the intended dose and three of the patients died from radiation poisoning. These accidents highlighted the dangers of software control of safety-critical systems. The investigation subsequently highlighted the poor software design of the system and the poor software development practices employed.

[4]The process maturity models focus mainly on the management, engineering and organizational practices required in software engineering. The models focus on what needs to be done rather how it should be done.

19.2   What Is Software Engineering?

Software engineering involves multi-person construction of multi-version programs. The IEEE 610.12 definition states that:

Definition 19.1 (Software Engineering) *Software engineering is the application of
a systematic, disciplined, quantifiable approach to the development, operation and
maintenance of software; that is, the application of engineering to software, and the
study of such approaches.*

Software engineering includes:

1. Methodologies to determine requirements, design, develop, implement and test
   software to meet customers' needs.
2. The philosophy of engineering: i.e. an engineering approach to developing
   software is adopted. That is, products are properly designed, developed, tested,
   with quality and safety properly addressed.
3. Mathematics[5] may be employed to assist with the design and verification of
   software products. The level of mathematics to be employed will depend on the
   safety-critical nature of the product, as systematic peer reviews and testing are
   often sufficient.
4. Sound project and quality management practices are employed.

Software engineering requires the engineer to state precisely the requirements
that the software product is to satisfy, and then to produce designs that will meet
these requirements. Engineers provide a precise description of the problem to be
solved; they then proceed to producing a design and validating its correctness;
finally, the design is implemented and testing is performed to verify the correctness
of the implementation with respect to the requirements. The software requirements
needs to be unambiguous, and should clearly state what is and what is not required.

Classical engineers produce the product design, and then analyse their design
for correctness. They use mathematics in their analysis, as this is the basis of
confirming that the specifications are met. The level of mathematics employed will
depend on the particular application and calculations involved. The term *'engineer'*
is generally applied only to people who have attained the necessary education and
competence to be called engineers, and who base their practice on mathematical and
scientific principles. Often in computer science, the term engineer is employed
rather loosely to refer to anyone who builds things, rather than to an individual with
a core set of knowledge, experience and competence.

---

[5]There is no consensus at this time as to the appropriate role of mathematics in software
engineering. The use of mathematics is invaluable in the safety-critical and security-critical fields
as it provides an extra level of confidence in the correctness of the software.

Fig. 19.1 David Parnas

Parnas[6] (Fig. 19.1) is a strong advocate of the classical engineering approach, and he argues that computer scientists should have the right education to apply scientific and mathematical principles to their work. This includes mathematics and design, to enable them to be able to build high-quality and safe products. Baber has argued (Baber 2011) that 'mathematics is the language of engineering', and that students should be shown how to turn a specification into a program using mathematics.

Parnas advocates a solid engineering approach to the teaching of mathematics with an emphasis on its application to developing and analysing product designs. He argues that software engineers need education on engineering mathematics; specification and design; converting designs into programs; software inspections, and testing. The education should enable the software engineer to produce well-designed programs that will correctly implement the requirements.

He argues that software engineers have individual responsibilities as professional engineers.[7] They are responsible for designing and implementing high-quality and reliable software that is safe to use. They are also accountable for their own decisions and actions,[8] and have a responsibility to object to decisions

---

[6]Parnas has made important contributions to software engineering including information hiding which is used in the object-oriented world.

[7]The concept of accountability is not new; indeed the ancient Babylonians employed a code of laws c. 1750 B.C. known as the Hammurabi Code. This code included the law that if a house collapsed and killed the owner then the builder of the house would be executed.

[8]However, it is unlikely that an individual programmer would be subject to litigation in the case of a flaw in a program causing damage or loss of life. A comprehensive disclaimer of responsibility for problems rather than a guarantee of quality accompany most software products. Software engineering is a team-based activity involving several engineers in various parts of the project, and it could be potentially difficult for an outside party to prove that the cause of a particular problem is due to the professional negligence of a particular software engineer, as there are many others involved in the process such as reviewers of documentation and code and the various test groups. Companies are more likely to be subject to litigation, as a company is legally responsible for the actions of their employees in the workplace, and the fact that a company is a financially richer entity than one of its employees.

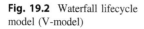

Fig. 19.2 Waterfall lifecycle
model (V-model)

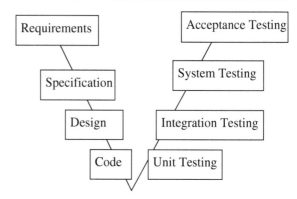

that violate professional standards. Professional engineers need to be honest about current capabilities, especially when asked to work on problems that have no appropriate technical solution. In other words, they should be honest and avoid accepting a contract for something that cannot be done.

The licensing of a professional engineer provides confidence that the engineer has the right education and experience to build safe and reliable products. Professional engineers are required to follow rules of good practice, and to object when the rules are violated.[9] The professional engineering body is responsible for enforcing standards and certification. The term 'engineer' is a title that is awarded on merit, but it also places responsibilities on its holder.

The approach used in traditional software engineering is to follow a well-defined software engineering process. The process includes activities such as project management, requirements gathering, requirements specification, architecture design, software design, coding and testing. Most companies use a set of templates for the various phases. The waterfall model (Royce 1970) and spiral model (Boehm 1988) are popular software development lifecycles.

The waterfall model (Fig. 19.2) starts with requirements, followed by specification, design, implementation and testing. It is typically used for projects where the requirements can be identified early in the process the left-hand side of the 'V' involves requirements, specification, design and coding, and the right-hand side is concerned with unit tests, integration tests, system tests and acceptance testing. Each phase has entry and exit criteria that must be satisfied before the next phase commences.

The spiral model (Fig. 19.3) is useful where the requirements are not fully known at project initiation. There is an evolution of the requirements during development which proceeds in a number of spirals, with each spiral typically involves updates to the requirements, design, code, testing and a user review of the particular iteration or spiral.

---

[9]Software companies that are following the CMMI or ISO 9000 will employ audits to verify that the rules and best practice have been followed. Auditors report their findings to management and the findings are addressed appropriately by the project team and affected individuals.

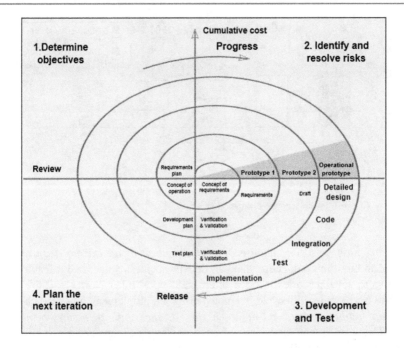

Fig. 19.3 SPIRAL lifecycle model

The spiral is, in effect, a reusable prototype and the customer examines the current iteration and provides feedback to the development team to be included in the next spiral. The approach is to partially implement the system. This leads to a better understanding of the requirements of the system and it then feeds into the next cycle in the spiral. The process repeats until the requirements and product are fully complete.

There has been a growth of popularity among software developers in lightweight methodologies such as *Agile*. This is a software development methodology that claims to be more responsive to customer needs than traditional methods such as the waterfall model. *The waterfall development model is similar to a wide and slow-moving value stream*, and halfway through the project 100% if the requirements are typically 50% done. *However, for agile development 50% of requirements are typically 100% done halfway through the project.*

Ongoing changes to requirements are considered normal in the Agile world, and it is believed to be more realistic to change requirements regularly throughout the project rather than attempting to define all of the requirements at the start of the project. The methodology includes controls to manage changes to the requirements, and good communication and early regular feedback is an essential part of the process.

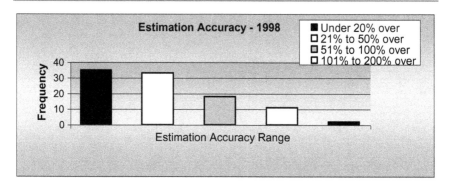

Fig. 19.4 Standish group report—estimation accuracy

A story may be a new feature or a modification to an existing feature. It is reduced to the minimum scope that can deliver business value, and a feature may give rise to several stories. Stories often build upon other stories and the entire software development lifecycle is employed for the implementation of each story. *Stories are either done or not done*: i.e. *there is such thing as a story being 80% done.* The story is complete only when it passes its acceptance tests. For more details on Agile see (O'Regan 2017, Beck 2000).

The challenge in software engineering is to deliver high-quality software on time to customers. The Standish Group research (Fig. 19.4) on project cost overruns in the US during 1998 showed that 33% of projects are between 21 and 50% over-estimate, 18% are between 51 and 100% overestimate, and 11% of projects are between 101 and 200% overestimate.

The accurate estimation of project cost and effort are key challenges, and project managers need to determine how good their current estimation process actually is and to make improvements. Many companies today employ formal project management methodologies such as Prince 2 or Project Management Professional (PMP). These methodologies allow projects to be rigorously managed and include processes for initiating a project, planning a project, executing a project, monitoring and controlling a project and closing a project.

The Capability Maturity Model developed by the Software Engineering Institute (SEI) has become useful in software engineering. The SEI has collected empirical data to suggest that there is a close relationship between software process maturity and the quality and the reliability of the delivered software. The CMMI enables the organization to improve processes such as:

–   Developing and managing requirements,
–   Design activities,
–   Configuration Management,
–   Selection and management of suppliers,
–   Planning and managing projects,

- Building quality into the product with peer reviews,
- Performing rigorous testing and
- Performing independent audits.

The rest of this chapter is focused on mathematical techniques to support software engineering to improve software quality, and the chapter concludes with a short discussion on software inspections and testing and process maturity models. For a more detailed account of software engineering see O'Regan (2017).

19.3   Early Software Engineering Mathematics

Robert Floyd was born in New York in 1936, and he did pioneering work on software engineering from the 1960s (Fig. 19.5). He made important contributions to the theory of parsing; the semantics of programming languages; program verification; and methodologies for the creation of efficient and reliable software.

Mathematics and Computer Science were regarded as two completely separate disciplines in the 1960s, and software development was based on the assumption that the completed code would always contain defects. It was therefore better and more productive to write the code as quickly as possible, and to then perform debugging to find the defects. Programmers then corrected the defects, made patches and retested and found more defects. This continued until they could no longer find defects. Of course, there was always the danger that defects remained in the code that could give rise to software failures.

Floyd believed that there was a way to construct a rigorous proof of the correctness of the programs using mathematics. He showed that mathematics could be used for program verification, and he introduced the concept of *assertions* that provided a way to verify the correctness of programs.

Fig. 19.5 Robert Floyd

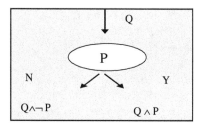

Fig. 19.6 Branch assertions in flowcharts

Flowcharts were employed in the 1960s to explain the sequence of basic steps for computer programs. Floyd's insight was to build upon flowcharts and to apply *an invariant assertion to each branch* in the flowchart. These assertions state the essential relations that exist between the variables at that point in the flowchart. An example relation is '$R = Z > 0$, $X = 1$, $Y = 0$'. He devised a general flowchart language to apply his method to programming languages. The language essentially contains boxes linked by flow of control arrows (Floyd 1967).

Consider the assertion Q that is true on entry to a branch where the condition at the branch is P. Then, the assertion on exit from the branch is $Q \land \neg P$ if P is false and $Q \land P$ otherwise (Fig. 19.6).

The use of assertions may be employed in an assignment statement. Suppose x represents a variable and v represents a vector consisting of all the variables in the program. Suppose $f(x,v)$ represents a function or expression of x and the other program variables represented by the vector v. Suppose the assertion $S(f(x,v), v)$ is true before the assignment $x = f(x,v)$. Then the assertion $S(x,v)$ is true after the assignment (Fig. 19.7). This is given by:

Floyd used flowchart symbols to represent entry and exit to the flowchart. This included entry and exit assertions to describe the program's entry and exit conditions.

Floyd's technique showed how a computer program is a sequence of logical assertions. Each assertion is true whenever control passes to it, and statements appear between the assertions. The initial assertion states the conditions that must be true for execution of the program to take place, and the exit assertion essentially describes what must be true when the program terminates.

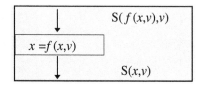

Fig. 19.7 Assignment assertions in flowcharts

Fig. 19.8 C. A. R. Hoare

Floyd's insight was his recognition that if it can be shown that the assertion immediately following each step is a consequence of the assertion immediately preceding it, then the assertion at the end of the program will be true, provided the appropriate assertion was true at the beginning of the program.

He published an influential paper, '*Assigning Meanings to Programs*', in 1967 (Floyd 1967), and this paper influenced Hoare's work on preconditions and post-conditions leading to Hoare logic (Hoare 1969). Floyd's paper also presented a formal grammar for flowcharts, together with rigorous methods for verifying the effects of basic actions like assignments.

Hoare logic is a formal system of logic used for programming semantics and for program verification. It was developed by C. A. R. Hoare (Fig. 19.8), and was originally published in Hoare's 1969 paper '*An axiomatic basis for computer programming*' (Hoare 1969). Hoare and others have subsequently refined it, and it provides a logical methodology for precise reasoning about the correctness of computer programs.

Hoare was influenced by Floyd's 1967 paper that applied assertions to flow-charts, and he recognized that this provided an effective method for proving the correctness of programs. He built upon Floyd's approach to cover the familiar constructs of high-level programming languages.

This led to the axiomatic approach to defining the semantics of every statement in a programming language, and the approach consists of axioms and proof rules. He introduced what has become known as the Hoare triple, and this describes how the execution of a fragment of code changes the state. A Hoare triple is of the form:

$$P\{Q\}R$$

where P and R are assertions and Q is a program or command. The predicate P is called the *precondition*, and the predicate R is called the *postcondition*.

Definition 4.2 (Partial Correctness) *The meaning of the Hoare triple above is that whenever the predicate P holds of the state before the execution of the command or program Q, then the predicate R will hold after the execution of Q. The brackets indicate partial correctness as if Q does not terminate then R can be any predicate. R may be chosen to be false to express that Q does not terminate.*

Total correctness requires Q to terminate, and at termination R is true. Termination needs to be proved separately. Hoare logic includes axioms and rules of inference rules for the constructs of imperative programming language.

Hoare and Dijkstra were of the view that the starting point of a program should always be the specification, and that the proof of the correctness of the program should be developed along with the program itself.

That is, the starting point is the mathematical specification of what a program is to do, and mathematical transformations are applied to the specification until it is turned into a program that can be executed. The resulting program is then known to be correct by construction.

19.4 Mathematics in Software Engineering

Mathematics plays a key role in classical engineering to assist with design and verification of products. It is therefore reasonable to apply appropriate mathematics in software engineering (especially for safety- and security-critical systems) to assure that the delivered systems conform to the requirements. The extent to which mathematics should be used is controversial with strong views in both camps. In many cases, peer reviews and testing will be sufficient to build quality into the software product. In other cases, and especially with safety- and security-critical applications, it is desirable to have the extra assurance that may be provided with mathematical techniques.

Mathematics allows a rigorous analysis to take place and avoids an over-reliance on intuition. The emphasis is on applying mathematics to solve practical problems and to develop products that are fit for use. Engineers are taught how to apply mathematics in their work, and the emphasis is always on the application of mathematics to solve practical problems.

Classical mathematics may be applied to software engineering and specialized mathematical methods and notations have also been developed. The classical mathematics employed includes sets, relations, functions, logic, number theory, algebra, graph theory, automata theory, model theory, matrix theory, probability and statistics, calculus and matrix theory. Specialized formal specification languages such as Z and VDM have been developed, and these allow the requirements to be formally specified in precise mathematical language.

The term *'formal methods'* refers to various mathematical techniques used in the safety-critical field for specification and formal development of software. Formal methods consist of formal specification languages or notations and employ a collection of tools to support the syntax checking of the specification, as well as the proof of properties about the specification. The term *'formal methods'* is used to describe a formal specification language and a method for the design and implementation of computer systems.

The mathematical analysis of the formal specification allows questions to be asked about what the system does, and these questions may be answered independently of the implementation. Mathematical notation is precise, and this helps to avoid the problem of ambiguity inherent in a natural language description of a system. The formal specification may be used to promote a common understanding for all stakeholders.

Formal methods have been applied to a diverse range of applications, including the safety-critical field; security-critical field; the railway sector; the nuclear field; microprocessor verification; the specification of standards, and the specification and verification of programs.

There are various tools to support formal methods including syntax checkers; specialized editors; tools to support refinement; automated code generators; theorem provers; and specification animation tools. Formal methods need to mature further before they will be used in mainstream software engineering, and they are described in more detail in Chap. 21.

19.5 Software Inspections and Testing

Software inspections play an important role in building quality into software products. The Fagan Inspection Methodology was developed by Michael Fagan at IBM in the mid-1970s (Fagan 1976). The methodology mandates that requirement documents, design documents, source code and test plans are all formally inspected. There are several *roles* defined in the process including the *moderator* who chairs the inspection; the *reader* who reads or paraphrases the particular deliverable; the *author* who is the creator of the deliverable; and the *tester* who is concerned with the testing viewpoint.

The process will consider whether a design is correct with respect to the requirements and whether the source code is correct with respect to the design. There are several stages in the Fagan inspection process, including planning, overview, preparation, inspection, process improvement, rework and follow-up.

Software testing plays a key role in verifying that a software product is of high quality and conforms to the customer's quality expectations. Testing is both a constructive activity in that it is verifying the correctness of functionality, and it is also a destructive activity in that the objective is to find as many defects as possible in the software. The testing verifies that the requirements are correctly implemented as well as identifying whether any defects are present in the software product.

There are various types of testing such as unit testing, integration testing, system testing, performance testing, usability testing, regression testing and customer acceptance testing. The testing needs to be planned and test cases prepared and executed. The results of testing are reported and any issues corrected and retested. The test cases will need to be appropriate to verify the correctness of the software. Software inspection and testing are described in more detail in O'Regan (2019).

19.6 Process Maturity Models

The Software Engineering Institute (SEI) developed the Capability Maturity Model (CMM) in the early 1990s as a framework to help software organizations to improve their software process maturity, and to implement best practice in software and systems engineering. The SEI believes that there is a close relationship between the maturity of software processes and the quality of the delivered software product.

The CMM applied the ideas of Deming (Deming 1986), Juran (Juran 2000) and Crosby (Crosby 1979) to the software field. These quality gurus were influential in transforming manufacturing companies with quality problems to effective quality-driven organizations with a reduced cost of poor quality.

Watt Humphries (Fig. 19.9) did early work on software process improvement at IBM (Humphry 1989). He moved to the SEI in the late 1980s and the first version of the CMM was released in 1991. It is now called the Capability Maturity Model Integration (CMMI®) (Chrissis et al. 2011).

The CMMI consists of five maturity levels with each maturity level (except level one) consisting of several process areas. Each process area consists of a set of goals that are implemented by practices related to that process area leading to an effective process.

The emphasis on level two of the CMMI is on maturing management practices. The emphasis on level three of the CMMI is to mature engineering and organization practices. Level four is concerned with ensuring that key processes are performing within strict quantitative limits, and adjusting processes, where necessary, to perform within these defined limits. Level five is concerned with continuous process improvement, which is quantitatively verified.

Fig. 19.9 Watts Humphrey.
Courtesy of Watts Humphrey

The CMMI allows organizations to benchmark themselves against other similar organizations. This is done by appraisals conducted by an authorized SCAMPI lead appraiser. An appraisal is useful in verifying that an organization has improved, and it enables the organization to prioritize improvements for the next improvement cycle. There is more detailed information on the CMMI and software process improvement in O'Regan (2010).

19.7 Review Questions

1. What is software engineering? Describe the difference between classical engineers and software engineers.
2. Describe the 'software crisis' of the late 1960s that led to the first software engineering conference in 1968.
3. Discuss the Standish Research Report and the level of success of IT projects today. In your view is there a crisis in software engineering today? Give reasons for your answer.
4. Discuss what the role of mathematics should be in current software engineering.
5. Describe the waterfall and spiral lifecycles. What are the similarities and differences between them?
6. Discuss the contributions of Floyd and Hoare.
7. Explain the difference between partial correctness and total correctness.
8. What are formal methods?
9. Discuss the process maturity models (including the CMMI). What are their advantages and disadvantages?
10. Discuss how software inspections and testing can assist in the delivery of high-quality software.

19.8 Summary

This chapter presented a short account of some important developments in software engineering. Its birth was at the Garmisch conference in 1968, and it was recognized that there was a crisis in the software field, and a need for sound methodologies to design, develop and maintain software to meet customer needs.

Classical engineering has a successful track record in building high-quality products that are safe for the public to use. It is therefore natural to consider using an engineering approach to developing software, and this involves identifying the customer requirements, carrying out a rigorous design to meet the requirements, developing and coding a solution to meet the design, and conducting appropriate inspections and testing to verify the correctness of the solution.

Mathematics plays a key role in classical engineering to assist with the design and verification of products. It makes sense to apply appropriate mathematics in software engineering (especially for safety-critical systems) to assure that the delivered systems conform to the requirements. The extent to which mathematics should be used remains controversial.

There is a lot more to the successful delivery of a project than just the use of mathematics or peer reviews and testing. Sound project management and quality management practices are essential, as a project that is not properly managed will suffer from schedule, budget or cost overruns as well as problems with quality.

Maturity models such as the CMMI can assist organizations in maturing key management and engineering practices, and may help companies in their goals to deliver high-quality software systems that are consistently delivered on time and budget.

References

Baber RL (2011) The language of mathematics. Utilizing math in practice. Wiley

Beck K (2000) Extreme programming explained. Embrace change. Addison Wesley

Boehm B (1988) A spiral model for software development and enhancement. Computer

Brooks F (1975) The mythical man month. Addison Wesley

Brooks F (1986) No silver bullet. Essence and accidents of software engineering. In: Information Processing. Elsevier, Amsterdam

Chrissis MB, Conrad M, Shrum S (2011) CMMI. Guidelines for process integration and product improvement, 3rd edn, SEI series in software engineering. Addison Wesley

Crosby P (1979) Quality is free. The art of making quality certain. McGraw Hill

Edwards Deming W (1986) Out of crisis. M.I.T. Press

Fagan M (1976) Design and code inspections to reduce errors in software development. IBM Syst J 15(3)

Floyd R (1967) Assigning meanings to programs. In: Proceedings of symposia in applied mathematics, no. 19, pp. 19–32

Hoare CAR (1969) An axiomatic basis for computer programming. Commun ACM 12(10):576–585

Humphry W (1989) Managing the software process. Addison Wesley

Juran J (2000) Juran's quality handbook, 5th edn. McGraw Hill

O'Regan G (2006) Mathematical approaches to software quality. Springer

O'Regan G (2010) Introduction to software process improvement. Springer

O'Regan G (2014) Introduction to software quality. Springer

O'Regan G (2017) Concise guide to software engineering. Springer

O'Regan G (2019) Concise guide to software testing. Springer

Royce W (1970) The software lifecycle model (waterfall model). In: Proceeding of WESTCON, August, 1970

Standish Group Research Note (1999) Estimating: art or science. Featuring Moritz cost expert

Software Reliability and Dependability

20

Key Topics

Software reliability
Software reliability model
System availability
Dependability
Computer security
Safety-critical systems
Cleanroom

20.1 Introduction

This chapter gives an introduction to the important area of software reliability and dependability, and it discusses important topics in software engineering such as software reliability; software availability; software reliability models; the cleanroom methodology; dependability and its various dimensions; security engineering; and safety-critical systems.

Software reliability is the probability that the program works without failure for a period of time, and it is usually expressed as the mean time to failure. It is different from hardware reliability, which is characterized by components that physically wear out, whereas software is intangible and software failures are due to design and implementation errors. In other words, software is either correct or incorrect when it is designed and developed, and it does not physically deteriorate with time.

© Springer Nature Switzerland AG 2020   319
G. O'Regan, *Mathematics in Computing*, Undergraduate Topics
in Computer Science, https://doi.org/10.1007/978-3-030-34209-8_20

Harlan Mills and others at IBM developed the cleanroom approach to software development, and the process is described in O' Regan (2006). It involves the application of statistical techniques to calculate a software reliability measure based on the expected usage of the software.[1] This involves executing tests chosen from the population of all possible uses of the software in accordance with the probability of its expected use. Statistical usage testing has been shown to be more effective in finding defects that lead to failure than coverage testing.

Models are simplifications of the reality and a good model allows accurate predictions of future behaviour to be made. A model is judged effective if there is good empirical evidence to support it, and a good software reliability model will have good theoretical foundations and realistic assumptions. The extent to which the software reliability model can be trusted depends on the accuracy of its predictions, and empirical data will need to be gathered to judge its accuracy. A good software reliability model will give good predictions of the reliability of the software.

It is essential that software that is widely used is dependable, which means that the software is available whenever required, and that it operates safely and reliably without any adverse side effects (e.g. the software problems with the Therac-25 radiography machine led to several patients receiving massive overdoses in radiation in the mid-1980s leading to serious injury and death of several patients).

Today, billions of computers are connected to the Internet, and this has led to a growth in attacks on computers. It is essential that computer security is carefully considered, and developers need to be aware of the threats facing a system and techniques to eliminate them. The developers need to be able to develop secure systems that are able to deal with and recover from external attacks.

20.2 Software Reliability

The design and development of high-quality software has become increasingly important for society. The hardware field has been very successful in developing sound reliability models, which allow useful predictions of how long a hardware component (or product) will function to be provided. This has led to a growing interest in the software field in the development of a sound software reliability model. Such a model would provide a sound mechanism to predict the reliability of the software prior to its deployment at the customer site, as well as confidence that the software is fit for purpose and safe to use.

Definition 20.1 (*Software Reliability*) Software reliability is the probability that the program works without failure for a specified length of time, and it is a statement of the future behaviour of the software. It is generally expressed in terms of the *Mean Time To Failure* (MTTF) or the *Mean Time Between Failure* (MTBF).

---

[1]The expected usage of the software (or operational profile) is a quantitative characterization (usually based on probability) of how the system will be used.

Statistical sampling techniques are often employed to predict the reliability of hardware, as it is not feasible to test all items in a production environment. The quality of the sample is then used to make inferences on the quality of the entire population, and this approach is effective in manufacturing environments where variations in the manufacturing process often lead to defects in the physical products.

There are similarities and differences between hardware and software reliability. A hardware failure generally arises due to a component wearing out due to its age, and often a replacement component is required. Many hardware components are expected to last for a certain period of time, and the variation in the failure rate of a hardware component is often due to variations in the manufacturing process, and to the operating environment of the component. Good hardware reliability predictors have been developed, and each hardware component has an expected mean time to failure. The reliability of a product may then be determined from the reliability of the individual components.

Software is an intellectual undertaking involving a team of designers and programmers. It does not physically wear out as such, and software failures manifest themselves from particular user inputs. Each copy of the software code is identical, and the software code is either correct or incorrect. That is, software failures are due to design and implementation errors, rather than to the software physically wearing out over time. The software community has not yet developed a sound software reliability predictor model.

The software population to be sampled consists of all possible execution paths of the software, and since this is potentially infinite it is generally not possible to perform exhaustive testing. The way in which the software is used (i.e. the inputs entered by the users) will impact upon its perceived reliability. Let I_f represent the fault set of inputs (i.e. $i_f \in I_f$ if and only if the input of i_f by the user leads to failure). The randomness of the time to software failure is due to the unpredictability in the selection of an input $i_f \in I_f$. It may be that the elements in I_f are inputs that are rarely used, and therefore the software will be perceived as reliable.

Statistical usage testing may be used to make predictions on the future performance and reliability of the software. This requires an understanding of the expected usage profile of the system, as well as the population of all possible usages of the software. The sampling is done in accordance with the expected usage profile, and a software reliability measure calculated.

20.2.1 Software Reliability and Defects

The release of an unreliable software product may result in damage to property or injury (including loss of life) to a third party. Consequently, companies need to be confident that their software products are fit for use prior to their release. The project team needs to conduct extensive inspections and testing of the software, as well as considering all associated risks prior to its release.

Objective product quality criteria may be set (e.g. 100% of tests performed and passed) that must be satisfied prior to the release of the product. This provides a degree of confidence that the software has the desired quality, and is fit for purpose. However, these results are historical in the sense that they are a statement of past and present quality. The question is whether the past behaviour and performance provides a sound indication of future behaviour.

Software reliability models are an attempt to predict the future reliability of the software and to assist in deciding on whether the software is ready for release. A defect does not always result in a failure, as it may occur on a rarely used execution path. Studies indicate that many observed failures arise from a small proportion of the existing defects.

Adam's 1984 case study of defects in IBM software (1984) indicate that over 33% of the defects led to an observed failure with mean time to failure greater than 5000 years, whereas less than 2% of defects led to an observed failure with a mean time to failure of less than 5 years. This suggests that a small proportion of defects often lead to almost all of the observed failures (Table 20.1).

The analysis shows that 61.6% of all fixes (Group 1 and 2) were for failures that will be observed less than once in 1580 years of expected use, and that these constitute only 2.9% of the failures observed by typical users. On the other hand, groups 7 and 8 constitute 53.7% of the failures observed by typical users and only 1.4% of fixes.

This case study indicates that *coverage testing* is not cost-effective in increasing MTTF. *Usage testing*, in contrast, would allocate 53.7% of the test effort to fixes that will occur 53.7% of the time for a typical user. Harlan Mills has argued (1990) that the data in the table shows that usage testing is 21 times more effective than coverage testing.

There is a need to be careful with *reliability growth models*, as there is no tangible growth in reliability unless the corrected defects are likely to manifest themselves as a failure.[2] Many existing software reliability growth models assume that all remaining defects in the software have an equal probability of failure, and that the correction of a defect leads to an increase in software reliability. These assumptions are questionable.

The defect count and defect density may be poor predictors of operational reliability, and an emphasis on removing a large number of defects from the software may not be sufficient to achieve high reliability.

The correction of defects in the software leads to a newer version of the software, and many software reliability models assume reliability growth, i.e. the new version is more reliable than the older version as several identified defects have been corrected. However, in some sectors such as the safety-critical field the view is that the new version of a program is a new entity, and that no inferences may be

---

[2]We are assuming that the defect has been corrected perfectly with no new defects introduced by the changes made.

Table 20.1 Adam's 1984 study of software failures of IBM products

	Rare					Frequent		
	1	2	3	4	5	6	7	8
MTTF (years)	5,000	1,580	500	158	50	15.8	5	1.58
Avg. % fixes	33.4	28.2	18.7	10.6	5.2	2.5	1.0	0.4
Prob. failure	0.008	0.021	0.044	0.079	0.123	0.187	0.237	0.300

Table 20.2 New and old version of software

Similarities and differences between new/old version
• The new version of the software is identical to the previous version except that the identified defects have been corrected
• The new version of the software is identical to the previous version, except that the identified defects have been corrected, but the developers have introduced some new defects
• No assumptions can be made about the behaviour of the new version of the software until further data is obtained

drawn until further investigation has been done. There are a number of ways to interpret the relationship between the new version of the software and the older version (Table 20.2).

The safety-critical industry (e.g. the nuclear power industry) takes the conservative viewpoint that any change to a program creates a new program. The new program is therefore required to demonstrate its reliability, and so extensive testing needs to be performed.

20.2.2 Cleanroom Methodology

Harlan Mills and others at IBM developed the cleanroom methodology as a way to develop high-quality software (1990). Cleanroom helps to ensure that the software is released only when it has achieved the desired quality level, and the probability of zero-defects is very high.

The way in which the software is used will impact on its perceived quality and reliability. Failures will manifest themselves on certain input sequences, and as the input sequences will vary among users, the result will be different perceptions of the reliability of the software among the users. The knowledge of how the software will be used allows the software testing to focus on verifying the correctness of common everyday tasks carried out by users.

Therefore, it is important to determine the operational profile of the users to enable effective software testing to be performed. This may be difficult to determine and could change over time, as users may potentially change their behaviour as their needs evolve. The determination of the operational profile involves identifying the common operations to be performed, and the probability of each operation being performed.

Cleanroom employs *statistical usage testing* rather than coverage testing, and this involves executing tests chosen from the population of all possible uses of the software in accordance with the probability of its expected use. The software reliability measure is calculated by statistical techniques based on the expected usage of the software, and cleanroom provides a certified mean time to failure of the software.

Coverage testing involves designing tests that cover every path through the program, and this type of testing is as likely to find a rare execution failure as well as a frequent execution failure. However, it is essential to find failures that occur on frequently used parts of the system.

The advantage of usage testing (that matches the actual execution profile of the software) is that it has a better chance of finding execution failures on frequently used parts of the system. This helps to maximize the expected mean time to failure of the software.

The cleanroom software development process and calculation of the software reliability measure is described in O' Regan (2006), and the cleanroom development process enables engineers to deliver high-quality software on time and on budget. Some of the benefits of the use of cleanroom on projects at IBM are described in Cobb and Mills (1990) and summarized in Table 20.3.

20.2.3 Software Reliability Models

Models are simplifications of the reality, and a good model allows accurate predictions of future behaviour to be made. It is important to determine the adequacy of the model, and this is done by model exploration, and determining the extent to which it explains the actual manifested behaviour, as well as the accuracy of its predictions.

Table 20.3 Cleanroom results in IBM

Project	Results
Flight control project (1987) 33KLOC	Completed ahead of schedule Error-fix effort reduced by factor of five 2.5 errors KLOC before any execution
Commercial product (1988)	Deployment failures of 0.1/KLOC Certification testing failures 3.4/KLOC Productivity 740 LOC/month
Satellite control (1989) 80 KLOC (partial cleanroom)	50% improvement in quality Certification testing failures of 3.3/KLOC Productivity 780 LOC/month 80% improvement in productivity
Research project (1990) 12 KLOC	Certified to 0.9978 with 989 test cases

Table 20.4 Characteristics of good software reliability model

Characteristics of good software reliability model
Good theoretical foundation
Realistic assumptions
Good empirical support
As simple as possible (Ockham's Razor)
Trustworthy and accurate

A model is judged effective if there is good empirical evidence to support it, and more accurate models are sought to replace inadequate models. Models are often modified (or replaced) over time, as further facts and observations lead to aberrations that cannot be explained with the current model. A good software reliability model will have the following characteristics (Table 20.4).

The underlying mathematics used in the calculation of software reliability (i.e. probability and statistics) is discussed in Chap. 25. There are several existing software reliability predictor models employed (Table 20.5) with varying degrees of success. Some of these models just compute defect counts rather than estimating software reliability in terms of mean time to failure. They may be categorized into:

• *Size and Complexity Metrics*

These are used to predict the number of defects that a system will reveal in operation or testing.

• *Operational Usage Profile*

These predict failure rates based on the expected operational usage profile of the system. The number of failures encountered is determined and the software reliability predicted (e.g. cleanroom and its prediction of the MTTF).

• *Quality of the Development Process*

These predict failure rates based on the process maturity of the software development process in the organization (e.g. CMMI maturity).

The extent to which the software reliability model can be trusted depends on the accuracy of its predictions, and empirical data will need to be gathered to make a judgment. It may be acceptable to have a little inaccuracy during the early stages of prediction, provided the predictions of operational reliability are close to the observations. A model that gives overly optimistic results is termed 'optimistic', whereas a model that gives overly pessimistic results is termed 'pessimistic'.

The assumptions in the reliability model need to be examined to determine whether they are realistic. Several software reliability models have questionable assumptions such as:

Table 20.5 Software reliability models

Model	Description	Comments
Jelinski/Moranda model	The failure rate is a Poisson process (The Poisson process is a widely used counting process, and especially in counting the occurrence of certain events that appear to happen at a certain rate but at random. A Poisson random variable is of the form $P\{X = i\} = e^{-\lambda} \lambda^i/i!$) and is proportional to the current defect content of program. The initial defect count is N; the initial failure rate is $N\varphi$; it decreases to $(N-1)\varphi$ after the first fault is detected and eliminated, and so on. The constant φ is termed the proportionality constant	Assumes defects corrected perfectly and no new defects are introduced Assumes each fault contributes the same amount to failure rate
Littlewood/Verrall model	Successive execution time between failures is independent exponentially distributed random variables. (The exponential distribution is used to model the time between the occurrence of events in an interval of time. The density function is given by $f(x) = \lambda e^{-\lambda x}$.) Software failures are the result of the particular inputs and faults introduced from the correction of defects	Does not assume perfect correction of defects
Seeding and Tagging	This is analogous to estimating the fish population of a lake (Mills). A known number of defects are inserted into a software program, and the proportion of these identified during testing determined Another approach (Hyman) is to regard the defects found by one tester as tagged, and then to determine the proportion of tagged defects found by a second independent tester	Estimate of the total number of defects in the software but not a not s/w reliability predictor Assumes all faults equally likely to be found and introduced faults representative of existing
Generalized Poisson model	The number of failures observed in ith time interval τ_i has a Poisson distribution with mean $\phi(N-M_{i-1}) \tau_i^\alpha$ where N is the initial number of faults; M_{i-1} is the total number of faults removed up to the end of the $(i-1)$th time interval; and ϕ is the proportionality constant	Assumes faults removed perfectly at end of time interval

- All defects are corrected perfectly.
- Defects are independent of one another.
- Failure rate decreases as defects are corrected.
- Each fault contributes the same amount to the failure rate.

20.3 Dependability

Software is ubiquitous and is important to all sections of society, and so it is essential that widely used software is dependable (or trustworthy). In other words, the software should be available whenever required, as well as operating properly, safely and reliably, without any adverse side effects or security concerns. It is essential that the software used in systems in the safety-critical and security-critical fields is dependable, as the consequence of failure (e.g. the failure of a nuclear power plant) could be massive damage leading to loss of life or endangering the lives of the public.

Dependability engineering is concerned with techniques to improve the dependability of systems, and it involves the use of a rigorous design and development process to minimize the number of defects in the software. A dependable system is generally designed for fault tolerance, where the system can deal with (and recover from) faults that occur during software execution. Such a system needs to be secure, and able to protect itself from accidental or deliberate external attacks. Table 20.6 lists several dimensions of dependability.

Modern software systems are subject to attack by malicious software such as viruses that change the behaviour of the software, or corrupt data causing the system to become unreliable. Other malicious attacks include a denial of service attack that negatively impacts the system's availability.

The design and development of dependable software needs to include protection measures that protect against external attacks that could compromise the availability and security of the system. Further, a dependable system needs to include recovery mechanisms to enable normal service to be restored as quickly as possible following an attack.

Dependability engineering is concerned with techniques to improve the dependability of systems, and in designing dependable systems. A dependable system will generally be developed using an explicitly defined repeatable process, and it may employ *redundancy* (spare capacity) and *diversity* (different types) to achieve reliability.

Table 20.6 Dimensions of dependability

Dimension	Description
Availability	System is available for use at any time
Reliability	The system operates correctly and is trustworthy
Safety	The system does not injure people or damage the environment
Security	The system prevents unauthorized intrusions

There is a trade-off between dependability and the performance of the system, as dependable systems often need to carry out extra checks to monitor themselves and to check for erroneous states, and to recover from faults before failure occurs. This inevitably leads to increased costs in the design and development of dependable systems.

Software availability is the percentage of the time that the software system is running, and is a measure of the uptime/downtime of the software during a particular time period. The downtime refers to a period of time when the software is unavailable for use (including planned and unplanned outages), and many companies aim to develop software that is available for use 99.999% of the time in the year (i.e. a downtime of less than 5 min per annum). This goal is known as *five nines*, and it is a common goal in the telecommunications sector.

Safety-critical systems are systems where it is essential that the system is safe for the public, and that people or the environment are not harmed in the event of system failure. These include aircraft control systems and process control systems for chemical and nuclear power plants. The failure of a safety-critical system could in some situations lead to loss of life or serious economic damage.

Formal methods are discussed in Chap. 21, and they provide a precise way of specifying the requirements of the proposed system, and demonstrating (using mathematics) that key properties are satisfied in the formal specification. Further, they may be used to show that the implemented program satisfies its specification. The use of formal methods generally leads to increased confidence in the correctness of safety-critical and security-critical systems.

The security of the system refers to its ability to protect itself from accidental or deliberate external attacks, which are common today since most computers are networked and connected to the Internet. There are various security threats in any networked system including threats to the confidentiality and integrity of the system and its data, and threats to the availability of the system.

Therefore, controls are required to enhance security and to ensure that attacks are unsuccessful. Encryption is one way to reduce system vulnerability, as encrypted data is unreadable to the attacker. There may be controls that detect and repel attacks, and these controls are used monitor the system and to take action to shut down parts of the system or restrict access in the event of an attack. There may be controls that limit exposure (e.g. insurance policies and automated backup strategies) that allow recovery from the problems introduced.

It is important to have a reasonable level of security as otherwise all of the other dimensions of dependability (reliability, availability and safety) are compromised. Security loopholes may be introduced in the development of the system, and so care needs to be taken to prevent hackers from exploiting security vulnerabilities.

Risk analysis plays a key role in the specification of security and dependability requirements, and this involves identifying risks that can result in serious incidents. This leads to the generation of specific security requirements as part of the system requirements to ensure that these risks do not materialize, or if they do materialize then serious incidents will not materialize.

20.4   Computer Security

The introduction of the world wide web in the early 1990s transformed the world of computing, and it led inexorably to more and more computers being connected to the Internet. This has subsequently led to an explosive growth in attacks on computers and systems, as hackers and malicious software seek to exploit known security vulnerabilities. It is therefore essential to develop secure systems that can deal with and recover from such external attacks.

Hackers will often attempt to steal confidential data and to disrupt the services being offered by a system. Security engineering is concerned with the development of systems that can prevent such malicious attacks, and recover from them. It has become an important part of software and system engineering, and software developers need to be aware of the threats facing a system, and develop solutions to eliminate them.

Hackers may probe parts of the system for weaknesses, and system vulnerabilities may lead to attackers gaining unauthorized access to the system. There is a need to conduct a risk assessment of the security threats facing a system early in the software development process, and this will lead to several security requirements for the system.

The system needs to be designed for security, as it is difficult to add security after it has been implemented. Security loopholes may be introduced in the development of the system, and so care needs to be taken to prevent these as well as preventing hackers from exploiting security vulnerabilities. There may be controls that detect and repel attacks, and these monitor the system and take appropriate action to restrict access in the event of an attack.

The choice of architecture and how the system is organized is fundamental to the security of the system, and different types of systems will require different technical solutions to provide an acceptable level of security to its users. There following guidelines for designing secure systems are described in (Sommerville 2011):

- Security decisions should be based on the security policy.
- A security-critical system should fail securely.
- A secure system should be designed for recoverability.
- A balance is needed between security and usability.
- A single point of failure should be avoided.
- A log of user actions should be maintained.
- Redundancy and diversity should be employed.
- Organization information in system into compartments.

It is important to have a reasonable level of security, as otherwise all of the other dimensions of dependability are compromised.

20.5 System Availability

System availability is the percentage of time that the software system is running without downtime, and robust systems will generally aim to achieve 5-nines availability (i.e. 99.999% availability). This is equivalent to approximately 5 min of downtime (including planned/unplanned outages) per year. The availability of a system is measured by its performance when a subsystem fails, and its ability to resume service in a state close to the original state. A fault-tolerant system continues to operate correctly (possibly at a reduced level) after some part of the system fails, and it aims to achieve 100% availability.

System availability and software reliability are related, with availability measuring the percentage of time that the system is operational, and reliability measuring the probability of failure-free operation over a period of time. The consequence of a system failure may be to freeze or crash the system, and system availability is measured by how long it takes to recover and restart after a failure. A system may be unreliable and yet have good availability metrics (fast restart after failure), or it may be highly reliable with poor availability metrics (taking a long time to recover after a failure).

Software that satisfies strict availability constraints is usually reliable. The downtime generally includes the time needed for activities such as re-booting a machine, upgrading to a new version of software, planned and unplanned outages. It is theoretically possible for software to be highly unreliable but yet to have good availability metrics or for software that is highly reliable to have poor availability metrics. Consequently, care is required before drawing conclusions between software reliability and software availability metrics.

20.6 Safety-Critical Systems

A safety-critical system is a system whose failure could result in significant economic damage or loss of life. There are many examples of safety-critical systems including aircraft flight control systems and missile systems. It is therefore essential to employ rigorous processes in their design and development, and testing alone is usually insufficient to verifying the correctness of a safety-critical system.

The safety-critical industry takes the view that any change to safety-critical software creates a new program. The new program is therefore required to demonstrate that it is reliable and safe to the public, and so extensive testing needs to be performed. Other techniques such as formal verification and model checking may be employed to provide an extra level of assurance in the correctness of the safety-critical system.

Safety-critical systems need to be dependable and available for use whenever required. Safety-critical software must operate correctly and reliably without any adverse side effects. The consequence of failure (e.g. the failure of a weapons system) could be massive damage, leading to loss of life or endangering the lives of the public.

Safety-critical systems are generally designed for fault tolerance, where the system can deal with (and recover from) faults that occur during execution. Fault tolerance is achieved by anticipating exceptional events, and in designing the system to handle them. A fault-tolerant system is designed to fail safely, and programs are designed to continue working (possibly at a reduced level of performance) rather than crashing after the occurrence of an error or exception. Many fault-tolerant systems mirror all operations, where each operation is performed on two or more duplicate systems, and so if one fails then the other system can take over.

The development of a safety-critical system needs to be rigorous, and subject to strict quality assurance to ensure that the system is safe to use and that the public will not be in danger. This involves rigorous design and development processes to minimize the number of defects in the software, as well as comprehensive testing to verify its correctness. Formal methods are often employed in the development of safety-critical systems (Chap. 21).

20.7   Review Questions

1. Explain the difference between software reliability and system availability.
2. What is software dependability?
3. Explain the significance of Adam's 1984 study of failures at IBM.
4. Describe the cleanroom methodology.
5. Describe the characteristics of a good software reliability model.
6. Explain the relevance of security engineering.
7. What is a safety-critical system?

20.8   Summary

This chapter gave an introduction to some important topics in software engineering including software reliability and the cleanroom methodology; dependability; availability; security; and safety-critical systems.

Software reliability is the probability that the program works without failure for a period of time, and it is usually expressed as the mean time to failure.

Cleanroom involves the application of statistical techniques to calculate software reliability, and it is based on the expected usage of the software.

It is essential that the software used in the safety- and security-critical fields is dependable, with the software available when required, as well as operating safely and reliably without any adverse side effects. Many of these systems are fault-tolerant and are designed to deal with (and recover) from faults that occur during execution.

Such a system needs to be secure and able to protect itself from external attacks and needs to include recovery mechanisms to enable normal service to be restored as quickly as possible. Another words, it is essential that if the system fails then it fails safely.

Today, billions of computers are connected to the Internet, and this has led to a growth in attacks on computers. It is essential that developers are aware of the threats facing a system and are familiar with techniques to eliminate them.

References

Adams E (1984) Optimizing preventive service of software products. IBM Res J 28(1):2–14
Cobb RH, Mills HD (1990) Engineering software under statistical quality control. IEEE Software
O' Regan G (2006) Mathematical approaches to software quality, Springer
Sommerville I (2011) Software engineering, 9th edn. Pearson

Overview of Formal Methods

21

Key Topics

Formal specification
Vienna development method
Z specification language
B-method
Model-oriented approach
Axiomatic approach
Process calculus
Refinement
Finite-state machines
Model checking
Usability of formal methods

21.1 Introduction

The term *'formal methods'* refer to various mathematical techniques used for the formal specification and development of software. They consist of a formal specification language and employ a collection of tools to support the syntax checking of the specification, as well as the proof of properties of the specification. They allow questions to be asked about what the system does independently of the implementation.

© Springer Nature Switzerland AG 2020
G. O'Regan, *Mathematics in Computing*, Undergraduate Topics
in Computer Science, https://doi.org/10.1007/978-3-030-34209-8_21

The use of mathematical notation avoids speculation about the meaning of phrases in an imprecisely worded natural language description of a system. Natural language is inherently ambiguous, whereas mathematics employs a precise rigorous notation. Spivey (1992) defines formal specification as:

Definition 21.1 (*Formal Specification*) Formal specification is the use of mathematical notation to describe in a precise way the properties that an information system must have, without unduly constraining the way in which these properties are achieved.

The formal specification thus becomes the key reference point for the different parties involved in the construction of the system. It may be used as the reference point for the requirements; program implementation; testing and program documentation. It thus promotes a common understanding for all those concerned with the system. The term '*formal methods*' is used to describe a formal specification language and a method for the design and implementation of a computer system. Formal methods may be employed at a number of levels:

- Formal specification only (program developed informally),
- Formal specification, refinement and verification (some proofs), and
- Formal specification, refinement and verification (with extensive theorem proving).

The specification is written in a mathematical language, and the implementation may be derived from the specification via stepwise refinement.[1] The refinement step makes the specification more concrete and closer to the actual implementation. There is an associated proof obligation to demonstrate that the refinement is valid and that the concrete state preserves the properties of the abstract state. Thus, assuming that the original specification is correct and the proof of correctness of each refinement step is valid, then there is a very high degree of confidence in the correctness of the implemented software.

Stepwise refinement is illustrated as follows: the initial specification S is the initial model M_0; it is then refined into the more concrete model M_1, and M_1 is then refined into M_2, and so on until the eventual implementation $M_n = E$ is produced.

$$S = M_0 \subseteq M_1 \subseteq M_2 \subseteq M_3 \subseteq, \ldots, \subseteq M_n = \mathrm{E}$$

Requirements are the foundation of the system to be built, and irrespective of the best design and development practices, the product will be incorrect if the requirements are incorrect. The objective of requirements validation is to ensure that the requirements reflect what is actually required by the customer (in order to

---

[1]It is questionable whether stepwise refinement is cost-effective in mainstream software engineering, as it involves rewriting a specification *ad nauseum*. It is time-consuming to proceed in refinement steps with significant time also required to prove that the refinement step is valid. It is more relevant to the safety-critical field. Others in the formal methods field may disagree with this position.

build the right system). Formal methods may be employed to model the requirements, and the model exploration yields further desirable or undesirable properties.

Formal methods provide the facility to prove that certain properties are true of the specification, and this is valuable, especially in safety-critical and security-critical applications. The properties are a logical consequence of the mathematical definition, and the requirements may be amended where appropriate. Thus, formal methods may be employed in a sense to debug the requirements during requirements validation.

The use of formal methods generally leads to more robust software and increased confidence in its correctness. Formal methods may be employed at different levels (e.g. just for specification with the program developed informally). The challenges involved in the deployment of formal methods in an organization include the education of staff in formal specification, as the use of these mathematical techniques, may be a culture shock to many staff.

Formal methods have been applied to a diverse range of applications, including the safety and security-critical fields to develop dependable software. The applications include the railway sector, microprocessor verification, the specification of standards and the specification and verification of programs. Parnas and others have criticized formal methods (Table 21.1).

However, formal methods are potentially quite useful and reasonably easy to use. The use of a formal method such as Z or VDM forces the software engineer to be precise and helps to avoid ambiguities present in natural language. Clearly, a formal specification should be subject to peer review to provide confidence in its correctness. New formalisms need to be intuitive to be usable by practitioners, and an advantage of the use of classical mathematics is that it is familiar to students.

21.2 Why Should We Use Formal Methods?

There is a strong motivation to use best practice in software engineering in order to produce software adhering to high-quality standards. Quality problems with software may cause minor irritations or major damage to a customer's business including loss of life. Formal methods are a leading-edge technology that may be of benefit to companies in reducing the occurrence of defects in software products. Brown (1990) argues that for the safety-critical field that:

Comment 21.1 (Missile Safety)

Missile systems must be presumed dangerous until shown to be safe, and that the absence of evidence for the existence of dangerous errors does not amount to evidence for the absence of danger.

This suggests that companies in the safety-critical field will need to demonstrate that every reasonable practice was taken to prevent the occurrence of defects. One such practice is the use of formal methods, and its exclusion may need to be justified in some domains. It is quite possible that a software company may be sued

Table 21.1 Criticisms of formal methods

No.	Criticism
1.	Often the formal specification is as difficult to read as the program. (Of course, others might reply by saying that some of Parnas's tables are not exactly intuitive, and that the notation he employs in some of his tables is quite unfriendly. The usability of all of the mathematical approaches needs to be enhanced if they are to be taken seriously by industrialists.)
2.	Many formal specifications are wrong. (Obviously, the formal specification must be analysed using mathematical reasoning and tools to provide confidence in its correctness. The validation of a formal specification can be carried out using mathematical proof of key properties of the specification; software inspections; or specification animation.)
3.	Formal methods are strong on syntax but provide little assistance in deciding on what technical information should be recorded using the syntax. (Approaches such as VDM include a method for software development as well as the specification language.)
4.	Formal specifications provide a model of the proposed system. However, a precise unambiguous mathematical statement of the requirements is what is needed. (Models are extremely valuable as they allow simplification of the reality. A mathematical study of the model demonstrates whether it is a suitable representation of the system. Models allow properties of the proposed requirements to be studied prior to implementation.)
5.	Stepwise refinement is unrealistic. (Stepwise refinement involves rewriting a specification with each refinement step producing a more concrete specification (that includes code and formal specification) until eventually the detailed code is produced. It is difficult and time-consuming but, tool support may make refinement easier.) It is like, for example, deriving a bridge from the description of a river and the expected traffic on the bridge. There is always a need for the creative step in design
6.	Many unnecessary mathematical formalisms have been developed rather than using the available classical mathematics. (Approaches such as VDM or Z are useful in that they add greater rigour to the software development process. They are reasonably easy to learn, and there have been some good results obtained by their use. Classical mathematics is familiar to students and therefore it is desirable that new formalisms are introduced only where absolutely necessary.)

for software which injures a third party, and this suggests that companies will need a rigorous quality assurance system to prevent the occurrence of defects.

There is some evidence to suggest that the use of formal methods provides savings in the cost of the project. For example, a 9% cost saving is attributed to the use of formal methods during the CICS project; the T800 project attributes a 12-month reduction in testing time to the use of formal methods. These are discussed in more detail in Chap. 1 of Hinchey and Bowen (1995).

The use of formal methods is mandatory in certain circumstances. The Ministry of Defence (MOD) in the United Kingdom issued two safety-critical standards in the early 1990s related to the use of formal methods in the software development lifecycle.

The first is Defence Standard 00-55, '*The Procurement of safety critical software in defense equipment*' (MOD 1991a) which makes it mandatory to employ formal methods in the development of safety-critical software in the UK. The standard mandates the use of formal proof that the most crucial programs correctly implement their specifications.

The other is Def. Stan 00-56 '*Hazard analysis and safety classification of the computer and programmable electronic system elements of defense equipment*' (MOD 1991b). The objective of this standard is to provide guidance to identify which systems or parts of systems being developed are safety-critical and thereby require the use of formal methods. This proposed system is subject to an initial hazard analysis to determine whether there are safety-critical parts.

The reaction to these defence standards 00-55 and 00-56 was quite hostile initially, as most suppliers were unlikely to meet the technical and organization requirements of the standard (Tierney 1991). The U.K. Defence Standards 0055 and 0056 were later revised to be less prescriptive on the use of formal methods.

21.3   Industrial Applications of Formal Methods

Formal methods have been employed in several domains such as the transport sector, the nuclear sector, the space sector, the defence sector, the semiconductor sector, the financial sector and the telecoms sector. The extent of the application of formal methods has varied from formal specification only, to specification with inspections, to proofs, to refinement, to test generation and to model checking. Formal methods is applicable to the regulated sector, and it has been applied to real-time applications in the nuclear industry, the aerospace industry, the security technology area and the railroad domain. These sectors are subject to stringent regulatory controls to ensure that safety and security are properly addressed.

Several organizations have piloted formal methods with varying degrees of success. IBM developed the VDM specification language at its laboratory in Vienna, and it piloted the Z and B formal specification languages on the CICS (Customer Information Control System) project at its plant in Hursley, England.

The mathematical techniques developed by Parnas (i.e. his requirements model and tabular expressions) were employed to specify the requirements of the A-7 aircraft (as part of a research project for the US Navy).[2] Tabular expressions were also employed for the software inspection of the automated shutdown software of the Darlington Nuclear power plant in Canada.[3] These are two successful uses of mathematical techniques in software engineering.

There are examples of the use of formal methods in the railway domain, with GEC Alsthom and RATP using B for the formal specification and verification of the computerized signalling system on the Paris Metro. Several examples dealing with the modelling and verification of a railroad gate controller and railway signalling

---

[2]However, the resulting software was never actually deployed on the A-7 aircraft.
[3]This was an impressive use of mathematical techniques and it has been acknowledged that formal methods must play an important role in future developments at Darlington. However, given the time and cost involved in the software inspection of the shutdown software some managers have less enthusiasm in shifting from hardware to software controllers Gerhart et al. (1994).

are described in Hinchey and Bowen (1995). Clearly, it is essential to verify safety-critical properties such as *'when the train goes through the level crossing then the gate is closed'*.

PVS is a mechanized environment for formal specification and verification, and it was developed at SRI in California. It includes a specification language integrated with support tools and an interactive theorem prover. The specification language is based on higher order logic, and the theorem prover is guided by the user in conducting proof. It has been applied to the verification of hardware and software, and PVS has been used for the formal specification and partial verification of the micro-code of the AAMP5 microprocessor.

A selection of applications of formal methods to industry is presented in Woodcock et al. (2009).

21.4 Industrial Tools for Formal Methods

Formal methods have been criticized for the limited availability of tools to support the software engineer in writing the formal specification and in conducting proof. Many of the early tools were criticized as not being of industrial strength. However, in recent years more advanced tools have become available to support the software engineer's work in formal specification and formal proof, and this is likely to continue in the coming years.

The tools include syntax checkers that determine whether the specification is syntactically correct; specialized editors which ensure that the written specification is syntactically correct; tools to support refinement; automated code generators that generate a high-level language corresponding to the specification; theorem provers to demonstrate the correctness of refinement steps and to identify and resolve proof obligations, as well proving the presence or absence of key properties; and specification animation tools where the execution of the specification can be simulated.

The B-Toolkit[4] from B-Core is an integrated set of tools that supports the B-Method. It provides functionality for syntax and type checking, specification animation, proof obligation generator, an auto-prover, a proof assistant and code generation. This, in theory, allows the complete formal development from the initial specification to the final implementation, with every proof obligation justified, leading to a provably correct program. There is also the Atelier B tool to support formal specification and development in B.

The IFAD Toolbox[5] is a support tool for the VDM-SL specification language, and it provides support for syntax and type checking, an interpreter and debugger to execute and debug the specification, and a code generator to convert from VDM-SL to C++. The Overture Integrated Development Environment (IDE) is an open-source tool for formal modelling and analysis of VDM-SL specifications.

---

[4]The source code for the B-Toolkit is now available.
[5]The IFAD Toolbox has been renamed to VDMTools as IFAD sold the VDM Tools to CSK in Japan. The CSK VDM tools are available for worldwide use.

There are various tools for model checking including Spin, Bandera, SMV and UppAal. These tools perform a systematic check on property P in all states and are applicable if the system generates a finite behavioural model. Spin is an open-source tool, and it checks finite-state systems with properties specified by linear temporal logic. It generates a counterexample trace if determines that a property is violated.

There are tools to support theorem provers (see Chap. 18) such as the Boyer–Moore Theorem prover (NQTHM) which was developed at the University of Texas in the late 1970s. It is far more automated than many other interactive theorem provers, but it requires detailed human guidance (with suggested lemmas) for difficult proofs. The user therefore needs to understand the proof being sought and the internals of the theorem prover. Many mathematical theorems have been proved including Gödel's incompleteness theorem.

The HOL system was developed at the University of Cambridge, and it is an environment for interactive theorem proving in a higher order logic. It requires skilled human guidance and has been used for the verification of microprocessor design. It is a widely used theorem prover.

21.5 Approaches to Formal Methods

There are two key approaches to formal methods: namely the *model-oriented approach* of VDM or Z, and the *algebraic* or *axiomatic approach* of the process calculi such as the calculus communicating systems (CCS) or communicating sequential processes (CSP).

21.5.1 Model-Oriented Approach

The model-oriented approach to specification is based on mathematical models, where a model is a simplification or abstraction of the real world that contains only the essential details. For example, the model of an aircraft will not include the colour of the aircraft, and the objective may be to model the aerodynamics of the aircraft. There are many models employed in the physical world, such as meteorological models that allow weather forecasts to be made.

The importance of models is that they serve to explain the behaviour of a particular entity and may also be used to predict future behaviour. Different models may vary in their ability to explain aspects of the entity under study. One model may be good at explaining some aspects of the behaviour, whereas another model might be good at explaining other aspects. The *adequacy* of a model is a key concept in modelling, and reflects the effectiveness of the model in representing the underlying behaviour, and in its ability to predict future behaviour. Model exploration consists of asking questions, and determining whether the model is able to give an effective answer to the particular question. A good model is chosen as a representation of the real world and is referred to whenever there are questions in relation to the aspect of the real world.

It is fundamental to explore the model to determine its adequacy, and to determine the extent to which it explains the underlying physical behaviour and allows accurate predictions of future behaviour to be made. There may be more than one possible model of a particular entity, for example, the Ptolemaic model and the Copernican model are different models of the solar system. This leads to the question as to which is the best or most appropriate model to use, and on the criteria to use to determine the most suitable model. The ability of the model to explain the behaviour, its simplicity and its elegance will be part of the criteria. The principle of 'Ockham's Razor' (law of parsimony) is often used in modelling, and it suggests that the simplest model with the least number of assumptions required should be selected.

The adequacy of the model will determine its acceptability as a representation of the physical world. Models that are ineffective will be replaced with models that offer a better explanation of the manifested physical behaviour. There are many examples in science of the replacement of one theory by a newer one. For example, the Copernican model of the universe replaced the older Ptolemaic model, and Newtonian physics was replaced by Einstein's theories of relativity. The structure of the revolutions that take place in science are described in Kuhn (1970).

Modelling can play a key role in computer science, as computer systems tend to be highly complex, whereas a model allows simplification or an abstraction of the underlying complexity, and it enables a richer understanding of the underlying reality to be gained. The model-oriented approach to software development involves defining an abstract model of the proposed software system, and the model is then explored to determine its suitability as a representation of the system. This takes the form of model interrogation, i.e. asking questions, and determining the extent to which the model can answer the questions. The modelling in formal methods is typically performed via elementary discrete mathematics, including set theory, sequences, functions and relations.

Various models have been applied to assist with the complexities in software development. These include the Capability Maturity Model (CMM), which is employed as a framework to enhance the capability of the organization in software development; UML, which has various graphical diagrams that are employed to model the requirements and design; and mathematical models that are employed for formal specification.

VDM and Z are model-oriented approaches to formal methods. VDM arose from work done at the IBM laboratory in Vienna in formalizing the semantics for the PL/1 compiler in the early 1970s, and it was later applied to the specification of software systems. The origin of the Z specification language is in work done at Oxford University in the early 1980s.

21.5.2 Axiomatic Approach

The axiomatic approach focuses on the properties that the proposed system is to satisfy, and there is no intention to produce an abstract model of the system. The required properties and behaviour of the system are stated in mathematical notation. The difference between the axiomatic specification and a model-based approach may be seen in the example of a stack.

The stack includes operators for pushing an element onto the stack and popping an element from the stack. The properties of *pop* and *push* are explicitly defined in the axiomatic approach. The model-oriented approach constructs an explicit model of the stack and the operations are defined in terms of the effect that they have on the model. The axiomatic specification of the *pop* operation on a stack is given by properties, for example, *pop(push(s, x)) = s*.

Comment 3.2 (Axiomatic Approach)

The property-oriented approach has the advantage that the implementer is not constrained to a particular choice of implementation, and the only constraint is that the implementation must satisfy the stipulated properties.

The emphasis is on specifying the required properties of the system, and implementation issues are avoided. The properties are typically stated using mathematical logic or higher order logics. Mechanized theorem-proving techniques may be employed to prove results.

One potential problem with the axiomatic approach is that the properties specified may not be satisfied in any implementation. Thus, whenever a 'formal axiomatic theory' is developed a corresponding 'model' of the theory must be identified, in order to ensure that the properties may be realized in practice. That is, when proposing a system that is to satisfy some set of properties, there is a need to prove that there is at least one system that will satisfy the set of properties.

21.6 Proof and Formal Methods

The nature of theorem proving was discussed in Chap. 18. A mathematical proof typically includes natural language and mathematical symbols, and often many of the tedious details of the proof are omitted. The proof may employ a '*divide and conquer*' technique; i.e. breaking the conjecture down into sub-goals and then attempting to prove each of the sub-goals.

Many proofs in formal methods are concerned with cross-checking the details of the specification, or checking the validity of the refinement steps, or checking that certain properties are satisfied by the specification. There are often many tedious lemmas to be proved, and theorem provers[6] play a key role in dealing with them. Machine proof is explicit and reliance on some brilliant insight is avoided. Proofs by hand are notorious for containing errors or jumps in reasoning, while machine proofs are explicit but are often extremely lengthy and unreadable. The infamous machine proof of the correctness of the VIPER microprocessor[7] consisted of several million formulae (Tierney 1991).

A formal mathematical proof consists of a sequence of formulae, where each element is either an axiom or derived from a previous element in the series by applying a fixed set of mechanical rules.

The application of formal methods in an industrial environment requires the use of machine-assisted proof, since thousands of proof obligations arise from a formal specification, and mechanized theorem provers are essential in resolving these efficiently. Automated theorem proving is difficult, as often mathematicians prove a theorem with an initial intuitive feeling that the theorem is true. Human intervention to provide guidance or intuition improves the effectiveness of the theorem prover.

The proof of various properties about a program increases confidence in its correctness. However, an absolute proof of correctness[8] is unlikely except for the most trivial of programs. A program may consist of legacy software that is assumed to work; a compiler that is assumed to work correctly creates it. Theorem provers are programs that are assumed to function correctly. The best that formal methods can claim is increased confidence in correctness of the software, rather than an absolute proof of correctness.

21.7   Mathematics in Software Engineering

The debate concerning the level of use of mathematics in software engineering is still ongoing. Many practitioners are against the use of mathematics and avoid its use. They tend to employ methodologies such as software inspections and testing to improve confidence in the correctness of the software. They argue that in the current competitive industrial environment where time to market is a key driver that the use of such formal mathematical techniques would seriously impact the market opportunity. Industrialists often need to balance conflicting needs such as quality, cost and delivering on time. They argue that the commercial necessities require methodologies and techniques that allow them to achieve their business goals effectively.

---

[6]Most existing theorem provers are difficult to use and are for specialist use only. There is a need to improve the usability of theorem provers.

[7]This verification was controversial with RSRE and Charter overselling VIPER as a chip design that conforms to its formal specification.

[8]This position is controversial with others arguing that if correctness is defined mathematically then the mathematical definition (i.e. formal specification) is a theorem, and the task is to prove that the program satisfies the theorem. They argue that the proofs for non-trivial programs exist, and that the reason why there are not many examples of such proofs is due to a lack of mathematical specifications.

The other camp argues that the use of mathematics is essential in the delivery of high-quality and reliable software, and that if a company does not place sufficient emphasis on quality it will pay the price in terms of poor quality and loss of reputation.

It is generally accepted that mathematics and formal methods must play a role in the safety-critical and security-critical fields. Apart from that the extent of the use of mathematics is a hotly disputed topic. The pace of change in the world is extraordinary, and companies face immense competitive forces in a global market place.

It is unrealistic to expect companies to deploy formal methods unless they have clear evidence that it will support them in delivering commercial products to the market place ahead of their competition, at the right price and with the right quality. Formal methods need to prove that it can do this if it wishes to be taken seriously in mainstream software engineering.

21.8   The Vienna Development Method

VDM was developed by a research team at the IBM research laboratory in Vienna. This group was specifying the semantics of the PL/1 programming language using an operational semantic approach. That is, the semantics of the language were defined in terms of a hypothetical machine which interprets the programs of that language (Bjørner and Jones 1978, 1982). Later work led to the Vienna Development Method (VDM) with its specification language, Meta IV. This was used to give the denotational semantics of programming languages; i.e. a mathematical object (set, function, etc.) is associated with each phrase of the language. The mathematical object is termed the *denotation* of the phrase (see Chap. 12).

VDM is a *model-oriented approach* and this means that an explicit model of the state of an abstract machine is given, and operations are defined in terms of the state. Operations may act on the system state, taking inputs and producing outputs as well as a new system state. Operations are defined in a precondition and post-condition style. Each operation has an associated proof obligation to ensure that if the precondition is true, then the operation preserves the system invariant. The initial state itself is, of course, required to satisfy the system invariant.

VDM uses keywords to distinguish different parts of the specification, e.g. preconditions, postconditions, as introduced by the keywords *pre* and *post*, respectively. In keeping with the philosophy that formal methods specifies *what* a system does as distinct from *how*, VDM employs postconditions to stipulate the effect of the operation on the state. The previous state is then distinguished by employing *hooked variables*, e.g. $v^{\overleftarrow{}}$ and the postcondition specifies the new state which is defined by a logical predicate relating the pre-state to the post-state.

VDM is more than its specification language VDM-SL, and is, in fact, a software development method, with rules to verify the steps of development. The rules enable the executable specification, i.e. the detailed code, to be obtained from the initial specification via refinement steps. Thus, we have a sequence $S = S_0, S_1, ...,$ $S_n = E$ of specifications, where S is the initial specification, and E is the final (executable) specification.

Retrieval functions enable a return from a more concrete specification to the more abstract specification. The initial specification consists of an initial state, a system state and a set of operations. The system state is a particular domain, where a domain is built out of primitive domains such as the set of natural numbers, integers, etc., or constructed from primitive domains using domain constructors such as Cartesian product, disjoint union, etc. A domain-invariant predicate may further constrain the domain, and a *type* in VDM reflects a domain obtained in this way. Thus, a type in VDM is more specific than the signature of the type, and thus represents values in the domain defined by the signature, which satisfy the domain-invariant. In view of this approach to types, it is clear that VDM types may not be 'statically type checked'.

VDM specifications are structured into modules, with a module containing the module name, parameters, types, operations, etc. Partial functions occur frequently in computer science as many functions may be undefined, or fail to terminate for some arguments in their domain. VDM addresses partial functions by employing non-standard logical operators, namely the logic of partial functions (LPFs), which was discussed in Chap. 17.

VDM has been used in industrial projects, and its tool support includes the IFAD Toolbox.[9] VDM is described in more detail in O'Regan (2017). There are several variants of VDM, including VDM++, the object-oriented extension of VDM, and the Irish school of the VDM, which is discussed in the next section.

21.9 VDM*, the Irish School of VDM

The Irish School of VDM is a variant of standard VDM and is characterized by its constructive approach, classical mathematical style and its terse notation (Mac An Airchinnigh 1990). This method aims to combine the *what* and *how* of formal methods in that its terse specification style stipulates in concise form *what* the system should do; furthermore, the fact that its specifications are constructive (or functional) means that the *how* is included with the *what*.

However, it is important to qualify this by stating that the how as presented by VDM* is not directly executable, as several of its mathematical data types have no corresponding structure in high-level programming languages or functional languages. Thus, a conversion or reification of the specification into a functional or higher level language must take place to ensure a successful execution. Further, the fact that a specification is constructive is no guarantee that it is a good implementation strategy, if the construction itself is naive.

---

[9]The VDM Tools are now available from the CSK Group in Japan.

The Irish school follows a similar development methodology as in standard VDM, and it is a model-oriented approach. The initial specification is presented, with the initial state and operations defined. The operations are presented with preconditions; however, no postcondition is necessary as the operation is 'functionally' (i.e. explicitly) constructed.

There are proof obligations to demonstrate that the operations preserve the invariant. That is, if the precondition for the operation is true, and the operation is performed, then the system invariant remains true after the operation. The philosophy is to exhibit existence *constructively* rather than providing a theoretical proof of existence that demonstrates the existence of a solution without presenting an algorithm to construct the solution.

The school avoids the existential quantifier of predicate calculus, and reliance on logic in proof is kept to a minimum, with emphasis instead placed on equational reasoning. Structures with nice algebraic properties are sought, and one nice algebraic structure employed is the monoid, which has closure, associative and a unit element. The concept of isomorphism is powerful, reflecting that two structures are essentially identical, and thus we may choose to work with either, depending on which is more convenient for the task in hand.

The school has been influenced by the work of Polya and Lakatos. The former Polya (1957) advocated a style of problem-solving characterized by first considering an easier sub-problem, and considering several examples. This generally leads to a clearer insight into solving the main problem. Lakatos's approach to mathematical discovery (1976) is characterized by heuristic methods. A primitive conjecture is proposed and if global counterexamples to the statement of the conjecture are discovered, then the corresponding *hidden lemma* for which this global counterexample is a local counterexample is identified and added to the statement of the primitive conjecture. The process repeats until no more global counterexamples are found. A sceptical view of absolute truth or certainty is inherent in this.

Partial functions are the norm in VDM♣, and as in standard VDM, the problem is that functions may be undefined, or fail to terminate for several of the arguments in their domain. The logic of partial functions (LPFs) is avoided, and instead care is taken with recursive definitions to ensure termination is achieved for each argument. Academic and industrial projects have been conducted using VDM♣, but tool support is limited. The Irish school of VDM is discussed in more detail in [ORg17b].

21.10 The *Z* Specification Language

Z is a formal specification language founded on Zermelo set theory, and it was developed by Abrial at Oxford University in the early 1980s. It is used for the formal specification of software and is a model-oriented approach. An explicit model of the state of an abstract machine is given, and the operations are defined in terms of the effect on the state. It includes a mathematical notation that is similar to

VDM and the visually striking schema calculus. The latter consists essentially of boxes (or schemas), and these are used to describe operations and states. The schema calculus enables schemas to be used as building blocks and combined with other schemas. The Z specification language was published as an ISO standard (ISO/IEC 13568:2002) in 2002.

The schema calculus is a powerful means of decomposing a specification into smaller pieces or schemas. This helps to make Z specification highly readable, as each individual schema is small in size and self-contained. Exception handling is done by defining schemas for the exceptional cases, and these are then combined with the original operation schema. Mathematical data types are used to model the data in a system and these data types obey mathematical laws. These laws enable simplification of expressions and are useful with proofs.

Operations are defined in a precondition/postcondition style. However, the precondition is implicitly defined within the operation; i.e. it is not separated out as in standard VDM. Each operation has an associated proof obligation to ensure that if the precondition is true, then the operation preserves the system invariant. The initial state itself is, of course, required to satisfy the system invariant. Postconditions employ a logical predicate which relates the pre-state to the post-state, and the post-state of a variable v is given by priming, e.g. v'. Various conventions are employed, e.g. $v?$ indicates that v is an input variable and $v!$ indicates that v is an output variable. The symbol Ξ Op operation indicates that this operation does not affect the state, whereas Δ Op indicates that this operation affects the state.

Many data types employed in Z have no counterpart in standard programming languages. It is therefore important to identify and describe the concrete data structures that will ultimately represent the abstract mathematical structures. The operations on the abstract data structures may need to be refined to yield operations on the concrete data structure that yield equivalent results. For simple systems, direct refinement (i.e. one step from abstract specification to implementation) may be possible; in more complex systems, deferred refinement is employed, where a sequence of increasingly concrete specifications are produced to eventually yield the executable specification.

Z has been successfully applied in industry, and one of its well-known successes is the CICS project at IBM Hursley in England. Z is described in more detail in Chap. 22.

21.11   The *B*-Method

The *B-Technologies* (McDonnell 1994) consist of three components: a method for software development, namely the *B*-Method; a supporting set of tools, namely, the *B*-Toolkit; and a generic program for symbol manipulation, namely, the *B*-Tool (from which the *B*-Toolkit is derived). The *B*-Method is a model-oriented approach and is closely related to the *Z* specification language. Abrial developed the B specification language, and every construct in the language has a set-theoretic counterpart, and the method is founded on Zermelo set theory. Each operation has an explicit precondition.

A key role of the *abstract machine* in the *B*-Method is to provide encapsulation of variables representing the state of the machine and operations that manipulate the state. Machines may refer to other machines, and a machine may be introduced as a refinement of another machine. The abstract machines are specification machines, refinement machines or implementable machines. The *B*-Method adopts a layered approach to design where the design is gradually made more concrete by a sequence of design layers. Each design layer is a refinement that involves a more detailed implementation in terms of the abstract machines of the previous layer. The design refinement ends when the final layer is implemented purely in terms of library machines. Any refinement of a machine by another has associated proof obligations, and proof is required to verify the validity of the refinement step.

Specification animation of the Abstract Machine Notation (AMN) specification is possible with the *B*-Toolkit, and this enables typical usage scenarios to be explored for requirements validation. This is, in effect, an early form of testing, and it may be used to demonstrate the presence or absence of desirable or undesirable behaviour. Verification takes the form of a proof to demonstrate that the invariant is preserved when the operation is executed within its precondition, and this is performed on the AMN specification with the *B*-Toolkit.

The *B*-Toolkit provides several tools that support the *B*-Method, and these include syntax and type checking; specification animation, proof obligation generator, auto prover, proof assistor and code generation. Thus, in theory, a complete formal development from initial specification to final implementation may be achieved, with every proof obligation justified, leading to a provably correct program.

The *B*-Method and toolkit have been successfully applied in industrial applications, including the CICS project at IBM Hursley in the United Kingdom (Hoare 1995). The automated support provided has been cited as a major benefit of the application of the *B*-Method and the *B*-Toolkit.

21.12   Predicate Transformers and Weakest Preconditions

The precondition of a program S is a predicate, i.e. a statement that may be true or false, and it is usually required to prove that if the precondition Q is true then execution of S is guaranteed to terminate in a finite amount of time in a state satisfying R. This is written as $\{Q\}\ S\ \{R\}$.

The weakest precondition of a command S with respect to a postcondition R (Gries 1981) represents the set of all states such that if execution begins in any one of these states, then execution will terminate in a finite amount of time in a state with R true. These set of states may be represented by a predicate Q', so that wp (S, R)= wp_S (R) = Q', and so wp_S is a predicate transformer: i.e. it may be regarded as a function on predicates. The weakest precondition is the precondition that places the fewest constraints on the state than all of the other preconditions of (S,R). That is, all of the other preconditions are stronger than the weakest precondition.

The notation $Q\{S\}R$ is used to denote partial correctness, and indicates that if execution of S commences in any state satisfying Q, and if execution terminates, then the final state will satisfy R. Often, a predicate Q which is stronger than the weakest precondition wp (S,R) is employed, especially where the calculation of the weakest precondition is non-trivial. Thus, a stronger predicate Q such that $Q \Rightarrow wp$ (S,R) is often employed.

There are many properties associated with the weakest preconditions, and these may be used to simplify expressions involving weakest preconditions, and in determining the weakest preconditions of various program commands such as assignments, iterations, etc. Weakest preconditions may be used in developing a proof of correctness of a program in parallel with its development (O' Regan 2006).

An imperative program may be regarded as a predicate transformer. This is since a predicate P characterizes the set of states in which the predicate P is true, and an imperative program may be regarded as a binary relation on states, which, leads to the Hoare triple $P\{F\}Q$. That is, the program F acts as a predicate transformer with the predicate P regarded as an input assertion, i.e. a Boolean expression that must be true before the program F is executed, and the predicate Q is the output assertion, which is true if the program F terminates (where F commenced in a state satisfying P).

21.13 The Process Calculi

The objectives of the process calculi (Hoare 1985) are to provide mathematical models which provide insight into the diverse issues involved in the specification, design and implementation of computer systems which continuously act and interact with their environment. These systems may be decomposed into sub-systems that interact with each other and their environment.

The basic building block is the *process*, which is a mathematical abstraction of the interactions between a system and its environment. A process that lasts indefinitely may be specified recursively. Processes may be assembled into systems; they may execute concurrently, or communicate with each other. Process communication may be synchronized, and this takes the form of one process outputting a message simultaneously to another process inputting a message. Resources may be shared among several processes. Process calculi such as CSP (Hoare 1985) and CCS (Milner 1989) have been developed and they enrich the understanding of communication and concurrency, and they obey several mathematical laws.

The expression $(a \mathrel{?} P)$ in CSP describes a process which first engages in event a, and then behaves as process P. A recursive definition is written as $(\mu X) \bullet F(X)$ and an example of a simple chocolate vending machine is:

$$\text{VMS} = \mu X : \{\text{coin}, \text{choc}\} \cdot (\text{coin} \mathrel{?} (\text{choc} \mathrel{?} X))$$

The simple vending machine has an alphabet of two symbols, namely, *coin* and *choc*. The behaviour of the machine is that a coin is entered into the machine, and then a chocolate selected and provided, and the machine is ready for further use. CSP processes use channels to communicate values with their environment, and input on channel c is denoted by $(c\mathrel{?}.x\, P_x)$. This describes a process that accepts any value x on channel c, and then behaves as process P_x. In contrast, $(c\mathord{!}e\ P)$ defines a process which outputs the expression e on channel c and then behaves as process P.

The π-calculus is a process calculus based on names. Communication between processes takes place between known channels, and the name of a channel may be passed over a channel. There is no distinction between channel names and data values in the π-calculus. The output of a value v on channel a is given by $\bar{a}v$; i.e. output is a negative prefix. Input on a channel a is given by $a(x)$ and is a positive prefix. Private links or restrictions are denoted by $(x)P$.

21.14   Finite-State Machines

Warren McCulloch and Walter Pitts published early work on finite-state automata in 1943. Moore and Mealy developed this work further, and these machines are referred to as the '*Moore machine*' and the '*Mealy machine*'.

A Finite-State Machine (FSM) is an abstract mathematical machine that consists of a finite number of states. It includes a start state q_0 in which the machine is in initially; a finite set of states Q; an input alphabet Σ; a state transition function δ; and a set of final accepting states F (where $F \subseteq Q$).

The state transition function takes the current state and an input and returns the next state. It provides rules that define the action of the machine for each input, and it may be extended to provide output as well as a state transition. State diagrams are used to represent finite-state machines, and each state accepts a finite number of inputs.

A finite-state machine may be deterministic or non-deterministic, and a *deterministic machine* (Fig. 21.1) changes to exactly one state for each input transition, whereas a *non-deterministic machine* may have a choice of states to move to for a particular input.

Finite-state automata can compute only very primitive functions and are not an adequate model for computing. There are more powerful automata such as the *Turing machine* that is essentially a finite automaton with an infinite storage (memory). Anything that is computable is computable by a Turing machine.

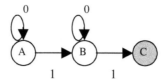

Fig. 21.1 Deterministic finite-state machine

The Turing machine provides a mathematical abstraction of computer execution and storage, as well as providing a mathematical definition of an algorithm. Automata theory is discussed in more detail in Chap. 23.

21.15   The Parnas Way

Parnas has been influential in the computing field, and his ideas on the specification, design, implementation, maintenance and documentation of computer software remain important. He advocates a solid engineering approach and argues that the role of the engineer is to apply scientific principles and mathematics to design and develop products. He argues that computer scientists need to be educated as engineers to ensure that they have the appropriate background to build software correctly.

His tabular expressions were used for the specification of the requirements of the A-7 aircraft for the US Navy, and his mathematical inspections were used to verify the correctness of the shutdown software at the Darlington Nuclear power plant in Canada. His contributions to software engineering include (Table 21.2).

21.16   Model Checking

Model checking is an automated technique such that given a finite-state model of a system and a formal property, (expressed in temporal logic) then it systematically checks whether the property is true or false in a given state in the model. It is an effective techniques to identify potential design errors, and it increases the confidence in the correctness of the system design. Model checking is a highly effective verification technology and is widely used in the hardware and software fields. It has been employed in the verification of microprocessors; in security protocols; in the transportation sector (trains); and in the verification of software in the space sector.

Model checking is a formal verification technique based on graph algorithms and formal logic. It allows the desired behaviour (specification) of a system to be verified, and its approach is to employ a suitable model of the system and to carry

Table 21.2 Parnas's contributions to software engineering

Area	Contribution
Tabular expressions	These are mathematical tables for specifying requirements and enable complex predicate logic expressions to be represented in a simpler form
Mathematical documentation	He advocates the use of precise mathematical documentation for requirements and design
Requirements specification	He advocates the use of mathematical relations to specify the requirements precisely
Software design	He developed *information hiding* that is used in object-oriented design (it is surprising that many in the object-oriented world seem unaware that information hiding goes back to the early 1970s and many have never heard of Parnas) and allows software to be designed for change. Every information-hiding module has an interface that provides the only means to access the services provided by the modules. The interface hides the module's implementation
Software inspections	His approach requires the reviewers to take an active part in the inspection. They are provided with a list of questions by the author and their analysis involves the production of mathematical table to justify the answers
Predicate logic	He developed an extension of the predicate calculus to deal with partial functions, and it preserves the classical two-valued logic when dealing with undefined values

out a systematic and exhaustive inspection of all states of the model to verify that the desired properties are satisfied. These properties are generally safety properties such as the absence of deadlock, request-response properties and invariants. Its systematic search determines whether a given system model truly satisfies a particular property or not. Model checking is discussed in more detail in Chap. 24.

21.17 Usability of Formal Methods

There are practical difficulties associated with the industrial use of formal methods. It seems to be assumed that programmers and customers are willing to become familiar with the mathematics used in formal methods, but this is true in only some domains.[10] It is usually possible to get a developer to learn a formal method, as a programmer has some experience of mathematics and logic. However, it is more difficult to get a customer to learn a formal method, and this makes it more difficult to perform a rigorous validation of the formal specification.

This often means that often a formal specification of the requirements and an informal definition of the requirements using a natural language are maintained. It is essential that both of these are consistent and that there is a rigorous validation of

---

[10]The domain in which the software is being used will influence the willingness or otherwise of the customers to become familiar with the mathematics required. There appears to be little interest in mainstream software engineering, and their perception is that formal methods are unusable. However, in there is a greater interest in the mathematical approach in the safety-critical field.

the formal specification. Otherwise, if the programmer proves the correctness of the code with respect to the formal specification, and the formal specification is incorrect, then the formal development of the software is incorrect. There are several techniques to validate a formal specification including:

– proof that the formal specification satisfies key properties,
– software inspections to compare formal specification and informal set of requirements and
– specification animation to validate the formal specification.

Formal methods are perceived as being difficult to use, and of providing limited value in mainstream software engineering. Programmers receive education in mathematics as part of their studies, but many never use formal methods again once they take an industrial position. Some of the reasons for this are:

– the notation is not intuitive,
– It is difficult to write a formal specification,
– Validation of a formal specification is difficult,
– Refinement and proof are difficult and
– Limited tool support.

It is important to investigate ways by which formal methods can be made more usable to software engineers, and to design more usable notations and better tools to support the process. Practical training and coaching to employees can help. Some of the characteristics of a usable formal method are:

– A formal method should be intuitive.
– It should have tool support.
– A formal method should be teachable.
– It should be able to adapt to change.
– The technology transfer path should be defined.
– A formal method should be cost-effective.

21.18 Review Questions

1. What are formal methods and describe their potential benefits? How essential is tool support?
2. What is stepwise refinement and how realistic is it in mainstream software engineering?

3. Discuss Parnas's criticisms of formal methods and discuss whether his views are valid.
4. Discuss the industrial applications of formal methods and which areas have benefited most from their use? What problems have arisen?
5. Describe a technology transfer path for the deployment of formal methods in an organization.
6. Explain the difference between the model-oriented approach and the axiomatic approach.
7. Discuss the nature of proof in formal methods and tools to support proof.
8. Discuss the Vienna Development Method and explain the difference between standard VDM and VDM$^{\clubsuit}$.
9. Discuss Z and B. Describe the tools in the B-Toolkit.
10. Discuss process calculi such as CSP, CCS or π–calculus.

21.19 Summary

This chapter discussed formal methods that offer a mathematical approach to the development of high-quality software. They are useful in the safety-critical field. They consist of a formal specification language; a methodology for formal software development; and a set of tools to support the syntax checking of the specification, as well as the proof of properties of the specification.

Formal methods may be model-oriented or axiomatic-oriented. The model-oriented approach includes formal methods such as VDM, Z and B. The axiomatic approach includes the process calculi such as CSP, CCS and the π calculus. VDM was developed at the IBM lab in Vienna and has been used in academia and industry. CSP was developed by Hoare and CCS by Milner.

Formal methods allow questions to be asked and answered about what the system does independently of the implementation. They offer a way to debug the requirements and to show that certain desirable properties are true of the specification, whereas certain undesirable properties are absent.

The use of formal methods generally leads to more robust software and to increased confidence in its correctness. There are challenges involved in the deployment of formal methods, as mathematical techniques may be a culture shock to staff. The usability of existing formal methods was considered, and reasons for their perceived difficulty considered. The characteristics of a usable formal method were explored.

There are various tools to support formal methods including syntax checkers; specialized editors; tools to support refinement; automated code generators to generate a high-level language corresponding to the specification; theorem provers; and specification animation tools for simulation of the specification.

References

Bjørner D, Jones C (1978) The vienna development method. The Meta language. Lecture notes in computer science (61), Springer

Bjørner D, Jones C (1982) Formal specification and software development. International series in computer science, Prentice Hall

Brown MJD (1990) Rational for the development of the U.K. defence standards for safety critical software. In: Compass Conference

Gerhart S, Craighen D, Ralston T (1994) Experience with formal methods in safety critical systems. IEEE Software 11(1): 21–28

Gries D (1981) The Science of programming, Springer, Berlin

Hinchey M, Bowen J (1995) Applications of formal methods. International series in computer science, Prentice Hall

Hoare CAR (1985) Communicating sequential processes. International series in computer science, Prentice Hall

Hoare JP (1995) Application of the B method to CICS. In: Applications of formal methods. International series in computer science, Prentice Hall

Kuhn T (1970) The structure of scientific revolutions, University of Chicago Press

Lakatos I (1976) Proof and refutations. The logic of mathematical discovery, Cambridge University Press

Mac An Airchinnigh M (1990) Computation models and computing. PhD thesis, Department of computer science, Trinity college Dublin

McDonnell E (1994) MSc. thesis, Department of computer science, Trinity college, Dublin

Milner R et al (1989) A calculus of mobile processes. Part 1. LFCS Report Series. ECS-LFCS-89-85, Department of computer science, University of Edinburgh

MOD (1991a) 00-55 (Part 1)/Issue 1. The procurement of safety critical software in defence equipment. Part 1: requirements. Ministry of Defence, Interim Defence Standard, UK

MOD UK (1991b) 00-55 (Part 2)/Issue 1. The procurement of safety critical software in defence equipment. Part 2: guidance. Ministry of Defence, Interim Defence Standard, UK

O' Regan G (2006) Mathematical approaches to software quality, Springer

O'Regan (2017) Concise guide to formal methods, Springer

Polya G (1957) How to solve it. A new aspect of mathematical method, Princeton University Press

Spivey JM (1992) The Z notation. A reference manual. International series in computer science, Prentice Hall

Tierney M (1991) The evolution of Def Stan 00-55 and 00-56. An intensification of the formal methods debate in the UK, Research Centre for Social Sciences, University of Edinburgh

Woodcock J, Larsen PG, Bicarregui J, Fitzgerald J (2009) Formal methods: practice and experience, ACM Computer Surveys

Z Formal Specification Language

22

Key Topics

Sets
Relations and functions
Bags and sequences
Data reification
Refinement
Schema calculus
Proof in Z

22.1 Introduction

Z is a formal specification language based on Zermelo set theory. It was developed at the Programming Research Group at Oxford University in the early 1980s (Diller 1990) and became an ISO standard in 2002. Z specifications are mathematical and employ a classical two-valued logic. The use of mathematics ensures precision and allows inconsistencies and gaps in the specification to be identified. Theorem provers may be employed to demonstrate that the software implementation meets its specification.

Z is a '*model-oriented*' approach with an explicit model of the state of an abstract machine given, and operations are defined in terms of this state. Its mathematical notation is used for formal specification, and the schema calculus is used to structure the specifications. The schema calculus is visually striking and

© Springer Nature Switzerland AG 2020

G. O'Regan, *Mathematics in Computing*, Undergraduate Topics
in Computer Science, https://doi.org/10.1007/978-3-030-34209-8_22

$$\begin{array}{|l}
\text{--}SqRoot\text{------} \\
\text{num?, } root! : \mathbb{R} \\
\hline
num? \geq 0 \\
root!^2 = num? \\
root! \geq 0 \\
\hline
\end{array}$$

Fig. 22.1 Specification of positive square root

consists essentially of boxes, with these boxes or schemas used to describe operations and states. The schemas may be used as building blocks and combined with other schemas. The simple schema in Fig. 22.1 is the specification of the positive square root of a real number.

The schema calculus is a powerful means of decomposing a specification into smaller pieces or schemas. This helps to make Z specifications highly readable, as each individual schema is small in size and self-contained. Exception handling is addressed by defining schemas for the exception cases. These are then combined with the original operation schema. Mathematical data types are used to model the data in a system, these data types obey mathematical laws. These laws enable simplification of expressions and are useful with proofs.

Operations are defined in a precondition/postcondition style. A precondition must be true before the operation is executed, and the postcondition must be true after the operation has executed. The precondition is implicitly defined within the operation. Each operation has an associated proof obligation to ensure that if the precondition is true, then the operation preserves the system invariant. The system invariant is a property of the system that must be true at all times. The initial state itself is, of course, required to satisfy the system invariant.

The precondition for the specification of the square root function in Fig. 22.1 is that $num? \geq 0$; i.e. the function $SqRoot$ may be applied to positive real numbers only. The postcondition for the square root function is $root!^2 = num?$ and $root! \geq 0$. That is, the square root of a number is positive and its square gives the number. Postconditions employ a logical predicate which relates the pre-state to the post-state, with the post-state of a variable being distinguished by priming the variable, e.g. v'.

Z is a typed language and whenever a variable is introduced its type must be given. A type is simply a collection of objects, and there are several standard types in Z. These include the natural numbers $\mathbb{N}$, the integers $\mathbb{Z}$ and the real numbers $\mathbb{R}$. The declaration of a variable x of type X is written x: X. It is also possible to create your own types in Z.

Various conventions are employed within Z specification, for example, $v?$ indicates that v is an input variable; $v!$ indicates that v is an output variable. The variable $num?$ is an input variable and $root!$ is an output variable for the square root example in Fig. 22.1. The notation Ξ in a schema indicates that the operation Op does not affect the state, whereas the notation Δ in the schema indicates that Op is an operation that affects the state.

-Library————
on-shelf, missing, borrowed : $\mathbb{P}$ Bkd-Id
————
on-shelf $\cap$ missing $= \emptyset$
on-shelf $\cap$ borrowed $= \emptyset$
borrowed $\cap$ missing $= \emptyset$

Fig. 22.2 Specification of a library system

Many of the data types employed in Z have no counterpart in standard programming languages. It is therefore important to identify and describe the concrete data structures that ultimately will represent the abstract mathematical structures. As the concrete structures may differ from the abstract, the operations on the abstract data structures may need to be refined to yield operations on the concrete data that yield equivalent results. For simple systems, direct refinement (i.e. one step from abstract specification to implementation) may be possible; in more complex systems, deferred refinement[1] is employed, where a sequence of increasingly concrete specifications are produced to yield the executable specification. The schema calculus is employed for combining schemas to make larger specifications and is discussed later in the chapter.

Example 22.1 The following is a Z specification to borrow a book from a library system. The library is made up of books that are on the shelf; books that are borrowed; and books that are missing (Fig. 22.2). The specification models a library with sets representing books on the shelf, on loan or missing. These are three mutually disjoint subsets of the set of books Bkd-Id.

The system state is defined in the *Library* schema in Fig. 22.2, and operations such as *Borrow* and *Return* affect the state. The *Borrow* operation is specified in Fig. 22.3.

The notation $\mathbb{P}$Bkd-Id is used to represent the power set of Bkd-Id (i.e. the set of all subsets of Bkd-Id). The disjointness condition for the library is expressed by the requirement that the pairwise intersection of the subsets *on-shelf, borrowed, missing* is the empty set.

The precondition for the *Borrow* operation is that the book must be available on the shelf to borrow. The postcondition is that the borrowed book is added to the set of borrowed books and is removed from the books on the shelf.

---

[1]Step-wise refinement involves producing a sequence of increasingly more concrete specifications until eventually the executable code is produced. Each refinement step has associated proof obligations to prove that it is valid.

―Borrow―――――
Δ Library
b? :Bkd-Id
⎯⎯⎯⎯⎯
b? ∈ on-shelf
on-shelf' = on-shelf \ {b?}
borrowed' = borrowed ∪ {b?}

Fig. 22.3 Specification of borrow operation

Z has been successfully applied in industry including the CICS project at IBM Hursley in the UK.[2] Next, we describe key parts of Z including sets, relations, functions, sequences and bags.

22.2 Sets

Sets were discussed in Chap. 3 and this section focuses on their use in Z. Sets may be enumerated by listing all of their elements. Thus, the set of all even natural numbers less than or equal to 10 is

$$\{2, 4, 6, 8, 10\}$$

Sets may be created from other sets using set comprehension, i.e. stating the properties that its members must satisfy. For example, the set of even natural numbers less than 10 is given by set comprehension as

$$\{n : \mathbb{N} \mid n \neq 0 \wedge n < 10 \wedge n \bmod 2 = 0 \cdot n\}$$

There are three main parts to the set comprehension above. The first part is the signature of the set and this is given by $n: \mathbb{N}$ above. The first part is separated from the second part by a vertical line. The second part is given by a predicate, and for this example, the predicate is $n \neq 0 \wedge n < 10 \wedge n \bmod 2 = 0$. The second part is separated from the third part by a bullet. The third part is a term, and for this example, it is simply n. The term is often a more complex expression, e.g. $\log(n^2)$.

In mathematics, there is just one empty set. However, since Z is a typed set theory, there is an empty set for each type of set. Hence, there are an infinite number of empty sets in Z. The empty set is written as ∅ [X] where X is the type of the empty set. In practice, X is omitted when the type is clear.

---

[2]This project claimed a 9% increase in productivity attributed to the use of formal methods.

Various operations on sets such as union, intersection, set difference and symmetric difference are employed in Z. The power set of a set X is the set of all subsets of X and is denoted by $\mathbb{P}\,X$. The set of non-empty subsets of X is denoted by $\mathbb{P}_1 X$ where

$$\mathbb{P}_1 X == \{U : \mathbb{P}\,X \mid U \neq \emptyset[X]\}$$

A finite set of elements of type X (denoted by $F\,X$) is a subset of X that cannot be put into a one to one correspondence with a proper subset of itself. This is defined formally as

$$F\,X == \{U : \mathbb{P}\,X \mid \neg \exists V: \mathbb{P}\,U \bullet V \neq U \wedge (\exists f: V \rightarrowtail U)\}$$

The expression $f: V \rightarrowtail U$ denotes that f is a bijection from U to V and injective, surjective and bijective functions were discussed in Chap. 3.

The fact that Z is a typed language means that whenever a variable is introduced (e.g. in quantification with $\forall$ and $\exists$) it is first declared. For example, $\forall j{:}J \cdot P \Rightarrow Q$. There is also the unique existential quantifier $\exists_1\, j{:}J \mid P$ which states that there is exactly one j of type J that has property P.

22.3 Relations

Relations were discussed in Chap. 3 and are used extensively in Z. A relation R between X and Y is any subset of the Cartesian product of X and Y; i.e. $R \subseteq (X \times Y)$, and a relation in Z is denoted by $R: X \leftrightarrow Y$. The notation $x \mapsto y$ indicates that the pair $(x, y) \in R$.

Consider, the relation *home_owner: Person $\leftrightarrow$ Home* that exists between people and their homes. An entry *daphne $\mapsto$ mandalay $\in$ home_owner* if *daphne* is the owner of *mandalay*. It is possible for a person to own more than one home:

$$rebecca \mapsto nirvana \in home\_owner$$
$$rebecca \mapsto tivoli \in home\_owner$$

It is possible for two people to share ownership of a home:

$$rebecca \mapsto nirvana \in home\_owner$$
$$lawrence \mapsto nirvana \in home\_owner$$

There may be some people who do not own a home, and there is no entry for these people in the relation *home_owner*. The type *Person* includes every possible person, and the type *Home* includes every possible home. The domain of the relation *home_owner* is given by

$$x \in \text{dom } home\_owner \Leftrightarrow \exists h : Home \cdot x \mapsto h \in home\_owner$$

The range of the relation *home_owner* is given by

$$h \in \text{ran } home\_owner \Leftrightarrow \exists x : Person \cdot x \mapsto h \in home\_owner$$

The composition of two relations *home_owner: Person* $\leftrightarrow$ *Home* and *home_-value: Home* $\leftrightarrow$ *Value* yields the relation *owner_wealth: Person* $\leftrightarrow$ *Value* and is given by the relational composition *home_owner; home_value* where

$$p \mapsto v \in home\_owner; home\_value \Leftrightarrow$$
$$(\exists h : Home \cdot p \mapsto h \in home\_owner \wedge h \mapsto v \in home\_value)$$

The relational composition may also be expressed as

$$owner\_wealth = home\_value \text{ o } home\_owner$$

The union of two relations often arises in practice. Suppose a new entry *aisling* $\mapsto$ *muckross* is to be added. Then this is given by

$$home\_owner' = home\_owner \cup \{aisling \mapsto muckross\}$$

Suppose that we are interested in knowing all females who are house owners. Then we restrict the relation *home_owner* so that the first element of all ordered pairs have to be female. Consider *female*: $\mathbb{P}$ *Person* with $\{aisling, rebecca\} \subseteq female$.

$$home\_owner = \{aisling \mapsto muckross, rebecca \mapsto nirvana,$$
$$lawrence \mapsto nirvana\}$$

$$female \lhd home\_owner = \{aisling \mapsto muckross, rebecca \mapsto nirvana\}$$

That is, *female* $\lhd$ *home_owner* is a relation that is a subset of *home_owner*, and the first element of each ordered pair in the relation is female. The operation $\lhd$ is termed domain restriction and its fundamental property is

$$x \mapsto y \in U \lhd R \Leftrightarrow (x \in U \wedge x \mapsto y \in R\}$$

where R: X $\leftrightarrow$ Y and U: $\mathbb{P}$ X.

There is also a domain anti-restriction (subtraction) operation and its fundamental property is

$$x \mapsto y \in U \lhd\!\!\!- R \Leftrightarrow (x \notin U \wedge x \mapsto y \in R\}$$

where R: X $\leftrightarrow$ Y and U: $\mathbb{P}$X.

There are also range restriction (the $\triangleright$ operator) and the range anti-restriction operator (the $\triangleright\!\!\!\!-$ operator). These are discussed in Diller (1990).

22.4 Functions

A function (Diller 1990) is an association between objects of some type X and objects of another type Y such that given an object of type X, there exists only one object in Y associated with that object. A function is a set of ordered pairs where the first element of the ordered pair has at most one element associated with it. A function is therefore a special type of relation, and a function may be *total* or *partial*.

A total function has exactly one element in Y associated with each element of X, whereas a partial function has at most one element of Y associated with each element of X (there may be elements of X that have no element of Y associated with them).

A partial function from X to $Y (f : X \nrightarrow Y)$ is a relation $f : X \leftrightarrow Y$ such that:

$$\forall x : X; y, z : Y \cdot (x \mapsto y \in f \wedge x \mapsto z \in f \Rightarrow y = z)$$

The association between x and y is denoted by $f(x) = y$, and this indicates that the value of the partial function f at x is y. A total function from X to Y (denoted $f: X \rightarrow Y$) is a partial function such that every element in X is associated with some value of Y.

$$f : X \rightarrow Y \Leftrightarrow f : X \nrightarrow Y \wedge \operatorname{dom} f = X$$

Clearly, every total function is a partial function but not vice versa.

One operation that arises quite frequently in specifications is the function override operation. Consider the following specification of a temperature map:

$$
\begin{array}{|l}
\hline
\text{–TempMap––––––} \\
CityList : \mathbb{P}City \\
temp : City \nrightarrow Z \\
\hline
\operatorname{dom} temp = CityList \\
\hline
\end{array}
$$

Suppose the temperature map is given by $temp = \{Cork \mapsto 17, Dublin \mapsto 19, London \mapsto 15\}$. Then consider the problem of updating the temperature map if a new temperature reading is made in Cork, e.g. $\{Cork \mapsto 18\}$. Then the new temperature chart is obtained from the old temperature chart by function override to yield $\{Cork \mapsto 18, Dublin \mapsto 19, London \mapsto 15\}$. This is written as

$$temp' = temp \oplus \{Cork \mapsto 18\}$$

The function override operation combines two functions of the same type to give a new function of the same type. The effect of the override operation is that the entry $\{Cork \mapsto 17\}$ is removed from the temperature chart and replaced with the entry $\{Cork \mapsto 18\}$.

Suppose $f, g : X \nrightarrow Y$ are partial functions then $f \oplus g$ is defined, and indicates that f is overridden by g. It is defined as follows:

$$(f \oplus g)(x) = g(x) \text{ where } x \in \text{dom } g$$
$$(f \oplus g)(x) = f(x) \text{ where } x \notin \text{dom } g \wedge x \in \text{dom} f$$

This may also be expressed (using domain anti-restriction) as

$$f \oplus g = ((\text{dom } g) \ntriangleleft f) \cup g$$

There is notation in Z for injective, surjective and bijective functions. An injective function is one to one, i.e.

$$f(x) = f(y) \Rightarrow x = y$$

A surjective function is onto, i.e.

$$\text{Given } y \in Y, \exists x \in X \text{ such that} f(x) = y$$

A bijective function is one to one and onto, and it indicates that the sets X and Y can be put into one to one correspondence with one another. Z includes lambda calculus notation to define functions (λ-calculus was discussed in Chap. 12). For example, the function cube $== \lambda x : \mathbf{N} \cdot x * x * x$. Function composition $f; g$ is similar to relational composition.

22.5   Sequences

The type of all sequences of elements drawn from a set X is denoted by seq X. Sequences are written as $\langle x_1, x_2, \ldots . x_n \rangle$ and the empty sequence is denoted by $\langle \rangle$. Sequences may be used to specify the changing state of a variable over time, with each element of the sequence representing the value of the variable at a discrete time instance.

Sequences are functions and a sequence of elements drawn from a set X is a finite function from the set of natural numbers to X. A partial finite function f from X to Y is denoted by $f : X \twoheadrightarrow Y$. A finite sequence of elements of X is given by a finite function $f : \mathbf{N} \twoheadrightarrow X$, and the domain of the function consists of all numbers between 1 and $\# f$ (where #f is the cardinality of f). It is defined formally as

$$\text{seq } X == \{f : \mathbf{N} \twoheadrightarrow X \mid \text{dom } f = 1..\#f \bullet f\}$$

The sequence $\langle x_1, x_2, \ldots x_n \rangle$ above is given by

$$\{1 \mapsto x_1, 2 \mapsto x_2, \ldots \ldots n \mapsto x_n\}$$

There are various functions to manipulate sequences. These include the sequence concatenation operation. Suppose $\sigma = \langle x_1, x_2, \ldots x_n \rangle$ and $\tau = \langle y_1, y_2, \ldots y_m \rangle$ then:

$$\sigma ^\frown \tau = \langle x_1, x_2, \ldots x_n, y_1, y_2, \ldots y_m \rangle$$

The head of a non-empty sequence gives the first element of the sequence.

$$\text{head } \sigma = \text{head} \langle x_1, x_2, \ldots x_n \rangle = x_1$$

The tail of a non-empty sequence is the same sequence except that the first element of the sequence is removed.

$$\text{tail } \sigma = \text{tail} \langle x_1, x_2, \ldots x_n \rangle = \langle x_2, \ldots x_n \rangle$$

Suppose $f: X \rightarrow Y$ and a sequence σ: seq X then the function map applies f to each element of σ:

$$\text{map} f \, \sigma = \text{map} f \langle x_1, x_2, \ldots x_n \rangle = \langle f(x_1), f(x_2), \ldots f(x_n) \rangle$$

The map function may also be expressed via function composition as

$$\text{map} f \, \sigma = \sigma; f$$

The reverse order of a sequence is given by the rev function:

$$\text{rev } \sigma = \text{rev} \langle x_1, x_2, \ldots x_n \rangle = \langle x_n, \ldots x_2, x_1 \rangle$$

22.6 Bags

A bag is similar to a set except that there may be multiple occurrences of each element in the bag. A bag of elements of type X is defined as a partial function from the type of the elements of the bag to positive whole numbers. The definition of a bag of type X is:

$$\text{bag } X == X \nrightarrow \mathbb{N}_1$$

For example, a bag of marbles may contain three blue marbles, two red marbles and one green marble. This is denoted by $B = [b, b, b, g, , r, r]$. The bag of marbles is thus denoted by

$$
\begin{array}{|l}
\hline
-\Delta Vending\ Machine\text{------} \\
stock : \text{bag}\ Good \\
price : Good \to \mathbb{N}_1 \\
\hline
\text{dom } stock \subseteq \text{dom } price \\
\\
\hline
\end{array}
$$

Fig. 22.4 Specification of vending machine using bags

$$\text{bag } Marble == Marble \nrightarrow \mathbb{N}_1$$

The function count determines the number of occurrences of an element in a bag. For the example above, count $Marble\ b = 3$, and count $Marble\ y = 0$ since there are no yellow marbles in the bag. This is defined formally as

$$
\begin{aligned}
&\text{count bag } X\ y = 0 &&y \notin \text{bag } X \\
&\text{count bag } X\ y = (\text{bag } X)(y) &&y \in \text{bag } X
\end{aligned}
$$

An element y is in bag X if and only if y is in the domain of bag X.

$$y \text{ in bag } X \Leftrightarrow y \in \text{dom}\,(\text{bag } X)$$

The union of two bags of marbles $B_1 = [b, b, b, g, , r, r]$ and $B_2 = [b, g, , r, y]$ is given by $B_1 \uplus B_2 = [b, b, b, b, g, g, r, r, r, y]$. It is defined formally as

$$
\begin{aligned}
&(B_1 \uplus B_2)(y) = B_2(y) &&y \notin \text{dom } B_1 \wedge y \in \text{dom } B_2 \\
&(B_1 \uplus B_2)(y) = B_1(y) &&y \in \text{dom } B_1 \wedge y \notin \text{dom } B_2 \\
&(B_1 \uplus B_2)(y) = B_1(y) + B_2(y) &&y \in \text{dom } B_1 \wedge y \in \text{dom } B_2
\end{aligned}
$$

A bag may be used to record the number of occurrences of each product in a warehouse as part of an inventory system. It may model the number of items remaining for each product in a vending machine (Fig. 22.4).

The operation of a vending machine would require other operations such as identifying the set of acceptable coins, checking that the customer has entered sufficient coins to cover the cost of the good, returning change to the customer and updating the quantity on hand of each good after a purchase [see (Diller 1990)].

22.7 Schemas and Schema Composition

The schemas in Z are visually striking and the specification is presented in two-dimensional graphic boxes. Schemas are used for specifying states and state transitions, and they employ notation to represent the before and after state (e.g.

s and s' where s' represents the after state of s). The schemas group all relevant information that belongs to a state description.

There are a number of useful schema operations such as schema inclusion, schema composition and the use of propositional connectives to link schemas together. The Δ convention indicates that the operation affects the state, whereas the Ξ convention indicates that the state is not affected. These operations and conventions allow complex operations to be specified concisely and assist with the readability of the specification. Schema composition is analogous to relational composition and allows new schemas to be derived from existing schemas.

A schema name S_1 may be included in the declaration part of another schema S_2. The effect of the inclusion is that the declarations in S_1 are now part of S_2, and the predicates of S_1 are S_2 are joined together by conjunction. If the same variable is defined in both S_1 and S_2, then it must be of the same type.

$$
\begin{array}{|l}
\hline
-S_1\text{------} \\
x,y : \mathbb{N} \\
\hline
x + y > 2 \\
\hline
\end{array}
\qquad
\begin{array}{|l}
\hline
-S_2\text{------} \\
S_1 ; z : \mathbb{N} \\
\hline
z = x + y \\
\hline
\end{array}
$$

The result is that S_2 includes the declarations and predicates of S_1 (Fig. 22.5).

Two schemas may be linked by propositional connectives such as $S_1 \wedge S_2$, $S_1 \vee S_2$, $S_1 \rightarrow S_2$, and $S_1 \leftrightarrow S_2$. The schema $S_1 \vee S_2$ is formed by merging the declaration parts of S_1 and S_2, and then combining their predicates by the logical $\vee$ operator. For example, $S = S_1 \vee S_2$ yields (Fig. 22.6).

Schema inclusion and the linking of schemas use normalization to convert sub-types to maximal types, and predicates are employed to restrict the maximal type to the sub-type. This involves replacing declarations of variables (e.g. $u: 1\ldots35$

$$
\begin{array}{|l}
\hline
-S_2\text{------} \\
x,y : \mathbb{N} \\
z : \mathbb{N} \\
\hline
x + y > 2 \\
z = x + y \\
\hline
\end{array}
$$

Fig. 22.5 Schema inclusion

$$
\begin{array}{|l}
\hline
-S\text{------} \\
x,y : \mathbb{N} \\
z : \mathbb{N} \\
\hline
x + y > 2 \vee z = x + y \\
\hline
\end{array}
$$

Fig. 22.6 Merging schemas ($S_1 \vee S_2$)

with u: Z, and adding the predicate $u > 0$ and $u < 36$ to the predicate part of the schema).

The Δ and Ξ conventions are used extensively, and the notation Δ *TempMap* is used in the specification of schemas that involve a change of state. The notation Δ *TempMap* represents:

$$\Delta\,TempMap = TempMap \land TempMap'$$

The longer form of Δ *TempMap* is written as

$$
\begin{array}{|l}
\hline
-\Delta TempMap\text{———} \\
CityList,\ CityList' : \mathbb{P}\ \ City \\
temp,\ temp' : City \nrightarrow Z \\
\hline
\mathrm{dom}\ temp = CityList \\
\mathrm{dom}\ temp' = CityList' \\
\hline
\end{array}
$$

The notation Ξ *TempMap* is used in the specification of operations that do not involve a change to the state.

$$
\begin{array}{|l}
\hline
-\Xi\ TempMap\text{———} \\
\Delta TempMap \\
\hline
CityList = CityList' \\
temp = temp' \\
\hline
\end{array}
$$

Schema composition is analogous to relational composition, and it allows new specifications to be built from existing ones. It allows the after state variables of one schema to be related with the before variables of another schema. The composition of two schemas S and T (S; T) is described in detail in Diller (1990) and involves four steps (Table 22.1).

The example in Fig. 22.7 should make schema composition clearer. Consider the composition of S and T where S and T are defined as follows:

Table 22.1 Schema composition

Step	Procedure
1.	Rename all *after* state variables in S to something new: $S[s^+/s']$
2.	Rename all *before* state variables in T to the same new thing, i.e. $T[s^+/s]$
3.	Form the conjunction of the two new schemas $S[s^+/s'] \land T[s^+/s]$
4.	Hide the variable introduced in steps 1 and 2 $S;T = (S[s^+/s'] \land T[s^+/s])\backslash(s^+)$

$$\left|\begin{array}{l} -S_1 \wedge T_1 \text{------} \\ x,x^+,x',y? : \mathbb{N} \\[4pt] \hline x^+ = y? - 2 \\ x' = x^+ + 1 \\ \hline \end{array}\right. \qquad \left|\begin{array}{l} -S ; T \text{------} \\ x, x', y? : \mathbb{N} \\[4pt] \hline \exists x^+ : \mathbb{N} \bullet \\ \quad (x^+ = y? - 2 \\ \quad\ \ x' = x^+ + 1) \\ \hline \end{array}\right.$$

Fig. 22.7 Schema composition

$$\left|\begin{array}{l} -S \text{------} \\ x,x',y? : \mathbb{N} \\[4pt] \hline x' = y? \text{ - } 2 \\ \hline \end{array}\right. \qquad \left|\begin{array}{l} -T \text{------} \\ x,x' : \mathbb{N} \\[4pt] \hline x' = x + 1 \\ \hline \end{array}\right.$$

$$\left|\begin{array}{l} -S_1 \text{------} \\ x,x^+,y? : \mathbb{N} \\[4pt] \hline x^+ = y? \text{ - } 2 \\ \hline \end{array}\right. \qquad \left|\begin{array}{l} -T_1 \text{------} \\ x^+,x' : \mathbb{N} \\[4pt] \hline x' = x^+ + 1 \\ \hline \end{array}\right.$$

S_1 and T_1 represent the results of step 1 and step 2, with x' renamed to x^+ in S, and x renamed to x^+ in T. Step 3 and step 4 yield (Fig. 22.7).

Schema composition is useful as it allows new specifications to be created from existing ones.

22.8 Reification and Decomposition

A Z specification involves defining the state of the system and then specifying the required operations. The Z specification language employs many constructs that are not part of conventional programming languages, and a Z specification is therefore not directly executable on a computer. A programmer implements the formal specification, and mathematical proof may be employed to prove that a program meets its specification.

Often, there is a need to write an intermediate specification that is between the original Z specification and the eventual program code. This intermediate specification is more algorithmic and uses less abstract data types than the Z specification. The intermediate specification is termed the design and the design needs to be correct with respect to the specification, and the program needs to be correct with respect to the design. The design is a refinement (reification) of the state of the specification, and the operations of the specification have been decomposed into those of the design.

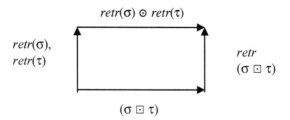

Fig. 22.8 Refinement commuting diagram

The representation of an abstract data type such as a set by a sequence is termed data reification, and data reification is concerned with the process of transforming an abstract data type into a concrete data type. The abstract and concrete data types are related by the retrieve function, and the retrieve function maps the concrete data type to the abstract data type. There are typically several possible concrete data types for a particular abstract data type (i.e. refinement is a relation), whereas there is one abstract data type for a concrete data type (i.e. retrieval is a function). For example, sets are often reified to unique sequences; and clearly, more than one unique sequence can represent a set, whereas a unique sequence represents exactly one set.

The operations defined on the concrete data type are related to the operations defined on the abstract data type. That is, the commuting diagram property is required to hold (Fig. 22.8). That is, for an operation $\square$ on the concrete data type to correctly model the operation $\odot$ on the abstract data type the following diagram must commute, and the commuting diagram property requires proof. That is, it is required to prove that:

$$ret(\sigma\square\tau) = (ret\ \sigma) \odot (ret\ \tau)$$

In Z, the refinement and decomposition is done with schemas. It is required to prove that the concrete schema is a valid refinement of the abstract schema, and this gives rise to a number of proof obligations. It needs to be proved that the initial states correspond to one another, and that each operation in the concrete schema is correct with respect to the operation in the abstract schema, and also that it is applicable (i.e. whenever the abstract operation may be performed the concrete operation may also be performed).

22.9 Proof in Z

We discussed the nature of theorem proving in Chap. 18. Mathematicians perform rigorous proof of theorems using technical and natural language, whereas logicians employ formal proofs using propositional and predicate calculus. Formal proofs generally involve a long chain of reasoning with every step of the proof justified, whereas rigorous mathematical proofs involve precise reasoning using a mixture of natural and mathematical language. Rigorous proofs (Diller 1990) have been described as being analogous to high-level programming languages, whereas formal proofs are analogous to machine language.

A mathematical proof includes natural language and mathematical symbols, and often many of the tedious details of the proof are omitted. Many proofs in formal methods such as Z are concerned with cross-checking on the details of the specification, or on the validity of the refinement step, or proofs that certain properties are satisfied by the specification. There are often many tedious lemmas to be proved, and tool support is essential as proof by hand often contain errors or jumps in reasoning. Machine proofs provide extra confidence as every step in the proof is justified, and the proof of various properties about the programs increases confidence in its correctness.

22.10   Industrial Applications of *Z*

The Z specification language is one of the more popular formal methods, and it has been employed for the formal specification and verification of safety-critical software. IBM piloted the Z formal specification language on the CICS (Customer Information Control System) project at its plant in Hursley, England.

Rolls Royce and Associates (RRA) developed a lifecycle suitable for the development of safety-critical software, and the safety-critical lifecycle used Z for the formal specification and the CADiZ tool provided support for specification, and Ada was the target implementation language.

Logica employed Z for the formal verification of a smartcard-based electronic cash system (the Mondex smartcard) in the early 1990s. The smartcard had an 8-bit microprocessor, and the objective was to formally specify both the high-level abstract security policy model and the lower level concrete architectural design in Z, and to provide a formal proof of correspondence between the two.

Computer Management Group (CMG) employed Z for modelling data and operations as part of the formal specification of a movable barrier (the Maeslantkering) in the mid-1990s, which is used to protect the port of Rotterdam from flooding. The decisions on opening and closing of the barrier are based on meteorological data provided by the computer system, and the focus of the application of formal methods was to the decision-making subsystem and its interfaces to the environment.

22.11   Review Questions

1. Describe the main features of the Z specification language.
2. Explain the difference between $\mathbb{P}_1 X$, $\mathbb{P} X$ and *FX*.
3. Give an example of a set derived from another set using set comprehension. Explain the three main parts of set comprehension in Z.
4. Discuss the applications of Z and which areas have benefited most from their use? What problems have arisen?
5. Give examples to illustrate the use of domain and range restriction operators and domain and range anti-restriction operators with relations in Z.
6. Give examples to illustrate relational composition.
7. Explain the difference between a partial and total function, and give examples to illustrate function override.
8. Give examples to illustrate the various operations on sequences including concatenation, head, tail, map and reverse operations.
9. Give examples to illustrate the various operations on bags.
10. Discuss the nature of proof in Z and tools to support proof.
11. Explain the process of refining an abstract schema to a more concrete representation, the proof obligations that are generated, and the commuting diagram property.

22.12   Summary

Z is a formal specification language that was developed in the early 1980s at Oxford University in England. It has been employed in both industry and academia, and it was used successfully on the IBM's CICS project. Its specifications are mathematical, and this leads to more rigorous software development. Its mathematical approach allows properties to be proved about the specification, and any gaps or inconsistencies in the specification may be identified.

Z is a model-oriented approach, and an explicit model of the state of an abstract machine is given, and the operations are defined in terms of their effect on the state. Its main features include a mathematical notation that is similar to VDM and the schema calculus. The latter consists essentially of boxes and is used to describe operations and states.

The schema calculus enables schemas to be used as building blocks to form larger specifications. It is a powerful means of decomposing a specification into smaller pieces, and helps with the readability of Z specifications, as each schema is small in size and self-contained.

Z is a highly expressive specification language, and it includes notation for sets, functions, relations, bags, sequences, predicate calculus and schema calculus. Z specifications are not directly executable as many of its data types and constructs are not part of modern programming languages. Therefore, there is a need to refine the Z specification into a more concrete representation and prove that the refinement is valid.

Reference

Diller A (1990) Z. An introduction to formal methods. Wiley, England

Automata Theory

23

Key Topics

Finite-state automata
State transition table
Deterministic FSA
Non-deterministic FSA
Pushdown automata
Turing machine

23.1 Introduction

Automata theory is the branch of computer science that is concerned with the study of abstract machines and automata. These include finite-state machines, pushdown automata and Turing machines. Finite-state machines are abstract machines that may be in one of a finite number of states. These machines are in only one state at a time (current state), and the input symbol causes a transition from the current state to the next state. Finite-state machines have limited computational power due to memory and state constraints, but they have been applied to a number of fields including communication protocols, neurological systems and linguistics.

Pushdown automata have greater computational power than finite-state machines, and they contain extra memory in the form of a stack from which symbols may be pushed or popped. The state transition is determined from the current state of the machine, the input symbol and the element on the top of the stack. The action may be to change the state and/or push/pop an element from the stack.

© Springer Nature Switzerland AG 2020
G. O'Regan, *Mathematics in Computing*, Undergraduate Topics
in Computer Science, https://doi.org/10.1007/978-3-030-34209-8_23

The Turing machine is the most powerful model for computation, and this theoretical machine is equivalent to an actual computer in the sense that it can compute exactly the same set of functions. The memory of the Turing machine is a tape that consists of a potentially infinite number of one-dimensional cells. It provides a mathematical abstraction of computer execution and storage, as well as providing a mathematical definition of an algorithm. However, Turing machines are not suitable for programming, and therefore, they do not provide a good basis for studying programming and programming languages.

23.2 Finite-State Machines

Warren McCulloch and Walter Pitts (two neurophysiologists) published early work on finite-state automata in 1943. They were interested in modelling the thought process for humans and machines. Moore and Mealy developed this work further in the mid-1950s, and their finite-state machines are referred to as the '*Mealy machine*' and the '*Moore machine*'. The Mealy machine determines its outputs from the current state and the input, whereas the output of Moore's machine is based upon the current state alone.

Definition 23.1 (*Finite-State Machine*) A Finite-State Machine (FSM) is an abstract mathematical machine that consists of a finite number of states. It includes a start state q_0 in which the machine is in initially; a finite set of states Q; an input alphabet Σ; a state transition function δ; and a set of final accepting states F (where $F \subseteq Q$).

The state transition function δ takes the current state and an input symbol and returns the next state. That is, the transition function is of the form:

$$\delta : Q \times \Sigma \rightarrow Q$$

The transition function provides rules that define the action of the machine for each input symbol, and its definition may be extended to provide output as well as a transition of the state. State diagrams are used to represent finite-state machines, and each state accepts a finite number of inputs. A finite-state machine (Fig. 23.1) may be deterministic or non-deterministic, and a *deterministic machine* changes to exactly (or at most)[1] one state for each input transition, whereas a *non-deterministic machine* may have a choice of states to move to for a particular input symbol.

---

[1] The transition function may be undefined for a particular input symbol and state.

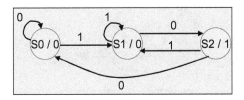

Fig. 23.1 Finite-state machine with output

Finite-state automata can compute only very primitive functions, and so they are not adequate as a model for computing. There are more powerful automata such as the Turing machine that is essentially a finite automaton with a potentially infinite storage (memory). Anything that is computable is computable by a Turing machine.

A finite-state machine can model a system that has a finite number of states, and a finite number of inputs/events that trigger transitions between states. The behaviour of the system at a point in time is determined from its current state and input, with behaviour defined for the possible input to that state. The system starts in an initial state.

A finite-state machine (also known as finite-state automata) is a quintuple $(\Sigma, Q, \delta, q_0, F)$. The alphabet of the FSM is given by Σ; the set of states is given by Q; the transition function is defined by $\delta: Q \times \Sigma \to Q$; the initial state is given by q_0; and the set of accepting states is given by F (where F is a subset of Q). A string is given by a sequence of alphabet symbols: i.e. $s \in \Sigma^*$, and the transition function δ can be extended to $\delta^*: Q \times \Sigma^* \to Q$.

A string $s \in \Sigma^*$ is accepted by the finite-state machine if $\delta^*(q_0, s) = q_f$ where $q_f \in F$, and the set of all strings accepted by a finite-state machine is the language generated by the machine. A finite-state machine is termed *deterministic* (Fig. 23.2) if the transition function δ is a function,[2] and otherwise (where it is a relation) it is said to be *non-deterministic*. A non-deterministic automaton is one for which the next state is not uniquely determined from the present state and input symbol, and the transition may be to a set of states rather than to a single state.

For the example above, the input alphabet is given by $\Sigma = \{0, 1\}$; the set of states by {A, B, C}; the start state by A; the accepting states by {C}; and the transition function is given by the state transition table in Table 23.1. The language accepted by the automata is the set of all binary strings that end with a one that contain exactly two ones.

A *non-deterministic* automaton (NFA) or non-deterministic finite-state machine is a finite-state machine where from each state of the machine and any given input, the machine may go to several possible next states. However, a non-deterministic

---

[2]It may be a total or a partial function (as discussed in Chap. 4).

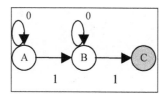

Fig. 23.2 Deterministic FSM

Table 23.1 State transition table

State	0	1
A	A	B
B	B	C
C	–	–

Fig. 23.3 Non-deterministic finite-state machine

automaton (Fig. 23.3) is equivalent to a deterministic automaton, in that they both recognize the same formal language (i.e. regular languages as defined in Chomsky's classification in Chap. 12). For any non-deterministic automaton, it is possible to construct the equivalent deterministic automaton using power set construction.

NFAs were introduced by Scott and Rabin in 1959, and a NFA is defined formally as a 5-tuple $(Q, \Sigma, \delta, q_o, F)$ as in the definition of a deterministic automaton, and the only difference is in the transition function δ.

$$\delta : Q \times \Sigma \rightarrow \mathbb{P}Q$$

The non-deterministic finite-state machine $M_1 = (Q, \Sigma, \delta, q_o, F)$ may be converted to the equivalent deterministic machine $M_2 = (Q', \Sigma, \delta', q_o', F')$ where

$Q' = \mathbb{P}Q$ (the set of all subsets of Q),
$q_o' = \{q_o\}$,
$F' = \{q \in Q' \text{ and } q \cap F \neq \varnothing\}$ and
$\delta'(q, \sigma) = \cup_{p \in q} \delta(p, \sigma)$ for each state $q \in Q'$ and $\sigma \in \Sigma$.

The set of strings (or language) accepted by an automaton M is denoted $L(M)$. That is, $L(M) = \{s: | \delta^*(q_0, s) = q_f \text{ for some } q_f \in F\}$. A language is termed regular if it is accepted by some finite-state machine. Regular sets are closed under union, intersection, concatenation, complement and transitive closure. That is, for regular sets A, B $\subseteq \Sigma^*$ then:

- A $\cup$ B and A $\cap$ B are regular.
- $\Sigma^* \setminus$ A (i.e. A^c) is regular.
- AB and A* is regular.

The proof of these properties is demonstrated by constructing finite-state machines to accept these languages. The proof for A $\cap$ B is to construct a machine $M_{A \cap B}$ that mimics the execution of M_A and M_B and is in a final state if and only if both M_A and M_B are in a final state. Finite-state machines are useful in designing systems that process sequences of data.

23.3   Pushdown Automata

A Pushdown Automaton (PDA) is essentially a finite-state machine with a stack, and its three components (Fig. 23.4) are an input tape; a control unit; and a potentially infinite stack. The stack head scans the top symbol of the stack, and two operations (push or pop) may be performed on the stack. The *push* operation adds a new symbol to the top of the stack, whereas the *pop* operation reads and removes an element from the top of the stack.

A pushdown automaton may remember a potentially infinite amount of information, whereas a finite-state automaton remembers only a finite amount of information. A PDA also differs from a FSM in that it may use the top of the stack to decide on which transition to take, and it may manipulate the stack as part of performing a transition. The input and current state determine the transition of a finite-state machine, and the FSM has no stack to work with.

A pushdown automaton is defined formally as a 7-tuple $(\Sigma, Q, \Gamma, \delta, q_0, Z, F)$. The set Σ is a finite set which is called the input alphabet; the set Q is a finite set of states; Γ is the set of stack symbols; δ, is the transition function which maps $Q \times \{\Sigma \cup \{\varepsilon\}\}^3 \times \Gamma$ into finite subsets of $Q \times \Gamma^{*4}$; q_0 is the initial state; Z is the initial stack top symbol on the stack (i.e. $Z \in \Gamma$); and F is the set of accepting states (i.e. $F \subseteq Q$).

---

[3] The use of $\{\Sigma \cup \{\varepsilon\}\}$ is to formalize that the PDA can either read a letter from the input, or proceed leaving the input untouched.

[4] This could also be written as $\delta:Q \times \{\Sigma \cup \{\varepsilon\}\}^4 \times \Gamma \to \mathbb{P}(Q \times \Gamma^*)$. It may also be described as a transition relation.

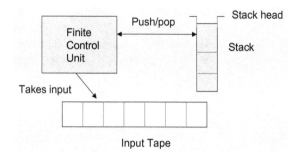

Fig. 23.4 Components of pushdown automata

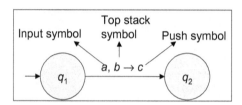

Fig. 23.5 Transition in pushdown automata

Figure 23.5 shows a transition from state q_1 to q_2, which is labelled as $a, b \rightarrow c$. This means that if the input symbol a occurs in state q_1, and the symbol on the top of the stack is b, then b is popped from the stack and c is pushed onto the stack. The new state is then q_2.

In general, a pushdown automaton has several transitions for a given input symbol, and so pushdown automata are mainly *non-deterministic*. If a pushdown automaton has at most one transition for the same combination of state, input symbol and top of stack symbol, it is said to be a *Deterministic PDA* (DPDA). The set of strings (or language) accepted by a pushdown automaton M is denoted $L(M)$.

The class of languages accepted by pushdown automata is the context-free languages, and every context-free grammar can be transformed into an equivalent non-deterministic pushdown automaton. There is more detailed information on the classification of languages in Chap. 12.

Example (Pushdown Automata) Construct a non-deterministic pushdown automaton which recognizes the language $\{0^n 1^n \mid n \geq 0\}$.

Solution We construct a pushdown automaton $M = (\Sigma, Q, \Gamma, \delta, q_0, Z, F)$ where $\Sigma = \{0, 1\}$; $Q = \{q_0, q_1, q_f\}$; $\Gamma = \{A, Z\}$; q_0 is the start state; the start stack symbol is Z; and the set of accepting states is given by $\{q_f\}$. The transition function (relation) δ is defined by

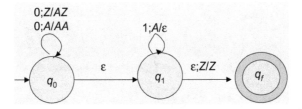

Fig. 23.6 Transition function for pushdown automata M

1. $(q_0, 0, Z) \to (q_0, AZ)$,
2. $(q_0, 0, A) \to (q_0, AA)$,
3. $(q_0, \varepsilon, Z) \to (q_1, Z)$,
4. $(q_0, \varepsilon, A) \to (q_1, A)$,
5. $(q_1, 1, A) \to (q_1, \varepsilon)$ and
6. $(q_1, \varepsilon, Z) \to (q_f, Z)$.

The transition function (Fig. 23.6) essentially says that whenever the value 0 occurs in state q_0 then A is pushed onto the stack. Parts (3) and (4) of the transition function essentially states that the automaton may move from state q_0 to state q_1 at any moment. Part (5) states when the input symbol is 1 in state q_1 then one symbol A is popped from the stack. Finally, part (6) states the automaton may move from state q_1 to the accepting state q_f only when the stack consists of the single stack symbol Z.

For example, it is easy to see that the string 0011 is accepted by the automaton, and the sequence of transitions is given by

$$(q_0, 0011, Z) \vdash (q_0, 011, AZ) \vdash (q_0, 11, AAZ)$$
$$\vdash (q_1, 11, AAZ) \vdash (q_1, 1, AZ) \vdash (q_1, \varepsilon, Z) \vdash (q_\xi, Z)$$

23.4 Turing Machines

Turing introduced the theoretical Turing Machine in 1936, and this abstract mathematical machine consists of a head and a potentially infinite tape that is divided into frames (Fig. 23.7). Each frame may be either blank or printed with a symbol from a finite alphabet of symbols. The input tape may initially be blank or have a finite number of frames containing symbols. At any step, the head can read the contents of a frame; the head may erase a symbol on the tape, leave it

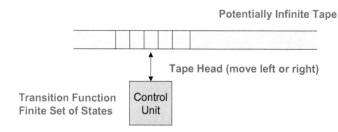

Fig. 23.7 Turing machine

unchanged or replace it with another symbol. It may then move one position to the right, one position to the left or not at all. If the frame is blank, the head can either leave the frame blank or print one of the symbols.

Turing believed that a human with finite equipment and with an unlimited supply of paper to write on could do every calculation. The unlimited supply of paper is formalized in the Turing machine by a paper tape marked off in squares, and the tape is potentially infinite in both directions. The tape may be used for intermediate calculations as well as input and output. The finite number of con-figurations of the Turing machine was intended to represent the finite states of mind of a human calculator.

The transition function determines for each state and the tape symbol what the next state to move to and what should be written on the tape, and where to move the tape head. The Turing Machine is defined formally as follows:

Definition 23.2 (*Turing Machine*) A Turing machine $M = (Q, \Gamma, b, \Sigma, \delta, q_0, F)$ is a 7-tuple is defined as follows in Hopcroft and Ullman (1979):

- Q is a finite set of *states*.
- Γ is a finite set of the *tape alphabet/symbols*.
- $b \in \Gamma$ is the *blank symbol* (This is the only symbol that is allowed to occur infinitely often on the tape during each step of the computation).
- Σ is the set of input symbols and is a subset of Γ (i.e. $\Gamma = \Sigma \cup \{b\}$).
- $\delta: Q \times \Gamma \rightarrow Q \times \Gamma \times \{L, R\}^5$ is the transition function. This is a partial function where L is left shift and R is right shift.
- $q_0 \in Q$ is the initial state.
- $F \subseteq Q$ is the set of final or accepting states.

---

[5]We may also allow no movement of the tape head to be represented by adding the symbol 'N' to the set.

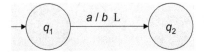

Fig. 23.8 Transition on Turing machine

The Turing machine is a simple machine that is equivalent to an actual physical computer in the sense that it can compute exactly the same set of functions. It is much easier to analyse and prove things about than a real computer, but it is not suitable for programming and does not provide a good basis for studying programming and programming languages.

Figure 23.8 illustrates the behaviour when the machine is in state q_1 and the symbol under the tape head is a, where b is written to the tape and the tape head moves to the left and the state changes to q_2.

A Turing machine is essentially a finite-state machine (FSM) with an unbounded tape. The tape is potentially infinite and unbounded, whereas real computers have a large but finite store. The machine may read from and write to the tape. The FSM is essentially the control unit of the machine, and the tape is essentially the store. However, the store in a real computer may be extended with backing tapes and discs, and in a sense may be regarded as unbounded. However, the maximum amount of tape that may be read or written within n steps is n.

A Turing machine has an associated set of rules that defines its behaviour. Its actions are defined by the transition function. It may be programmed to solve any problem for which there is an algorithm. However, if the problem is unsolvable then the machine will either stop or compute forever. The solvability of a problem may not be determined beforehand. There is, of course, some answer (i.e. either the machine halts or it computes forever). The applications of the Turing machine to computability and decidability were discussed in Chap. 13.

Turing also introduced the concept of a Universal Turing Machine and this machine is able to simulate any other Turing machine. For more detailed information on automata theory see Hopcroft and Ullman (1979).

23.5 Review Questions

1. What is a finite-state machine?
2. Explain the difference between a deterministic and non-deterministic finite-state machine.
3. Show how to convert the non-deterministic finite-state automaton in Fig. 7.3 to a deterministic automaton.
4. What is a pushdown automaton?

5. What is a Turing machine?
6. Explain what is meant by the language accepted by an automaton.
7. Give an example of a language accepted by a pushdown automaton but not by a finite-state machine.
8. Describe the applications of the Turing machine to computability and decidability.

23.6   Summary

Automata theory is concerned with the study of abstract machines and automata. These include finite-state machines, pushdown automata and Turing machines. Finite-state machines are abstract machines that may be in one of a finite number of states. These machines are in only one state at a time (current state), and the state transition function determines the new state from the current state and the input symbol. Finite-state machines have limited computational power due to memory and state constraints, but they have been applied to a number of fields including communication protocols and linguistics.

Pushdown automata have greater computational power than finite-state machines, and they contain extra memory in the form of a stack from which symbols may be pushed or popped. The state transition is determined from the current state of the machine, the input symbol and the element on the top of the stack. The action may be to change the state and/or push/pop an element from the stack.

The Turing machine is the most powerful model for computation, and it is equivalent to an actual computer in the sense that it can compute exactly the same set of functions. The Turing machine provides a mathematical abstraction of computer execution and storage, as well as providing a mathematical definition of an algorithm.

Reference

Hopcroft JE, Ullman JD (1979) Introduction to automata theory, languages and computation. Addison-Wesley, Boston

Model Checking

24

Key Topics

Concurrent systems
Temporal logic
State explosion
Safety and liveness properties
Fairness properties
Linear temporal logic
Computational tree logic

24.1 Introduction

Model checking is an automated technique such that given a finite-state model of a system and a formal property (expressed in temporal logic), then it systematically checks whether the property is true or false in a given state in the model. It is an effective technique to identify potential design errors, and it increases the confidence in the correctness of the system design. Model checking is a highly effective verification technology and is widely used in the hardware and software fields. It has been employed in the verification of microprocessors; in security protocols; in the transportation sector (trains); and in the verification of software in the space sector.

Early work on model checking commenced in the early 1980s (especially in checking the presence of properties such as mutual exclusion and the absence of deadlocks), and the term 'model checking' was coined by Clarke and Emerson (1981) who combined the state exploration approach and temporal logic in an

© Springer Nature Switzerland AG 2020
G. O'Regan, *Mathematics in Computing*, Undergraduate Topics
in Computer Science, https://doi.org/10.1007/978-3-030-34209-8_24

efficient manner. Clarke and Emerson received the ACM Turing Award in 2007 for their role in developing model checking into a highly effective verification technology.

Model checking is a formal verification technique based on graph algorithms and formal logic. It allows the desired behaviour (specification) of a system to be verified, and its approach is to employ a suitable model of the system and to carry out a systematic and exhaustive inspection of all states of the model to verify that the desired properties are satisfied. These properties are generally safety properties such as the absence of deadlock, request-response properties and invariants. The systematic search shows whether a given system model truly satisfies a particular property or not.

The phases in the model checking process include the modelling, running and analysis phases (Table 24.1).

The model-based techniques use mathematical models to describe the required system behaviour in precise mathematical language, and the system models have associated algorithms that allow all states of the model to be systematically explored. Model checking is used for formally verifying finite-state concurrent systems (typically modelled by automata), where the specification of the system is expressed in temporal logic, and efficient algorithms are used to traverse the model defined by the system (in its entirety) to check if the specification holds or not. *Of course, any verification using model-based techniques is only as good as the underlying model of the system.*

Model checking is an automated technique such that given a finite-state model of a system and a formal property, then a systematic search may be conducted to determine if the property holds for a given state in the model. The set of all possible states is called the model's state space, and when a system has a finite-state space it is then feasible to apply model checking algorithms to automate the demonstration of properties, with a counterexample exhibited if the property is not valid.

The properties to be validated are generally obtained from the system specification, and they may be quite elementary, e.g. a deadlock scenario should never arise (i.e. the system should never be able to reach a situation where no further progress is possible). The formal specification describes what the system should do,

Table 24.1 Model checking process

Phase	Description
Modelling phase	Model the system under consideration Formalize the property to be checked
Running phase	Run the model checker to determine the validity of the property in the model
Analysis phase	Is the property satisfied? If applicable, check next property If the property is violated then 1. Analyse generated counterexample 2. Refine model, design or property If out of space try alternative approach (e.g. abstraction of system model)

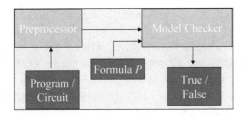

Fig. 24.1 Concept of model checking

whereas the model description (often automatically generated) is an accurate and unambiguous description of how the system actually behaves. The model is often expressed in a finite-state machine consisting of a finite set of states and a finite set of transitions.

Figure 24.1 shows the structure of a typical model checking system where a preprocessor extracts a state transition graph from a program or circuit. The model checking engine then takes the state transition graph and a temporal formula P and determines whether the formula is true or not in the model.

The properties need to be expressed precisely and unambiguously (usually in temporal logic) to enable rigorous verification to take place. Model checking extracts a finite model from a system and then checks some property of that model. The model checker performs an exhaustive state search, which involves checking all system states to determine whether they satisfy the desired property or not (Fig. 24.2).

If a state that violates the desired property is determined (i.e. a defect has been found once it is shown that the system does not fulfil one of its specified properties), then the model checker provides a counterexample indicating how the model can reach this undesired state. The system is considered to be correct if it satisfies all of the specified properties. In the cases of where the model is too large to fit within the physical memory of the computer (state explosion problem), then other approaches

Fig. 24.2 Model checking

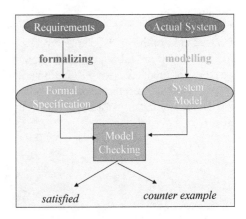

such as abstraction of the system model or probabilistic verification may be employed.

There may be several causes of a state violating the desired property. It may be due to a modelling error (i.e. the model does not reflect the design of the system, and the model may need to be corrected and the model checking restarted). Alternatively, it may be due to a design error with improvements needed to the design, or it may be due to an error in the statement of the property with a modification to the property required and the model checking needs to be restarted.

Model checking is expressed formally by showing that a desired property P (expressed as a temporal logic formula) and a model M with initial state s, that P is always true in any state derivable from s (i.e. $M, s \models P$). We discussed temporal logic briefly in Chap. 17, and model checking is concerned with verifying that linear time properties such as *safety*, *liveness* and *fairness* properties are always satisfied, and it employs *linear temporal logic* and *branching temporal logic*. Computational tree logic is a branching temporal logic where the model of time is a tree-like structure, with many different paths in the future, one of which might be an actual path which is realized.

One problem with model checking is the state space explosion problem, where the transition graph grows exponentially on the size of the system, which makes the exploration of the state space difficult or impractical. Abstraction is one technique that aims to deal with the state explosion problem, and it involves creating a simplified version of the model (the abstract model). The abstract model may be explored in a reasonable period of time, and the abstract model must respect the original model with respect to key properties such that if the property is valid in the abstract model it is valid in the original model.

Model checking has been applied to areas such as the verification of hardware designs, embedded systems, protocol verification and software engineering. Its algorithms have improved over the years, and today model checking is a mature technology for verification and debugging with many successful industrial applications.

The advantages of model theory include the fact that the user of the model checker does not need to construct a correctness proof (as in automated theorem proving or proof checking). Essentially, all the user needs to do is to input a description of the program or circuit to be verified and the specification to be checked, and to then press the return key. The checking process is then automatic and fast, and it provides a counterexample if the specification is not satisfied. One weakness of model checking is that it verifies an actual model rather than the actual system, and so its results are only as good as the underlying model. Model checking is described in detail in Baier and Katoen (2008).

24.2   Modelling Concurrent Systems

Concurrency is a form of computing in which multiple computations (processes) are executed during the same period of time. *Parallel computing* allows execution to occur in the same time instant (on separate processors of a multiprocessor machine), whereas *concurrent computing* consists of process lifetimes overlapping and where execution need not happen at the same time instant.

Concurrency employs *interleaving* where the execution steps of each process employs time-sharing slices so that only one process runs at a time, and if it does not complete within its time slice it is paused; another process begins or resumes; and then later the original process is resumed. In other words, only one process is running at a given time instant, whereas multiple processes are part of the way through execution.

It is important to identify concurrency-specific errors such as deadlock and livelock. A *deadlock* is a situation in which the system has reached a state in which no further progress can be made, and at least one process needs to complete its tasks. *Livelock* refers to a situation where the processes in a system are stuck in a repetitive task and are making no progress towards their functional goals.

It is essential that safety properties such as *mutual exclusion* (at most one process is in its critical section at any given time) are not violated. In other words, something bad (e.g. a deadlock situation) should never happen; *liveness properties* (a desired event or something good eventually happens) are satisfied; and *invariants* (properties that are true all the time) are never violated. These behaviour errors may be mechanically detected if the systems are properly modelled and analysed.

Transition systems (Fig. 24.3) are often used as models to describe the behaviour of systems, and these are directed graphs with nodes representing *states* and edges represent *state transitions*. A state describes information about a system at a certain moment of time. For example, the state of a sequential computer consists of the values of all program variables and the current value of the program counter (pointer to next program instruction to be executed).

A transition describes the conditions under which a system moves from one state to another. Transition systems are expressive in that programs are transition systems; communicating processes are transition systems; and hardware circuits are transition systems.

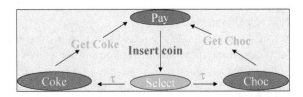

Fig. 24.3 Simple transition system

The transitions are associated with action labels that indicate the actions that cause the transition. For example, in Fig. 24.3 the *Insert coin* is a user action, whereas the *Get coke* and *Get choc* are actions that are performed by the machine. The activity τ represents an internal activity of the vending machine that is not of interest to the modeller. Formally, a transition system TS is a tuple $(S, Act, \rightarrow, I, AP, L)$ such that:

S is the set of states.
Act is the set of actions.
$\rightarrow S \times Act \times S$ is the transition relation (source state, action and target state).
$I \subseteq S$ is the set of initial states.
AP is a set of atomic propositions.
$L : S \rightarrow \mathbb{P} \, AP$ (power set of AP) is a labelling function.

The transition (s, a, s') is written as $s \xrightarrow{a} s'$
$L(s)$ are the atomic propositions in AP that are satisfied in state s.
A concurrent system consists of multiple processes executing concurrently. If a concurrent system consists of n processes where each process $proc_i$ is modelled by a transition system TS_i, then the concurrent system may be modelled by a transition system ($\|$ is the parallel composition operator):

$$TS = TS_1 \| TS_2 \| \ldots \| TS_n$$

There are various operators used in modelling concurrency with transition systems, including operators for interleaving, communication via shared variables, handshaking and channel systems.

24.3   Linear Temporal Logic

Temporal logic was discussed in Chap. 17 and is concerned with the expression of properties that have time dependencies. The existing temporal logics allow facts about the past, present and future to be expressed. Temporal logic has been applied to specify temporal properties of natural language, as well as the specification and verification of program and system behaviour. It provides a language to encode temporal knowledge in artificial intelligence applications, and it plays a useful role in the formal specification and verification of temporal properties (e.g. *liveness* and *fairness*) in safety-critical systems.

The statements made in temporal logic can have a truth-value that varies over time. In other words, sometimes the statement is true and sometimes it is false, but it is never true or false at the same time. The two main types of temporal logics are *linear time logics* (reason about a single timeline), and *branching-time logics* (reason about multiple timelines).

Table 24.2 Basic temporal operators

Operator	Description
Fp	p holds sometime in the future
Gp	p holds globally in the future
Xp	p holds in next time instant
p U q	p holds until q is true

Linear Temporal Logic (LTL) is a modal temporal logic that can encode formulae about the future of paths (e.g. a condition that will eventually be true). The basic linear temporal operators that are often employed (p is an atomic proposition below) are listed in Table 24.2 and illustrated in Fig. 24.4.

For example, consider how the sentence '*This microwave does not heat up until the door is closed*' is expressed in temporal logic. This is naturally expressed with the until operator p U q as follows:

$$\neg\text{Heatup U Door Closed}$$

24.4 Computational Tree Logic

In linear logic, we look at the execution paths individually, whereas in branching-time logics we view the computation as a tree. Computational Tree Logic (CTL) is a branching-time logic, which means that its model of time is a tree-like structure in which the future is not determined, and so there are many paths in the future such that any of them could be an actual path that is realized. CTL was first proposed by Clarke and Emerson in the early 1980s.

Computational tree logic can express many properties of finite-state concurrent systems. Each operator of the logic has two parts namely the path quantifier (A —'every path', E—'there exists a path'), and the state quantifier (F, P, X, U as explained in Table 24.3). The operators in CTL logic are given by.

For example, the following is a valid CTL formula that states that it is always possible to get to the restart state from any state:

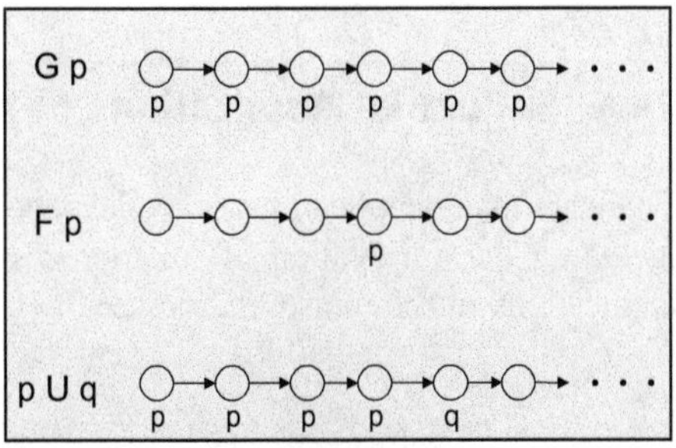

Fig. 24.4 LTL operators

Table 24.3 CTL temporal operators

Operator	Description
Aφ (*all*)	φ holds on *all* paths starting from the current state
Eφ (*exists*)	φ holds on at least one path starting from the current state
Xφ (*next*)	φ holds in the next state
Gφ (*global*)	φ has to hold on the entire subsequent path
Fφ (*finally*)	φ eventually has to hold (somewhere on the subsequent path)
φ U ψ (*until*)	φ has to hold until at some position ψ holds
φ W ψ (weak *until*)	φ has to hold until ψ holds (no guarantee ψ will ever hold)

$$AG\,(EF\,restart)$$

24.5   Tools for Model Checking

There are various tools for model checking including Spin, Bandera, SMV and UppAal. These tools perform a systematic check on property P in all states and are applicable if the system generates a finite behavioural model. Model checking tools use a model-based approach rather than a proof-rule-based approach, and the goal is to determine whether the concurrent program satisfies a given logical property.

Spin is a popular open-source tool that is used for the verification of distributed software systems (especially concurrent protocols), and it checks finite-state systems with properties specified by linear temporal logic. It generates a counterexample trace if determines that a property is violated.

Spin has its own input specification language (PROMELA), and so the system to be verified needs to be translated to the language of the model checker. The properties are specified using LTL.

Bandera is a tool for model checking Java source code, and it automates the extraction of a finite-state model from the Java source code. It then translates into an existing model checker's input language. The properties to be verified are specified in the Bandera Specification Language (BSL), which supports pre- and postconditions and temporal properties.

24.6   Industrial Applications of Model Checking

There are many applications of model checking in the hardware and software fields, including the verification of microprocessors and security protocols, as well as applications in the transportation sector (trains) and in the space sector.

The Mars Science Laboratory (MSL) mission used model checking as part of the verification of the critical software for the landing of Curiosity (a large rover) on its mission to Mars. The hardware and software of a spacecraft must be designed for a high degree of reliability, as an error can lead to a loss of the mission. The Spin model checker was employed for the model verification, and the rover was launched in Nov 2011 and landed safely on Mars in Aug 2012.

CMG employed formal methods as part of the specification and verification of the software for a movable flood barrier in the mid-1990s. This is used to protect the port of Rotterdam from flooding, and Z was employed for modelling data and operations and Spin/PROMELA was used for model checking.

Lucent's PathStar Access Server was developed in the late 1990s, and this system is capable of sending voice and data over the Internet. The automated verification techniques applied to PathStar consist of generating an abstract model from the implemented C code, and then defining the formal requirements that the application is satisfy. Finally, the model checker is employed to perform the verification.

24.7 Review Questions

1. What is model checking?
2. Explain the state explosion problem.
3. Explain the difference between parallel processing and concurrency.
4. Describe the basic temporal operators.
5. Describe the temporal operators in CTL.
6. Explain the difference between liveness and fairness properties.
7. What is a transition system?
8. Explain the difference between linear temporal logic and branching temporal logic.
9. Investigate tools to support model checking.

24.8 Summary

Model checking is a formal verification technique which allows the desired behaviours of a system to be verified. Its approach is to employ a suitable model of the system and to carry out a systematic inspection of all states of the model to verify that the required properties are satisfied in each state. The properties to be validated are generally obtained from the system specification, and a defect is found once it is shown that the system does not fulfil one of its specified properties. The system is considered to be correct if it satisfies all of the specified properties.

The desired behaviour (specification) of the system is verified by employing a suitable model of the system, and then carrying out a systematic exhaustive inspection of all states of the model to verify that the desired properties are satisfied. These properties are generally properties such as the absence of deadlock and invariants. The systematic search shows whether a given system model truly satisfies a particular property or not.

The model-based techniques use mathematical models to describe the required system behaviour in precise mathematical language, and the system models have associated algorithms that allow all states of the model to be systematically explored. The specification of the system is expressed in temporal logic, and efficient algorithms are used to traverse the model defined by the system (in its entirety) to check if the specification holds or not. Model-based techniques are only as good as the underlying model of the system.

References

Baier C, Katoen JP (2008) Principles of model checking. MIT Press, Cambridge, MA
Clarke EM, Emerson EA (1981) Design and synthesis of synchronization skeletons using branching time temporal logic. In Logic of programs: work-shop, Yorktown Heights, NY, May 1981, volume 131 of LNCS. Springer

Probability and Statistics

25

Key Topics

Sample spaces
Random variables
Mean, mode and median
Variance
Normal distributions
Histograms
Hypothesis testing
Software reliability models

25.1 Introduction

Statistics is an empirical science that is concerned with the collection, organization, analysis, interpretation and presentation of data. The data collection needs to be planned and may include surveys and experiments. Statistics are widely used by government and industrial organizations, and they may be employed for forecasting as well as for presenting trends. They allow the behaviour of a population to be studied and inferences to be made about the population. These inferences may be tested (*hypothesis testing*) to ensure their validity.

The analysis of statistical data allows an organization to understand its performance in key areas, and to identify problematic areas. Organizations will often examine performance trends over time, and will devise appropriate plans and actions to address problematic areas. The effectiveness of the actions taken will be judged by improvements in performance trends over time.

© Springer Nature Switzerland AG 2020
G. O'Regan, *Mathematics in Computing*, Undergraduate Topics
in Computer Science, https://doi.org/10.1007/978-3-030-34209-8_25

It is often not possible to study the entire population, and instead, a representative subset or sample of the population is chosen. This *random sample* is used to make inferences regarding the entire population, and it is essential that the sample chosen is indeed random and representative of the entire population. Otherwise, the inferences made regarding the entire population will be invalid.[1]

A statistical experiment is a causality study that aims to draw a conclusion regarding values of a *predictor variable*(s) on a *response variable*(s). For example, a statistical experiment in the medical field may be conducted to determine if there is a causal relationship between the use of a particular drug and the treatment of a medical condition such as lowering of cholesterol in the population. A statistical experiment involves:

- Planning the research,
- Designing the experiment,
- Performing the experiment,
- Analysing the results and
- Presenting the results.

Probability is a way of expressing the likelihood of a particular event occurring. It is normal to distinguish between the frequency interpretation and the subjective interpretation of probability (Ross 1987). For example, if a geologist states that 'there is a 70% chance of finding gas in a certain region' then this statement is usually interpreted in two ways:

- The geologist is of the view that over the long run 70% of the regions whose environment conditions are very similar to the region under consideration have gas [*Frequency Interpretation*].
- The geologist is of the view that it is likely that the region contains gas, and that 0.7 is a measure of the geologist's belief in this hypothesis [*Personal Interpretation*].

However, the mathematics of probability is the same for both the frequency and personal interpretation.

25.2 Probability Theory

Probability theory provides a mathematical indication of the likelihood of an event occurring, and the probability of an event is a numerical value between 0 and 1. A probability of 0 indicates that the event cannot occur whereas a probability of 1 indicates that the event is guaranteed to occur. If the probability of an event is

---

[1]The random sample leads to predictions for the entire population that may be inaccurate. For example, consider the opinion polls on the 2016 British Referendum on EU membership and the 2016 presidential election in the United States.

greater than 0.5 then this indicates that the event is more likely to occur than not to occur.

A *sample space* is the set of all possible outcomes of an experiment, and an *event* E is a subset of the sample space. For example, the sample space for the experiment of tossing a coin is the set of all possible outcomes of this experiment: i.e. head or tails. The event that the toss results a tail is a subset of the sample space.

$$S = \{h, t\} \quad E = \{t\}$$

Similarly, the sample space for the gender of a newborn baby is the set of outcomes: i.e. the newborn baby is a boy or a girl. The event that the baby is a girl is a subset of the sample space.

$$S = \{b, g\} \quad E = \{g\}$$

For any two events E and F of a sample space S we can also consider the union and intersection of these events. That is,

- $E \cup F$ consists of all outcomes that are in E or F or both.
- $E \cap F$ (normally written as EF) consists of all outcomes that are in both E and F.
- E^c denotes the complement of E with respect to S and represents the outcomes of S that are not in E (i.e. S\E).

If $EF = \varnothing$ then there are no outcomes in both E and F, and so the two events E and F are mutually exclusive. The union and intersection of two events can be extended to the union and intersection of a family of events $E_1, E_2, \ldots, E_n$ (i.e. $\cup_{i=1}^{n} E_i$ and $\cap_{i=1}^{n} E_i$).

25.2.1 Laws of Probability

The laws of probability essentially state that the probability of an event is between 0 and 1, and that the probability of the union of a mutually disjoint set of events is the sum of their individual probabilities.

(i) $P(S) = 1$
(ii) $P(\varnothing) = 0$
(iii) $0 \le P(E) \le 1$
(iv) For any sequence of mutually exclusive events $E_1, E_2, \ldots, E_n$. (i.e. $E_i E_j = \varnothing$ where $i \ne j$) then the probability of the union of these events is the sum of their individual probabilities: i.e.

$$P\left(\bigcup_{i=1}^{n} E_i\right) = \sum_{i=1}^{n} P(E_i)$$

The probability of the union of two events (not necessarily disjoint) is given by

$$P(E \cup F) = P(E) + P(F) - P(EF)$$

The probability of an event E not occurring is denoted by E^c and is given by $1 - P(E)$. The probability of an event E occurring given that an event F has occurred is termed the *conditional probability* (denoted by P(E|F)) and is given by

$$P(E|F) = \frac{P(EF)}{P(F)} \quad \text{where } P(F) > 0$$

This formula allows us to deduce that

$$P(EF) = P(E|F)P(F)$$

Bayes formula enables the probability of an event E to be determined by a weighted average of the conditional probability of E given that the event F occurred and the conditional probability of E given that F has not occurred:

$$E = E \cap S = E \cap (F \cup F^c)$$
$$= EF \cup EF^c$$

$$P(E) = P(EF) + P(EF^c) \quad (\text{since } EF \cap EF^c = \varnothing)$$
$$= P(E|F)P(F) + P(E|F^c)P(F^c)$$
$$= P(E|F)P(F) + P(E|F^c(1 - P(F))$$

Two events E, F are *independent* if knowledge that F has occurred does not change the probability that E has occurred. That is, P(E|F) = P(E) and since P(E|F) = P(EF)/P(F) we have that two events E, F are independent if:

$$P(EF) = P(E)P(F)$$

Two events E and F that are not independent are said to be *dependent*.

25.2.2 Random Variables

Often, some numerical quantity determined by the result of the experiment is of interest rather than the result of the experiment itself. These numerical quantities are termed *random variables*. A random variable is termed *discrete* if it can take on a finite or countable number of values; otherwise, it is termed *continuous*.

The *distribution function* of a random variable is the probability that the random variable X takes on a value less than or equal to x. It is given by

$$F(x) = P\{X \leq x\}$$

All probability questions about X can be answered in terms of its distribution function F. For example, the computation of P $\{a < X < b\}$ is given by

$$P\{a < X < b\} = P\{X \leq b\} - P\{X \leq a\}$$
$$= F(b) - F(a)$$

The probability mass function for a discrete random variable X (denoted by $p(a)$) is the probability that it is a certain value. It is given by

$$p(a) = P\{X = a\}$$

Further, F(a) can also be expressed in terms of the probability mass function

$$F(a) = \sum_{\forall x \leq a} p(x)$$

X is a continuous random variable if there exists a non-negative function $f(x)$ termed the *probability density function* defined for all $x \in (-\infty, \infty)$ such that

$$P\{X \in B\} = \int_B f(x)dx$$

All probability statements about X can be answered in terms of its density function $f(x)$. For example,

$$P\{X \in (-\infty, \infty)\} = 1 = \int_{-\infty}^{\infty} f(x)dx$$

$$P\{a \leq X \leq b\} = \int_a^b f(x)dx$$

The function $f(x)$ is termed the probability density function and the probability distribution function F(a) is defined by

$$F(a) = P\{X \leq a\} = \int_{-\infty}^a f(x)dx$$

Further, the derivative of the probability distribution function yields the probability density function. That is,

$$\frac{d}{da}F(a) = f(a)$$

The expected value (i.e. the *mean*) of a discrete random variable X (denoted E[X]) is given by the weighted average of the possible values of X, and the expected value of a function of a random variable is given by E[g(X)]. These are defined as

$$E[X] = \sum_i x_i P\{X = x_i\}$$

$$E[g(X)] = \sum_i g(x_i)P\{X = x_i\}$$

The expected value (i.e. the *mean*) of a continuous random variable X, and the expected value of a function of a continuous random variable is given by

$$E(X) = \int_{-\infty}^{\infty} xf(x)dx$$

$$E(g(X)) = \int_{-\infty}^{\infty} g(x)f(x)dx$$

The *variance* of a random variable is a measure of the spread of values from the mean and is defined by

$$Var(X) = E[X^2] - (E[X])^2$$

The standard deviation σ is given by the square root of the variance. That is,

$$\sigma = \sqrt{Var(X)}$$

The *covariance* of two random variables is a measure of the relationship between two random variables X and Y, and indicates the extent to which they both change (in either similar or opposite ways) together. It is defined by

$$Cov(X, Y) = E[XY] - E[X]E[Y]$$

It follows from the definition that the covariance of two independent random variables is zero (and this would be expected as a change in one variable would not affect the other). Variance is a special case of covariance (when the two random variables are identical). This follows since $Cov(X, X) = E[X \cdot X] - (E[X])(E[X]) = E[X^2] - (E[X])^2 = Var(X)$.

A positive covariance $(Cov(X, Y) \geq 0)$ indicates that Y tends to increase as X does, whereas a negative covariance indicates that Y tends to decrease as X increases. The *correlation* of two random variables is an indication of the relationship between two variables X and Y. If the correlation is negative then Y tends to decrease as X increases, and if it is positive number then Y tends to increase as X increases. The correlation coefficient is a value that is between ± 1 and it is defined by

$$Corr(X, Y) = \frac{Cov(X, Y)}{\sqrt{Var(X)Var(Y)}}$$

Once the correlation between two variables has been calculated the probability that the observed correlation was due to chance can be computed. This is to ensure that the observed correlation is a real one and not due to a chance occurrence.

There are a number of special random variables, and these include the Bernoulli trial, where there are just two possible outcomes of an experiment: i.e. success or failure. The probability of success and failure is given by

$$P\{X = 0\} = 1 - p$$
$$P\{X = 1\} = p$$

The mean of the Bernoulli distribution is given by p and the variance by $p(1 - p)$. The *Binomial distribution* involves n Bernoulli trials, each of which results in success or failure. The probability of i successes from n trials is then given by

$$P\{X = i\} = \binom{n}{i} p^i (1 - p)^{n-i}$$

where the mean of the Binomial distribution given by np, and the variance is given by $np(1 - p)$.

The *Poisson distribution* may be used as an approximation to the Binomial Distribution when n is large and p is small. The probability of i successes is given by

$$P\{X = i\} = e^{-\lambda} \lambda^i / i!$$

and the mean and variance of the Poisson distribution is given by λ.

There are many other well-known distributions such as the *hypergeometric distribution* that describes the probability of i successes in n draws from a finite population without replacement; the *uniform distribution*; the *exponential distribution*, the *normal distribution* and the *gamma* distribution. Table 25.1 presents several important probability distributions including their mean and variance.

The reader is referred to Ross (1987) for a more detailed account of probability theory.

Table 25.1 Probability distributions

Distribution name	Density function	Mean/variance
Binomial	$P\{X = i\} = \binom{n}{i} p^i (1-p)^{n-i}$	np, $np(1-p)$
Poisson	$P\{X = i\} = e^{-\lambda} \lambda^i / i!$	λ, λ
Hypergeometric	$P\{X = i\} = \binom{N}{i}\binom{M}{n-i} / \binom{N+M}{n}$	$nN/N + M$, $np(1-p)[1 - (n-1)/N + M - 1]$
Uniform	$f(x) = 1/(\beta - \alpha) \alpha \le x \le \beta, 0$	$(\alpha + \beta)/2$, $(\beta - \alpha)^2/12$
Exponential	$f(x) = \lambda e^{-\lambda x}$	$1/\lambda$, $1/\lambda^2$
Normal	$f(x) = \frac{1}{\sqrt{2\pi}\sigma} e^{-(x-\mu)^2/2\sigma^2}$	μ, σ^2
Gamma	$f(x) = \lambda e^{-\lambda x} (\lambda x)^{\alpha-1}/\Gamma(\alpha)\ (x \ge 0)$	α/λ, α/λ^2

25.3 Statistics

The field of statistics is concerned with summarizing, digesting and extracting information from large quantities of data. Statistics provide a collection of methods for planning and conducting an experiment, and analysing the data to draw accurate conclusions. We distinguish between descriptive statistics and inferential statistics:

Descriptive Statistics

This is concerned with describing the information in a set of data elements in graphical format, or by describing its distribution.

Inferential Statistics

This is concerned with making inferences with respect to the population by using information gathered in the sample.

25.3.1 Abuse of Statistics

Statistics are extremely useful in drawing conclusions about a population. However, it is essential that the random sample is valid, and that the experiment is properly conducted to enable valid conclusions to be inferred. Further, the presentation of the statistics should not be misleading. Some examples of the abuse of statistics include:

–   The sample size may be too small to draw conclusions.
–   It may not be a genuine random sample of the population.
–   Graphs may be drawn to exaggerate small differences.
–   Area may be misused in representing proportions.
–   Misleading percentages may be used.

The quantitative data used in statistics may be discrete or continuous. *Discrete data* is numerical data that has a finite number of possible values, and *continuous data* is numerical data that has an infinite number of possible values.

25.3.2 Statistical Sampling

Statistical sampling is concerned with the methodology of choosing a random sample of a population, and the systematic study of the sample with the goal of drawing valid conclusions about the entire population.

The assumption is that if a genuine representative sample of the population is chosen, then the detailed study of the sample will provide insight into the whole population. This helps to avoid a lengthy expensive (and potentially infeasible) study of the entire population.

The sample chosen must be random (this can be difficult to achieve), and the sample size sufficiently large to enable valid conclusions to be made for the entire population.

Random Sample

A *random sample* is a sample of the population such that each member of the population has an equal chance of being chosen.

There are various ways of generating a random sample from the population including (Table 25.2).

Once the random sample group has been chosen the next step is to obtain the required information from the sample. This may be done by interviewing each member in the sample; phoning each member; conducting a mail survey, and so on (Table 25.3).

Table 25.2 Sampling techniques

Sampling technique	Description
Systematic sampling	Every kth member of the population is sampled
Stratified sampling	The population is divided into two or more strata, and each subpopulation (stratum) is then sampled. Each element in the subpopulation shares the same characteristics (e.g. age groups, gender)
Cluster sampling	A population is divided into clusters and a few of these clusters are exhaustively sampled (i.e. every element in the cluster is considered)
Convenience sampling	Sampling is done as convenient and often allows the element to choose whether or not it is sampled

Table 25.3 Types of survey

Survey type	Description
Direct measurement	This may involve the direct measurement of all entities in the sample (e.g. the height of students in a class)
Mail survey	This involves sending a mail survey to the sample. This may have a lower response rate and may thereby invalidate the findings
Phone survey	This is a reasonably efficient and cost-effective way to gather data. However, refusals or hang-ups may affect the outcome
Personal interview	This tends to be expensive and time-consuming, but it allows detailed information to be collected
Observational study	An observational study allows individuals to be studied, and the variables of interest to be measured
Experiment	An experiment imposes some treatment on individuals in order to study the response

25.3.3 Averages in a Sample

The term 'average' generally refers to the arithmetic *mean* of a sample, but it could also refer to the *mode* or *median* of the sample. These terms are defined below:

Mean

The *arithmetic mean* of a set of n numbers is defined to be the sum of the numbers divided by n. That is, the arithmetic mean for a sample of size n is given by

$$\bar{x} = \frac{\sum_{i=1}^{n} x_i}{n}$$

The actual mean of the population is denoted by μ, and it may differ from the sample mean $\bar{x}$.

Mode

The mode is the data element that occurs most frequency in the sample. It is possible that two elements occur with the same frequency, and if this is the case then we are dealing with a bi-modal or possibly a multimodal sample.

Median

The median is the middle element when the data set is arranged in increasing order of magnitude.

If there are an odd number of elements in the sample the median is the middle element. Otherwise, the median is the arithmetic mean of the two middle elements.

Mid Range

The midrange is the arithmetic mean of the highest and lowest data elements in the sample. That is, $(x_{max} + x_{min})/2$.

The *arithmetic mean* is the most widely used average in statistics.

25.3.4 Variance and Standard Deviation

An important characteristic of a sample is its distribution, and the spread of each element from some measure of central tendency (e.g. the mean). One elementary measure of dispersion is that of the sample *range*, and it is defined to be the difference between the maximum and minimum value in the sample. That is, the sample range is defined to be:

$$\text{range} = x_{\max} - x_{\min}.$$

The sample range is not a reliable measure of dispersion as only two elements in the sample are used, and extreme values in the sample can distort the range to be very large even if most of the elements are quite close to one another.

The standard deviation is the most common way to measure dispersion, and it gives the average distance of each element in the sample from the mean. The sample standard deviation is denoted by s and is defined by

$$s = \sqrt{\frac{\sum (x_i - \bar{x})^2}{n - 1}}$$

The population standard deviation is denoted by σ and is defined by

$$\sigma = \sqrt{\frac{\sum (x_i - \mu)^2}{N}}$$

Variance is another measure of dispersion and it is defined as the square of the standard deviation. The sample variance is given by

$$s^2 = \frac{\sum (x_i - \bar{x})^2}{n - 1}$$

The population variance is given by

$$\sigma^2 = \frac{\sum (x_i - \mu)^2}{N}$$

25.3.5 Bell-Shaped (Normal) Distribution

The German mathematician Gauss (Fig. 25.1) originally studied the *normal distribution*, and it is also known as the *Gaussian distribution*. It is shaped like a bell and so is popularly known as the *bell-shaped* distribution. The empirical frequencies of many natural populations exhibit a bell-shaped (normal) curve.

Fig. 25.1 Carl Friedrich
Gauss

The *normal distribution* N has mean μ and standard deviation σ. Its density function $f(x)$ (where $-\infty < x < \infty$) is given by

$$f(x) = \frac{1}{\sqrt{2\pi}\sigma} e^{-(x-\mu)^2/2\sigma^2}$$

The *unit* (or *standard*) normal distribution $Z(0, 1)$ has mean 0 and standard deviation of 1 (Fig. 25.2). Every normal distribution may be converted to the unit normal distribution by $Z = (X - \mu)/\sigma$, and every probability statement about X has an equivalent probability statement about Z. The unit normal density function is given by

Fig. 25.2 Standard unit
normal bell curve (Gaussian
distribution)

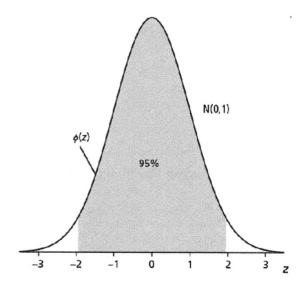

$$f(y) = \frac{1}{\sqrt{2\pi}} e^{-y^2/2}$$

For a normal distribution, 68.2% of the data elements lie within one standard deviation of the mean; 95.4% of the population lies within two standard deviations of the mean; and 99.7% of the data lies within three standard deviations of the mean. For example, the shaded area under the curve within two standard deviations of the mean represents 95% of the population.

A fundamental result in probability theory is the *Central Limit Theorem*, and this theorem essentially states that the sum of a large number of independent and identically distributed random variables has a distribution that is approximately normal. That is, suppose X_1, X_2, ..., X_n is a sequence of independent random variables each with mean μ and variance σ^2. Then for large n the distribution of

$$\frac{X_1 + X_2 + \cdots + X_n - n\mu}{\sigma \sqrt{n}}$$

is approximately that of a unit normal variable Z. One application of the central limit theorem is in relation to the binomial random variables, where a binomial random variable with parameters (n, p) represents the number of successes of n independent trials, where each trial has a probability of p of success. This may be expressed as

$$X = X_1 + X_2 + \cdots + X_n$$

where $X_i = 1$ if the *i*th trial is a success and is 0 otherwise. $E(X_i) = p$ and $\text{Var}(X_i) = p(1 - p)$, and then by applying the central limit theorem it follows that for large n

$$\frac{X - np}{\sqrt{np(1 - p)}}$$

will be approximately a unit normal variable (which becomes more normal as n becomes larger).

The sum of independent normal random variables is normally distributed, and it can be shown that the sample average of X_1, X_2, ..., X_n is normal, with a mean equal to the population mean but with a variance reduced by a factor of $1/n$.

$$E(\bar{X}) = \sum_{i=1}^{n} \frac{E(X_i)}{n} = \mu$$

$$\text{Var}(\bar{X}) = \frac{1}{n^2} \sum_{i=1}^{n} \text{Var}(X_i) = \frac{\sigma^2}{n}$$

It follows that from this that the following is a unit normal random variable:

Table 25.4 Frequency table
—salary

Profession	Salary	Frequency
Project manager	65,000	3
Architect	65,000	1
Programmer	50,000	8
Tester	45,000	2
Director	90,000	1

Table 25.5 Frequency table
—test results

Mark	Frequency
0–24	3
25–49	10
50–74	15
75–100	2

$$\sqrt{n}\frac{(X - \mu)}{\sigma}$$

The term *six-sigma* (6σ) is a methodology concerned with continuous process improvement and aims for very high quality (close to perfection). A 6σ process is one in which 99.9996% of the products are expected to be free from defects (3.4 defects per million).

25.3.6 Frequency Tables, Histograms and Pie Charts

A frequency table is used to present or summarize data (Tables 25.4 and 25.5). It lists the data classes (or categories) in one column and the frequency of the category in another column.

A histogram is a way to represent data in bar chart format (Fig. 25.3). The data is divided into intervals where an interval is a certain range of values. The horizontal

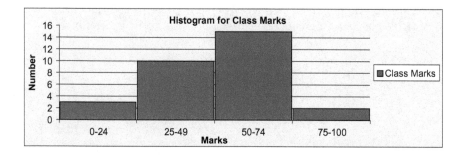

Fig. 25.3 Histogram test results

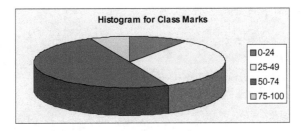

Fig. 25.4 Pie chart test results

axis of the histogram contains the intervals (also known as buckets) and the vertical axis shows the frequency (or relative frequency) of each interval. The bars represent the frequency and there is no space between the bars.

A histogram has an associated shape. For example, it may resemble a normal distribution, a bi-modal or multimodal distribution. It may be positively or negatively skewed. The construction of a histogram first involves the construction of a frequency table where the data is divided into disjoint classes and the frequency of each class is determined.

A pie chart (Fig. 25.4) offers an alternate way to histograms in the presentation of data. A frequency table is first constructed, and the pie chart presents a visual representation of the percentage in each data class.

25.3.7 Hypothesis Testing

The basic concept of inferential statistics is *hypothesis testing*, where a hypothesis is a statement about a particular population whose truth or falsity is unknown. Hypothesis testing is concerned with determining whether the values of the random sample from the population are consistent with the hypothesis. There are two mutually exclusive hypotheses: one of these is the null hypothesis H_0 and the other is the alternate research hypothesis H_1. The null hypothesis H_0 is what the researcher is hoping to reject, and the research hypothesis H_1 is what the researcher is hoping to accept.

Statistical testing is then employed to test the hypothesis, and the result of the test is that we either reject the null hypothesis (and therefore accept the alternative hypothesis), or that we fail to reject (i.e. we accept) the null hypothesis. The rejection of the null hypothesis means that the null hypothesis is highly unlikely to be true, and that the research hypothesis should be accepted.

Statistical testing is conducted at a certain level of significance, with the probability of the null hypothesis H_0 being rejected when it is true never greater than α. The value α is called the level of significance of the test, with α usually being 0.1, 0.05, 0.005. A significance level β may also be applied to with respect to accepting the null hypothesis H_0 when H_0 is false.

Table 25.6 Hypothesis testing

Action	H_0 true, H_1 false	H_0 false, H_1 true
Reject H_1	Correct	False positive—Type 2 error $P(\text{Accept } H_0 \mid H_0 \text{ false}) = \beta$
Reject H_0	False negative—type 1 error $P(\text{Reject } H_0 \mid H_0 \text{ true}) = \alpha$	Correct

The objective of a statistical test is not to determine whether or not H_0 is actually true, but rather to determine whether its validity is consistent with the observed data. That is, H_0 should only be rejected if the resultant data is very unlikely if H_0 is true.

The errors that can occur with hypothesis testing include type 1 and type 2 errors. Type 1 errors occur when we reject the null hypothesis when the null hypothesis is actually true. Type 2 errors occur when we accept the null hypothesis when the null hypothesis is false (Table 25.6).

For example, an example of a false positive is where the results of a blood test comes back positive to indicate that a person has a particular disease when in fact the person does not have the disease. Similarly, an example of a false negative is where a blood test is negative indicating that a person does not have a particular disease when in fact the person does. Both errors can potentially be very serious.

The terms α and β represent the level of significance that will be accepted, and usually $\alpha = \beta$. In other words, α is the probability that we will reject the null hypothesis when the null hypothesis is true, and β is the probability that we will accept the null hypothesis when the null hypothesis is false.

Testing a hypothesis at the $\alpha = 0.05$ level is equivalent to establishing a 95% confidence interval. For 99% confidence α will be 0.01, and for 99.999% confidence then α will be 0.00001.

The hypothesis may be concerned with testing a specific statement about the value of an unknown parameter θ of the population. This test is to be done at a certain level of significance, and the unknown parameter may, for example, be the mean or variance of the population. An estimator for the unknown parameter is determined, and the hypothesis that this is an accurate estimate is rejected if the random sample is not consistent with it. Otherwise, it is accepted.

The steps involved in hypothesis testing include:

1. Establish the null and alternative hypothesis.
2. Establish error levels (significance).
3. Compute the test statistics (often a t-test).
4. Decide on whether to accept or reject the null hypothesis.

The difference between the observed and expected test statistic, and whether the difference could be accounted for by normal sampling fluctuations is the key to the acceptance or rejection of the null hypothesis.

25.4 Review Questions

1. What is probability? What is statistics? Explain the difference between them.
2. Explain the laws of probability.
3. What is a sample space? What is an event?
4. Prove Boole's inequality $P\left(\cup_{i=1}^{n} E_i\right) \leq \sum_{i=1}^{n} P(E_i)$ where the E_i are not necessarily disjoint.
5. A couple has two children. What is the probability that both are girls if the eldest is a girl?
6. What is a random variable?
7. Explain the difference between the probability density function and the probability distribution function.
8. Explain expectation, variance, covariance and correlation.
9. Describe how statistics may be abused.
10. What is a random sample? Describe methods available to generate a random sample from a population. How may information be gained from a sample?
11. Explain how the average of a sample may be determined, and discuss the mean, mode and median of a sample.
12. Explain sample variance and sample standard deviation.
13. Describe the normal distribution and the central limit theorem.
14. Explain hypothesis testing and acceptance or rejection of the null hypothesis.

25.5 Summary

Statistics is an empirical science that is concerned with the collection, organization, analysis and interpretation and presentation of data. The data collection needs to be planned and this may include surveys and experiments. Statistics are widely used by government and industrial organizations, and they may be used for forecasting as well as for presenting trends. Statistical sampling allows the behaviour of a random sample to be studied, and inferences to be made about the population.

Probability theory provides a mathematical indication of the likelihood of an event occurring, and the probability is a numerical value between 0 and 1. A probability of 0 indicates that the event cannot occur, whereas a probability of 1 indicates that the event is guaranteed to occur. If the probability of an event is greater than 0.5, then this indicates that the event is more likely to occur than not to occur.

Statistical sampling is concerned with the methodology of choosing a random sample of a population and the systematic study of the sample with the goal of drawing valid conclusions about the entire population. Hypothesis testing is concerned with determining whether the values from a random sample from the population are consistent with a particular hypothesis. There are two mutually exclusive hypotheses: one of these is the null hypothesis H_0 and the other is the alternate research hypothesis H_1. The null hypothesis H_0 is what the researcher is hoping to reject, and the research hypothesis H_1 is what the researcher is hoping to accept.

Reference

Ross SM (1987) Introduction to probability and statistics for engineers and scientists. Wiley Publications, New York

Complex Numbers and Quaternions

26

26.1 Introduction

A complex number z is a number of the form $a + bi$ where a and b are real numbers and $i^2 = -1$. Cardona, who was a sixteenth-century Italian mathematician, introduced complex numbers, and he used them to solve cubic equations. The set of complex numbers is denoted by $\mathbb{C}$, and each complex number has two parts, namely, the real part $\text{Re}(z) = a$ and the imaginary part $\text{Im}(z) = b$. The set of complex numbers is an extension of the set of real numbers, and this is clear since every real number is a complex number with an imaginary part of zero. A complex number with a real part of zero (i.e. $a = 0$) is termed an imaginary number. Complex numbers have many applications in physics, engineering and applied mathematics.

A complex number may then be viewed as a point in a two-dimensional Cartesian coordinate system (called complex plane or Argand diagram), where the complex number $a + bi$ is represented by the point (a, b) on the complex plane (Fig. 26.1). The real part of the complex number is the horizontal component, and the imaginary part is the vertical component.

© Springer Nature Switzerland AG 2020   411
G. O'Regan, *Mathematics in Computing*, Undergraduate Topics
in Computer Science, https://doi.org/10.1007/978-3-030-34209-8_26

Fig. 26.1 Argand diagram

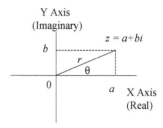

Quaternions are an extension of complex numbers. A quaternion number is a quadruple of the form $(a + bi + cj + dk)$ where $i^2 = j^2 = k^2 = ijk = -1$. The set of quaternions is denoted by $\mathbb{H}$, and the quaternions form an algebraic system known as a division ring. The multiplication of two quaternions is not commutative: i.e. given $q_1, q_2 \in \mathbb{H}$ then $q_1, q_2 \neq q_2, q_1$. Quaternions were one of the first non-commutative algebraic structures to be discovered, as matrix algebra came later.

The Irish mathematician, Sir William Rowan Hamilton,[1] discovered quaternions. Hamilton was trying to generalize complex numbers to triples without success. He had a moment of inspiration along the banks of the Royal Canal in Dublin, and he realized that if he used quadruples instead of triples that a generalization from the complex numbers was possible. He was so overcome with emotion at his discovery that he traced the famous quaternion formula[2] on Broom's Bridge in Dublin. This formula is given by

$$i^2 = j^2 = k^2 = ijk = -1$$

Quaternions have many applications in physics and quantum mechanics and are applicable to the computing field. They are useful and efficient in describing rotations and are therefore applicable to computer graphics, computer vision and robotics.

26.2 Complex Numbers

There are several operations on complex numbers such as addition, subtraction, multiplication, division and so on (Table 26.1). Consider two complex numbers $z_1 = a + bi$ and $z_2 = c + di$. Then,

---

[1]There is a possibility that the German mathematician, Gauss, discovered quaternions earlier.
[2]Eamonn De Valera, a former taoiseach and president of Ireland, was formerly a mathematics teacher, and his interests included maths, physics and quaternions. He is alleged to have carved the quaternion formula on the door of his cell while in prison in Lincoln Jail, England during the Irish struggle for independence from Britain.

Table 26.1 Operations on complex numbers

Operation	Definition
Addition	$z_1 + z_2 = (a + bi) + (c + di) = (a + c) + (b + d)i$ The addition of two complex numbers may be interpreted as the addition of two vectors
Subtraction	$z_1 - z_2 = (a + bi) - (c + di) = (a - c) + (b - d)i$
Multiplication	$z_1 z_2 = (a + bi)(c + di) = (ac - bd) + (ad + cb)i$
Division	This operation is defined for $z_2 \neq 0$ $\frac{z_1}{z_2} = \frac{a+bi}{c+di} = \frac{ac+bd}{c^2+d^2} + \frac{bc-ad}{c^2+d^2}i$
Conjugate	The conjugate of a complex number $z = a + bi$ is given by $z^* = a - bi$ Clearly, $z^{**} = z$ and $(z_1 + z_2)^* = z_1^* + z_2^*$. Further, $\mathrm{Re}(z) = z + z^*/2$ and $\mathrm{Im}(z) = z - z^*/2i$
Absolute value	The absolute value or modulus of a complex number $z = a + bi$ is given by $|z| = \sqrt{(a^2 + b^2)}$. Clearly, $z\,z^* = |z|^2$
Reciprocal	The reciprocal of a complex number z is defined for $z \neq 0$ and is given by $\frac{1}{z} = \frac{1}{a+bi} = \frac{a-bi}{a^2+b^2} = \frac{z^*}{|z|^2}$

Properties of Complex Numbers

The absolute value of a complex number z is denoted by $|z| = \sqrt{(a^2 + b^2)}$, and is just its distance from the origin. It has the following properties:

(i) $|z| \geq 0$ and $|z| = 0$ if and only if $z = 0$.
(ii) $|z| = |z^*|$.
(iii) $|z_1 + z_2| \leq |z_1| + |z_2|$ (this is known as the triangle inequality).
(iv) $|z_1 z_2| = |z_1|\,|z_2|$.
(v) $|1/z| = 1/|z|$.
(vi) $|z_1/z|_2 = |z_1|/|z_2|$.

Proof (iii)

$$\begin{aligned}
|z_1 + z_2|^2 &= (z_1 + z_2)(z_1 + z_2)^* \\
&= (z_1 + z_2)(z_1^* + z_2^*) \\
&= z_1 z_1^* + z_1 z_2^* + z_2 z_1^* + z_2 z_2^* \\
&= |z_1|^2 + z_1 z_2^* + z_2 z_1^* + |z_2|^2 \\
&= |z_1|^2 + z_1 z_2^* + (z_1 z_2^*)^* + |z_2|^2 \\
&= |z_1|^2 + 2\mathrm{Re}(z_1 z_2^*) + |z_2|^2 \\
&\leq |z_1|^2 + 2|z_1 z_2^*| + |z_2|^2 \\
&= |z_1|^2 + 2|z_1||z_2^*| + |z_2|^2 \\
&= |z_1|^2 + 2|z_1||z_2| + |z_2|^2 \\
&= (|z_1| + |z_2|)^2
\end{aligned}$$

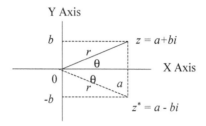

Fig. 26.2 Interpretation of complex conjugate

Therefore, $|z_1 + z_2| \leq |z_1| + |z_2|$ and so the triangle inequality is proved.

The modulus of z is used to define a distance function between two complex numbers, and $d(z_1, z_2) = |z_1 - z_2|$. This turns the complex numbers into a metric space.[3]

Interpretation of Complex Conjugate

The complex conjugate of the complex number $z = a + bi$ is defined as $z^* = a - bi$ and this is the reflection of z about the real axis in Fig. 26.2.

The modulus $|z|$ of the complex number z is the distance of the point z from the origin.

Polar Representation of Complex Numbers

The complex number $z = a + bi$ may also be represented in polar form (r, θ) in terms of its modulus $|z|$ and the argument θ.

$$\cos \theta = \frac{a}{\sqrt{a^2 + b^2}} = \frac{a}{|z|}$$

$$\sin \theta = \frac{b}{\sqrt{a^2 + b^2}} = \frac{b}{|z|}$$

Let r denote the modulus of z: i.e. $r = |z|$. Then, z may be represented by $(r \cos \theta + i r \sin \theta) = r (\cos \theta + i \sin \theta)$. Clearly, $\text{Re}(z) = r \cos \theta$ and $\text{Im}(z) = r \sin \theta$. Euler's formula (discussed below) states that $re^{i\theta} = r (\cos \theta + i \sin \theta)$.

The *principle argument* θ (denoted by Arg θ) is chosen so that $\theta \in [-\pi, \pi]$. There is, of course, more than one argument θ that will satisfy $z = r (\cos \theta + i \sin \theta)$. In fact, the full set of arguments (denoted by arg z) is given by arg $z = \theta + 2k\pi$, where $k \in \mathbb{Z}$ and satisfies $z = re^{i(\theta + 2k\pi)}$.

---

[3]A non-empty set X with a distance function d is a metric space if
(i) $d(x, y) \geq 0$ and $d(x, y) = 0 \Leftrightarrow x = y$
(ii) $d(z, y) = d(y, x)$
(iii) $d(x, y) \leq d(x, z) + d(z, y)$.

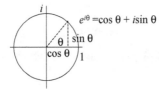

Fig. 26.3 Interpretation of Euler's formula

Euler's Formula

Euler's remarkable formula expresses the relationship between the exponential function for complex numbers and trigonometric functions. It is named after the eighteenth-century Swiss mathematician, Euler.

It may be interpreted as the function $e^{i\theta}$ traces out the unit circle in the complex plane as the angle θ ranges through the real numbers (Fig. 26.3). Euler's formula provides a way to convert between Cartesian coordinates and polar coordinates (r, θ). It states that

$$e^{i\theta} = \cos\theta + i\sin\theta$$

Further, the complex number $z = a + bi$ may be represented in polar coordinates as $z = r(\cos\theta + i\sin\theta) = re^{i\theta}$. Further,

$$e^z = e^a(\cos b + i\sin b) \quad \text{where } z = a + bi$$

Next, we prove Euler's formula: i.e. $e^{i\theta} = \cos\theta + i\sin\theta$.

Proof Recall the exponential expansion for e^x:

$$e^x = 1 + x + x^2/2! + x^3/3! + \cdots + x^r/r! + \cdots$$

The expansion of $e^{i\theta}$ is then given by

$$e^{i\theta} = 1 + i\theta + \frac{(i\theta)^2}{2!} + \frac{(i\theta)^3}{3!} + \cdots + \frac{(i\theta)^r}{r!} + \cdots$$

$$= 1 + i\theta - \frac{\theta^2}{2!} - \frac{i\theta^3}{3!} + \frac{\theta^4}{4!} + \frac{i\theta^5}{5!} + \cdots \frac{(i\theta)^r}{r!} + \cdots$$

$$= \left[1 - \frac{\theta^2}{2!} + \frac{\theta^4}{4!} - \frac{\theta^6}{6!} + \cdots\right] + i\left[\theta - \frac{\theta^3}{3!} + \frac{\theta^5}{5!} - \frac{\theta^7}{7!} + \cdots\right]$$

$$= \cos\theta + i\sin\theta$$

(This follows from the Taylor Series expansion of $\sin(\theta)e$ and $\cos(\theta)$).

De Moivre's Theorem

$(\cos \theta + i \sin \theta)^n = (\cos n\theta + i \sin n\theta)$ (where $n \in \mathbb{Z}$).

Proof This result is proved by mathematical induction and the result is clearly true for the base case $n = 1$.

Inductive Step:

The inductive step is to assume that the theorem is true for $n = k$ and to then show that it is true for $n = k + 1$. That is, we assume that

$$(\cos \theta + i \sin \theta)^k = (\cos k\theta + i \sin k\theta) \quad \text{(for some } k > 1)$$

We next show that the result is true for $n = k + 1$:

$$
\begin{aligned}
(\cos \theta + i \sin \theta)^{k+1} &= (\cos \theta + i \sin \theta)^k (\cos \theta + i \sin \theta) \\
&= (\cos k\theta + i \sin k\theta)(\cos \theta + i \sin \theta) \quad \text{(from inductive step)} \\
&= (\cos k\theta \cos \theta - \sin k\theta \sin \theta) + i(\cos k\theta \sin \theta + \sin k\theta \cos \theta) \\
&= \cos(k\theta + \theta) + i \sin(k\theta + \theta) \\
&= \cos(k+1)\theta + i \sin(k+1)\theta
\end{aligned}
$$

Therefore, we have shown that if the result is true for some value of n say $n = k$, then the result is true for $n = k + 1$. We have shown that the base case of $n = 1$ is true, and it therefore follows that the result is true for $n = 2, 3, \ldots$ and for all natural numbers. The result may also be shown to be true for the integers.

Complex Roots

Suppose that z is a non-zero complex number and that n is a positive integer. Then z has exactly n distinct n-th roots, and these roots are given in polar form by

$$\sqrt[n]{|z|} \left[\cos \left\{ \frac{\text{Arg } z + 2k\pi}{n} \right\} + i \sin \left\{ \frac{\text{Arg } z + 2k\pi}{n} \right\} \right]$$

for $k = 0, 1, 2, \ldots, n-1$.

Proof The objective is to find all complex numbers w such that $w^n = z$ where $w = |w| (\cos \Phi + i \sin \Phi)$. Using De Moivre's Theorem, this results in

$$|w|^n (\cos n\Phi + i \sin n\Phi) = |z|(\cos \theta + i \sin \theta)$$

Therefore, $|w| = \sqrt[n]{|z|}$ and $n\Phi = \theta + 2k\pi$ for some k. That is,

$$\Phi = (\theta + 2k\pi)/n = (\text{Arg} z + 2k\pi)/n$$

The choices $k = 0, 1, \ldots, n-1$ produce the distinct n-th roots of z.

Fundamental Theorem of Algebra

Every polynomial equation with complex coefficients has complex solutions, and the roots of a complex polynomial of degree n exist, and the n roots are all complex numbers.

Complex Derivatives

A function $f: A \rightarrow \mathbb{C}$ is said to be differentiable at a point z_0 if f is continuous at z_0

$$f'(z_0) = \lim_{z \to z_0} \frac{f(z) - f(z_0)}{z - z_0}$$

and if the limit below exists. The derivative at z_0 is denoted by $f'(z_0)$.

It is often written as

$$f'(z_0) = \lim_{h \to 0} \frac{f(z_0 + h) - f(z_0)}{h}$$

26.3   Quaternions

The Irish mathematician, Sir William Rowan Hamilton, discovered quaternions in the nineteenth-century (Fig. 26.4). Hamilton was born in Dublin in 1805 and attended Trinity College, Dublin. He was appointed professor of astronomy in 1827 while still an undergraduate. He made important contributions to optics, classical mechanics and mathematics.

He discovered quaternions in 1843 while he was walking with his wife from his home at Dunsink Observatory to the Royal Irish Academy in Dublin. This route followed the towpath of the Royal Canal, and Hamilton had a sudden flash of inspiration on the idea of quaternion algebra at Broom's Bridge. He was so

Fig. 26.4 William Rowan
Hamilton

Fig. 26.5 Plaque at Broom's Bridge

overcome with emotion at his discovery that he carved the quaternion formula into the stone on the bridge. Today, there is a plaque at Broom's Bridge that commemorates Hamilton's discovery (Fig. 26.5).

$$i^2 = j^2 = k^2 = ijk = -1$$

Quaternions are an extension of complex numbers and Hamilton had been trying to extend complex numbers to three-dimensional space without success. Complex numbers are numbers of the form $(a + bi)$ where $i^2 = -1$, and may be regarded as points on a two-dimensional plane. A quaternion number is of the form $(a + bi + cj + dk)$ where $i^2 = j^2 = k^2 = ijk = -1$, and may be regarded as points in four-dimensional space.

The set of quaternions is denoted $\mathbb{H}$ by and the quaternions form an algebraic system known as a division ring. The multiplication of two quaternions is not commutative: i.e. given $q_1, q_2 \in \mathbb{H}$ then $q_1, q_2 \neq q_2, q_1$. Quaternions were the first non-commutative algebraic structure to be discovered, and other non-commutative algebras (e.g. matrix algebra) were discovered in later years.

Quaternions have many applications in physics, quantum mechanics and theoretical and applied mathematics. Gibbs and Heaviside later developed vector analysis, and it replaced quaternions from the 1880s. Quaternions have become important in computing in recent years, as they are useful and efficient in describing rotations. They are applicable to computer graphics, computer vision and robotics.

26.3.1 Quaternion Algebra

Hamilton had been trying to extend the two-dimensional space of the complex numbers to a three-dimensional space of triples. He wanted to be able to add, multiply and divide triples of numbers but he was unable to make progress on the problem of the division of two triples.

The generalization of complex numbers to the four-dimensional quaternions rather than triples allows the division of two quaternions to take place. The quaternion is a number of the form $(a + bi + cj + dk)$ where $1, i, j, k$ are the basis elements (where 1 is the identity) and satisfy the following properties:

$$i^2 = j^2 = k^2 = ijk = -1 \quad \text{(Quaternion Formula)}$$

This formula leads to the following properties:

$$ij = k = -ji$$
$$jk = i = -kj$$
$$ki = j = -ik$$

These properties can be easily derived from the quaternion formula. For example:

$$ijk = -1$$
$$\Rightarrow ijkk = -k \quad \text{(Right multiplying by } k)$$
$$\Rightarrow ij(-1) = -k \quad \text{(since } k^2 = -1)$$
$$\Rightarrow ij = k$$

Similarly, from

$$ij = k$$
$$\Rightarrow i^2 j = ik \quad \text{(Left multiplying by } i)$$
$$\Rightarrow -j = ik \quad \text{(since } i^2 = -1)$$
$$\Rightarrow j = -ik$$

Table 26.2 represents the properties of quaternions under multiplication.

The quaternions ($\mathbb{H}$) are a four-dimensional vector space over the real numbers with three operations: addition, scalar multiplication and quaternion multiplication.

Addition and subtraction of Quaternions

The addition of two quaternions $q_1 = (a_1 + b_1 i + c_1 j + d_1 k)$ and $q_2 = (a_2 + b_2 i + c_2 j + d_2 k)$ is given by

$$q_1 + q_2 = (a_1 + a_2) + (b_1 + b_2)i + (c_1 + c_2)j + (d_1 + d_2)k$$
$$q_1 - q_2 = (a_1 - a_2) + (b_1 - b_2)i + (c_1 - c_2)j + (d_1 - d_2)k$$

Table 26.2 Basic quaternion multiplication

$\times$	1	i	j	k
1	1	i	j	k
i	i	-1	k	$-j$
j	j	$-k$	-1	i
k	k	j	$-i$	-1

Identity Element

The addition identity is given by the quaternion $(0 + 0i + 0j + 0k)$ and the multiplicative identity is given by $(1 + 0i + 0j + 0k)$.

Multiplication of Quaternions

The multiplication of two quaternions q_1 and q_2 is determined by the product of the basis elements and the distributive law. It yields:

$$
\begin{aligned}
q_1 \cdot q_2 = \; & a_1 a_2 + a_1 b_2 i + a_1 c_2 j + a_1 d_2 k \\
& + b_1 a_2 i + b_1 b_2 ii + b_1 c_2 ij + b_1 d_2 ik \\
& + c_1 a_2 j + c_1 b_2 ji + c_1 c_2 jj + c_1 d_2 jk \\
& + d_1 a_2 k + d_1 b_2 ki + d_1 c_2 kj + d_1 d_2 kk
\end{aligned}
$$

This may then be simplified to:

$$
\begin{aligned}
q_1 \cdot q_2 = \; & a_1 a_2 - b_1 b_2 - c_1 c_2 - d_1 d_2 \\
& + (a_1 b_2 + b_1 a_2 + c_1 d_2 - d_1 c_2)i \\
& + (a_1 c_2 - b_1 d_2 + c_1 a_2 + d_1 b_2)j \\
& + (a_1 d_2 + b_1 c_2 - c_1 b_2 + d_1 a_2)k
\end{aligned}
$$

The multiplication of two quaternions may be defined in terms of matrix multiplication. It is easy to see that the product of the two quaternions above is equivalent to:

$$
q_1 q_2 = \begin{pmatrix} a_1 & -b_1 & -c_1 & -d_1 \\ b_1 & a_1 & -d_1 & c_1 \\ c_1 & d_1 & a_1 & -b_1 \\ d_1 & -c_1 & b_1 & a_1 \end{pmatrix} \begin{pmatrix} a_2 \\ b_2 \\ c_2 \\ d_2 \end{pmatrix}
$$

This may also be written as:

$$
(a_2 b_2 c_2 d_2) \begin{pmatrix} a_1 & b_1 & c_1 & d_1 \\ -b_1 & a_1 & d_1 & -c_1 \\ -c_1 & -d_1 & a_1 & b_1 \\ -d_1 & c_1 & -b_1 & a_1 \end{pmatrix} = q_1 q_2
$$

Property of Quaternions under Multiplication

The quaternions are *not commutative* under multiplication. That is,

$$
q_1 q_2 \neq q_2 q_1
$$

The quaternions are associative under multiplication. That is,

$$q_1(q_2 q_3) = (q_1 q_2)q_3$$

Conjugation

The conjugation of a quaternion is analogous to the conjugation of a complex number. The conjugate of a complex number $z = (a + bi)$ is given by $z^* = (a - bi)$. Similarly, the conjugate of a quaternion is determined by reversing the sign of the vector part of the quaternion. That is, the conjugate of $q = (a + bi + cj + dk)$ (denoted by q^*) is given by $q^* = (a - bi - cj - dk)$.

Scalar and Vector Parts

A quaternion $(a + bi + cj + dk)$ consists of a *scalar* part a and a *vector* part $bi + cj + dk$. The scalar part is always real and the vector part is imaginary. That is, the quaternion q may be represented $q = (s, v)$ where s is the scalar part and v is the vector part. The scaler part of a quaternion is given by $q + q^*/2$ and the vector part is given by $q - q^*/2$. The *norm* of a quaternion q (denoted by $\|q\|$) is given by

$$\|q\| = \sqrt{qq^*} = \sqrt{q^*q} = \sqrt{a^2 + b^2 + c^2 + d^2}$$

A quaternion of norm one is termed a unit quaternion (i.e. $\|u\| = 1$). Any quaternion u where u is defined by $u = q/\|q\|$ is a unit quaternion. Given $\alpha \in \mathbb{R}$ then $\|\alpha q\| = |\alpha| \|q\|$. The inverse of a quaternion q is given by q^{-1} where

$$q^{-1} = q^*/\|q\|^2$$

and $qq^{-1} = q^{-1}q = 1$.

Given two quaternions p and q we have

$$\|pq\| = \|p\|\|q\|$$

The norm is used to define the distance between two quaternions, and the distance between two quaternions p and q (denoted by $d(p, q)$) is given by

$$\|p - q\|$$

Representing Quaternions with 2 × 2 Matrices over Complex Numbers

The quaternions have an interpretation under the 2 × 2 matrices where the basis elements i, j, k may be interpreted as matrices. Recall, that the multiplicative identity for 2 × 2 matrices is

$$1 = \begin{bmatrix} 1 & 0 \\ 0 & 1 \end{bmatrix} \quad -1 = \begin{bmatrix} -1 & 0 \\ 0 & -1 \end{bmatrix}$$

Consider then the quaternion basis elements defined as follows:

$$i = \begin{bmatrix} 0 & 1 \\ -1 & 0 \end{bmatrix} \quad j = \begin{bmatrix} 0 & i \\ i & 0 \end{bmatrix} \quad k = \begin{bmatrix} i & 0 \\ 0 & -i \end{bmatrix}$$

Then a simple calculation shows that

$$i^2 = j^2 = k^2 = ijk = \begin{bmatrix} -1 & 0 \\ 0 & -1 \end{bmatrix} = -1$$

Then the quaternion $q = (a + bi + cj + dk)$ may also be defined as

$$a \begin{bmatrix} 1 & 0 \\ 0 & 1 \end{bmatrix} + b \begin{bmatrix} 0 & 1 \\ -1 & 0 \end{bmatrix} + c \begin{bmatrix} 0 & i \\ i & 0 \end{bmatrix} + d \begin{bmatrix} i & 0 \\ 0 & -i \end{bmatrix}$$

This may be simplified to the complex matrix

$$q = \begin{bmatrix} a + di & b + ci \\ -b + ci & a - di \end{bmatrix}$$

and this is equivalent to:
where $u = a + di$ and $v = b + ci$.

The addition and multiplication of quaternions is then just the usual matrix addition

$$q = \begin{bmatrix} u & v \\ -v^* & u^* \end{bmatrix}$$

and multiplication. Quaternions may also be represented by 4×4 real matrices.

26.3.2   Quaternions and Rotations

Quaternions may be applied to computer graphics; computer vision and robotics, and unit quaternions provide an efficient mathematical way to represent rotations in three dimensions (Fig. 26.6). They offer an alternative to Euler angles and matrices.

The unit quaternion $q = (s, v)$ that computes the rotation about the unit vector u by an angle θ is given by

$$(\text{Cos}(\theta/2), u \, \text{Sin}(\theta/2))$$

The scalar part is given by $s = \text{Cos}(\theta/2)$ and the vector part is given by $v = u \, \text{Sin}(\theta/2)$.

A point p in space is represented by the quaternion $P = (0, p)$. The result of the rotation of p is given by

Fig. 26.6 Quaternions and
rotations

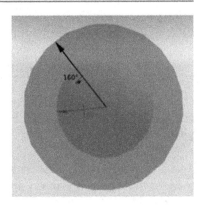

$$P_q = qPq^{-1}$$

Suppose we have two rotations represented by the unit quaternions q_1 and q_2, and that we first wish to perform q_1 followed by the rotation q_2. Then the composition of the two relations is given by applying q_2 to the result of applying q_1. This is given by the following:

$$P(q_2 \cdot q_1) = q_2(q_1 P q_1^{-1})q_2^{-1}$$
$$= q_2 q_1 P q_1^{-1} q_2^{-1}$$
$$= (q_2 q_1)P(q_2 q_1)^{-1}$$

26.4   Review Questions

1. What is a complex number?
2. Show how a complex number may be represented as a point in the complex plane.
3. Show that $|z_1 z_2| = |z_1| \, |z_2|$.
4. Evaluate the following
   (a) $(-1)^{1/4}$
   (b) $1^{1/5}$.
5. What is the fundamental theorem of algebra?
6. Show that $d/dz \, z^n = nz^{n-1}$.
7. What is a quaternion?

8. Investigate the application of quaternions to computer graphics, computer vision and robotics.

26.5 Summary

A complex number z is a number of the form $a + bi$ where a and b are real numbers and $i^2 = -1$. The set of complex numbers is denoted by $\mathbb{C}$, and a complex number has two parts, namely, its real part a and imaginary part b. The complex numbers are an extension of the set of real numbers, and complex numbers with a real part $a = 0$ are termed imaginary numbers. Complex numbers have many applications in physics, engineering and applied mathematics.

The Irish mathematician, Sir William Rowan Hamilton, discovered quaternions. Hamilton had been trying to extend the two-dimensional space of the complex numbers to a three-dimensional space of triples. He wanted to be able to add, multiply and divide triples of numbers, but he was unable to make progress on the problem of the division of two triples. His insight was that if he considered quadruples rather than triples that this structure would give him the desired mathematical properties. Hamilton also made important contributions to optics, classical mechanics and mathematics.

The generalization of complex numbers to the four-dimensional quaternions rather than triples allows the division of two quaternions to take place. The quaternion is a number of the form $(a + bi + cj + dk)$ where 1, i, j, k are the basis elements (where 1 is the identity) and satisfy the quaternion formula:

$$i^2 = j^2 = k^2 = ijk = -1$$

Quaternions have many applications in physics and quantum mechanics. Quaternions have become important in computing in recent years as they are useful and efficient in describing rotations. They are applicable to computer graphics, computer vision and robotics.

Calculus

27

Key Topics

Limit of a function
Mean value theorem
Taylor's theorem
Differentiation
Maxima and minima
Integration
Numerical analysis
Fourier series
Laplace transforms
Differential equations

27.1 Introduction

Newton and Leibniz independently developed calculus in the late seventeenth century[1]. It plays a key role in describing how rapidly things change and may be employed to calculate areas of regions under curves, volumes of figures and in

---

[1] The question of who invented the calculus led to a bitter controversy between Newton and Leibniz with the latter accused of plagiarizing Newton's work. Newton, an English mathematician and physicist, was the giant of the late seventeenth century, and Leibniz was a German mathematician and philosopher. Today, both Newton and Leibniz are credited with the independent development of the calculus.

© Springer Nature Switzerland AG 2020
G. O'Regan, *Mathematics in Computing*, Undergraduate Topics
in Computer Science, https://doi.org/10.1007/978-3-030-34209-8_27

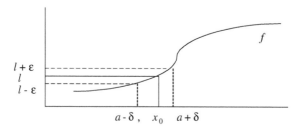

Fig. 27.1 Limit of a function

finding tangents to curves. It is an important branch of mathematics concerned with limits, continuity, derivatives and integrals of functions.

The concept of a *limit* is fundamental in the calculus. Let f be a function defined on the set of real numbers, then the limit of f at a is l (written as $\lim_{x \to a} f(x) = l$) if given any real number $\varepsilon > 0$ then there exists a real number $\delta > 0$ such that $|f(x) - l| < \varepsilon$ whenever $|x - a| < \delta$. The idea of a limit can be seen in Fig. 27.1.

The function f defined on the real numbers is *continuous* at a if $\lim_{x \to a} f(x) = f(a)$. The set of all continuous functions on the closed interval $[a, b]$ is denoted by C $[a, b]$.

If f is a function defined on an open interval containing x_0 then f is said to be *differentiable* at x_0 if the limit exists.

$$\lim_{x \to x_0} \frac{f(x) - f(x_0)}{x - x_0}$$

whenever this limit exists, it is denoted by $f''(x_0)$ and is *derivative* of f at x_0. Differential calculus is concerned with the properties of the derivative of a function. The derivative of f at x_0 is the slope of the tangent line to the graph of f at $(x_0, f(x_0))$ (Fig. 27.2).

It is easy to see that if a function f is differentiable at x_0, then f is continuous at x_0.

Rolle's Theorem

Suppose $f \in C[a, b]$ and f is differentiable on (a, b). If $f(a) = f(b)$ then there exists c such that $a < c < b$ with $f'(c) = 0$.

Fig. 27.2 Derivative as a tangent to curve

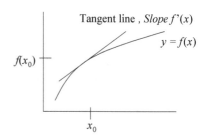

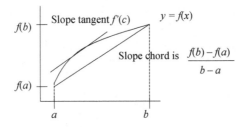

Fig. 27.3 Interpretation of mean value theorem

Mean Value Theorem

Suppose $f \in C[a, b]$ and f is differentiable on (a, b). Then there exists c such that $a < c < b$ with

$$f'(c) = \frac{f(b) - f(a)}{b - a}$$

Proof The mean value theorem is a special case of Rolle's theorem, and the proof involves defining the function $g(x) = f(x)-rx$ where $r = (f(b)-f(a))/(b-a)$.

It is easy to verify that $g(a) = g(b)$. Clearly, g is differentiable on (a, b) and so by Rolle's theorem, there is a c in (a, b) such that $g'(c) = 0$. Therefore, $f'(c)-r = 0$ and so $f'(c) = r = f(b)-f(a)/(b-a)$ as required.

Interpretation of the Mean Value Theorem

The mean value theorem essentially states that there is at least one point c on the curve $f(x)$ between a and b such that slope of the chord is the same as the tangent $f'(c)$ (Fig. 27.3).

Intermediate Value Theorem

Suppose $f \in C[a, b]$ and K is any real number between $f(a)$ and $f(b)$. Then there exists c in (a, b) for which $f(c) = K$.

Proof The proof of this relies on the completeness property of the real numbers. It involves considering the set S in $[a, b]$ such that $f(x) \leq K$ and noting that this set is non-empty since $a \in S$ and bounded above by b. Therefore, the supremum[2] sup $S = c$ exists and it is straightforward to show (using ε and δ arguments and the fact that f is continuous) that $f(c) = K$ (Fig. 27.4).

---

[2]The supremum is the least upper bound and the infimum is the greatest lower bound.

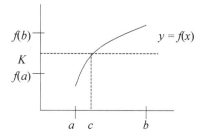

Fig. 27.4 Interpretation of intermediate value theorem

L'Hôpital's Rule

Suppose that $f(a) = g(a) = 0$ and that $f'(a)$ and $g'(a)$ exist and that $g'(a) \neq 0$. Then Then L'Hôpital's rule states that:

$$\lim_{x \to a} \frac{f(x)}{g(x)} = \frac{f'(a)}{g'(a)}$$

Proof Taylor's Theorem

$$\lim_{x \to a} \frac{f(x)}{g(x)} = \lim_{x \to a} \frac{f(x) - f(a)}{g(x) - g(a)}$$

$$= \lim_{x \to a} \frac{\frac{f(x) - f(a)}{x - a}}{\frac{g(x) - g(a)}{x - a}}$$

$$= \frac{\lim_{x \to a} \frac{f(x) - f(a)}{x - a}}{\lim_{x \to a} \frac{g(x) - g(a)}{x - a}}$$

$$= \frac{f'(a)}{g'(a)}$$

The Taylor series is concerned with the approximation to values of the function f near x_0. The approximation employs a polynomial (or power series) in powers of $(x - x_0)$ as well as the derivatives of f at $x = x_0$. There is an error term (or remainder) associated with the approximation.

Suppose $f \in C^n[a, b]$ and f^{n+1} exists on (a, b). Let $x_0 \in (a, b)$ then for every $x \in (a, b)$ there exists $\xi(x)$ between x_0 and x with

$$f(x) = P_n(x) + R_n(x)$$

where $P_n(x)$ is the nth Taylor polynomial for f about x_0, and $R_n(x)$ is the remainder term associated with $P_n(x)$. The infinite series obtained by taking the limit of $P_n(x)$ as $n \to \infty$ is termed the Taylor series for f about x_0.

$$P_n(x) = f(x_0) + f'(x_0)(x - x_0) + \frac{f''(x_0)}{2!}(x - x_0)^2 + \cdots + \frac{f^n(x_0)}{n!}(x - x_0)^n$$

The remainder term is give by

$$R_n(x) = \frac{f^{n+1}(\xi(x))(x - x_0)^{n+1}}{(n+1)!}.$$

27.2 Differentiation

Mathematicians of the seventeenth century were working on various problems concerned with motion. These included problems such as determining the motion or velocity of objects on or near the earth, as well as the motion of the planets. They were also interested in changes of motion: i.e. in the acceleration of these moving bodies.

Velocity is the rate at which distance changes with respect to time, and the average speed during a journey is the distance travelled divided by the elapsed time. However, since the speed of an object may be variable over a period of time, there is a need to be able to determine its velocity at a specific time instance. That is, there is a need to determine the rate of change of distance with respect to time at any time instant.

The direction in which an object is moving at any instant of its flight was also studied. For example, the direction in which a projectile is fired determines the horizontal and vertical components of its velocity. The direction in which an object is moving can vary from one instant to another.

The problem of finding the maximum and minimum values of a function was also studied: e.g. the problem to determine the maximum height that a bullet reaches when it is fired. Other problems studied including problems to determine the lengths of paths, the areas of figures and the volume of objects.

Newton and Leibniz (Figs. 27.5 and 27.6) showed that these problems could be solved by means of the concept of the derivative of a function: i.e. the rate of change of one variable with respect to another.

Rate of Change

The average rate of change and instantaneous rate of change are of practical interest. For example, if a motorist drives 200 miles in four hours then the average speed is 50 miles per hour: i.e. the distance travelled divided by the elapsed time.

Fig. 27.5 Isaac newton

Fig. 27.6 Wilhelm Gottfried
Leibniz

The actual speed during the journey may vary as if the driver stops for lunch then
the actual speed is zero for the duration of lunch.

The actual speed is the instantaneous rate of change of distance with respect to
time. This has practical implications as motorists are required to observe speed
limits, and a speed camera may record the actual speed of a vehicle with the driver
subject to a fine if the permitted speed limit has been exceeded. The actual speed is
relevant in a car crash as speed is a major factor in road fatalities.

In calculus, the term Δx means a change in x and Δy means the corresponding
change in y. The derivative of f at x is the instantaneous rate of change of f, and f is
said to be differentiable at x. It is defined as

$$\frac{dy}{dx} = \lim_{\Delta x \to 0} \frac{\Delta y}{\Delta x} = \lim_{\Delta x \to 0} \frac{f(x + \Delta x) - f(x)}{\Delta x}$$

In the formula, Δy is the increment $f(x + \Delta x) - f(x)$

The average velocity of a body moving along a line in the time interval t to $t + \Delta t$ where the body moves from position $s = f(t)$ to position $s + \Delta s$ is given by

$$V_{av} = \frac{\text{displacement}}{\text{Time travelled}} = \frac{\Delta s}{\Delta t} = \frac{f(t + \Delta t) - f(t)}{\Delta t}$$

The instantaneous velocity of a body moving along a line is the derivative of its position $s = f(t)$ with respect to t. It is given by

$$v = \frac{ds}{dt} = \lim_{\Delta t \to 0} \frac{\Delta s}{\Delta t} = f'(t)$$

27.2.1 Rules of Differentiation

(i)   The derivative of a constant is 0. That is, for $y = f(x) = c$ (a constant value) we have $dy/dx = 0$.

(ii) $d/dx(f + g) = df/dx + dg/dx$

(iii) The derivative of $y = f(x) = x^n$ is given by $dy/dx = nx^{n-1}$.

(iv) If c is a constant and u is a differentiable function of x then $dy/dx = c\, du/dx$ where $y = cu(x)$.

(v)   The product of two differentiable functions u and v is differentiable and $\frac{d}{dx}(uv) = v\frac{du}{dx} + u\frac{dv}{dx}$

(vi) The quotient of two differentiable functions u, v is differentiable (where $v \neq 0$) and $\frac{d}{dx}\left[\frac{u}{v}\right] = \frac{v\frac{du}{dx} - u\frac{dv}{dx}}{v^2}$

(vii) *Chain Rule.* Suppose $h = g \circ f$ is the composite of two differentiable functions $y = g(x)$ and $x = f(t)$. Then h is a differentiable function of t whose derivative at each value of t is: $h'(t) = (g \circ f)'(t) = g'(f(t))f'(t)$
      This may also be written as: $\frac{dy}{dt} = \frac{dy}{dx}\frac{dx}{dt}$

Derivatives of Well-Known Functions

The following are the derivatives of some well-known functions including basic trigonometric functions, the exponential function and the natural logarithm function:

(i) $d/dx \, \mathrm{Sin}x = \mathrm{Cos}x$

(ii) $d/dx \, \mathrm{Cos}x = -\mathrm{Sin}x$

(iii) $d/dx \, \mathrm{Tan}x = \mathrm{Sec}^2 x$

(iv) $d/dx \, e^x = e^x$

(v) $d/dx \, \ln x = 1/x$ where $x > 0$

(vi) $d/dx \, a^x = \ln(a)a^x$

(vii) $d/dx \, \log_a x = 1/x \ln(a)$

(viii) $d/dx \, \arcsin x = 1/\sqrt{(1-x^2)}$

(ix) $d/dx \arccos x = -1/\sqrt{(1-x^2)}$
(x) $d/dx \arctan x = 1/(1+x^2)$

Increasing and Decreasing Functions

Suppose that a function f has a derivative at every point x of an interval I. Then

(i) f increases on I if $f'(x) > 0$ for all x in I.
(ii) f decreases on I if $f'(x) < 0$ for all x in I.

The geometric interpretation of the first derivative test is that it states that differentiable functions increase on intervals where their graphs have positive slopes and decrease on intervals where their graphs have negative slopes.

If f' changes from positive to negative values as x passes from left to right through point c then the value of f at c is a *local maximum* value of f. Similarly, if f' changes from negative to positive values as x passes from left to right through point c then the value of f at c is a *local minimum* value of f (Fig. 27.7).

The graph of a differentiable function $y = f(x)$ is concave down in an interval where f' decreases and concave up in an interval where f' increases. This may be defined by the second interval test for concavity. In other words, the graph of $y = f(x)$ is concave down in an interval where $f' < 0$ and concave up in an interval where $f'' > 0$.

A point on the curve where the concavity changes is termed a point of inflection. That is, at a *point of inflection c* we have that f'' is positive on one side and negative on the other side. At the point of inflection c, the value of the second derivative is zero: i.e. $f''(c) = 0$.

27.3 Integration

The derivative is a functional operator that takes a function as an argument and returns a function as a result. The inverse operation involves determining the original function from the known derivative, and integral calculus is the branch of

Fig. 27.7 Local minima and maxima

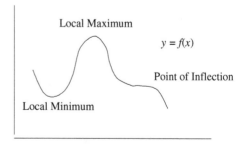

Local Maximum

$y = f(x)$

Point of Inflection

Local Minimum

the calculus concerned with this problem. The integral of a function consists of all those functions that have it as a derivative.

Integration is applicable to problems involving area and volume. It is the mathematical process that allows the area of a region with curved boundaries to be determined, and it also allows the volume of a solid to be determined.

The problem of finding functions whose derivatives is known involves finding a function $y = F(x)$ whose derivative is given by the differential equation:

$$\frac{dy}{dx} = f(x)$$

The solution to this differentiable equation over the interval I is F if F is differentiable at every point of I and for every x in I we have:

$$\frac{d}{dx}F(x) = f(x)$$

Clearly, if $F(x)$ is a particular solution to $d/dx \ F(x) = f(x)$ then the general solution is given by:

$$y = \int f(x)dx = F(x) + k$$

Since

$$d/dx(F(x) + k) = f(x) + 0 = f(x).$$

Rules of Integration

(i) $\int (u'(x)dx = u(x) + k$
(ii) $\int (u(x) + v(x))dx = \int u(x)dx + \int v(x)dx.$
(iii) $\int au(x)dx = a \int u(x)dx$ (where a is a constant)
(iv) $\int x^n dx = \frac{x^{n+1}}{n=1} + k$ (where $n \neq -1$)
(v) $\int \cos x \, dx = \sin x + k$
(vi) $\int \sin x \, dx = -\cos x + k$
(vii) $\int \sec^2 x \, dx = \tan x + k$
(viii) $\int e^x dx = e^x + k$

It is easy to check that the integration has been carried out correctly. This is done by computing the derivative of the function obtained and checking that it is the same as the function to be integrated.

Often, the goal will be to determine a particular solution satisfying certain conditions rather than the general solution. The general solution is first determined, and then the constant k that satisfies the particular solution is identified.

The *substitution method* of integration can often change an unfamiliar integral into one that is easier to evaluate. It is a useful method by which integrals are evaluated, and the procedure to evaluate $\int f(g(x))g'(x)dx$ where f', g' are continuous functions is as follows:

1. Substitute $u = g(x)$ and $du = g'(x)dx$ to obtain $\int f(u))du$.
2. Integrate with respect to u.
3. Replace u by $g(x)$ in the result.

The method of *integration by parts* is a rule of integration that transforms the integral of a product of functions into simpler integrals. It is a consequence of the product rule for differentiation.

$$\int udv = uv - \int vdu$$

$$\int f(x)g'(x)dx = f(x)g(x) - \int f'(x)g(x)dx$$

27.3.1 Definite Integrals

A definite integral defines the area under the curve $y = f(x)$, and the area of the region between the graph of a non-negative continuous function $y = f(x)$ for the interval $a \leq x \leq b$ of the x-axis is given by the definite integral.

The sum of the area of the rectangles approximates the area under the curve and the more rectangles that are used the better the approximation (Fig. 27.8).

The definition of the area of the region beneath the graph of $y = f(x)$ from a to b is defined to be the limit of the sum of the rectangle areas as the rectangles become smaller and smaller, and the number of rectangles used approaches infinity. The limit of the sum of the rectangle areas exists for any continuous function.

The approximation of the area under the graph $y = f(x)$ between $x = a$ and $x = b$ is done by dividing the region into n strips with each strip of uniform width given by $\Delta x = (b-a)/n$ and drawing lines perpendicular to the x-axis (Fig. 27.9).

Fig. 27.8 Area under the curve

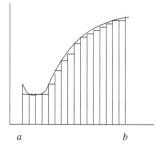

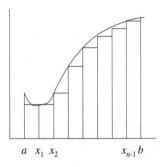

Fig. 27.9 Area under the curve—lower sum

Each strip is approximated with an inscribed rectangle where the base of the rectangle is on the x-axis to the lowest point on the curve above. We let c_k be a point in which f takes on its minimum value in the interval from x_{k-1} to x_k and the height of the rectangle is $f(c_k)$. The sum of these areas is the approximation of the area under the curve and is given by:

$$S_n = f(c_1)\Delta x + f(c_2)\Delta x + \cdots + f(c_n)\Delta x$$

The area under the graph of a non-negative continuous function f over the interval $[a, b]$ is the limit of the sums of the areas of inscribed rectangles of equal base length as n approaches infinity.

$$A = \lim_{n\to\infty} S_n$$

$$= \lim_{n\to\infty} f(c_1)\Delta x + f(c_2)\Delta x + \cdots + f(c_n)\Delta x$$

$$= \lim_{n\to\infty} \sum_{k=1}^{n} f(c_k)\Delta x$$

It is not essential that the division of $[a, b]$ into $a, x_1, x_2, \ldots, x_{n-1}, b$ gives equal subintervals $\Delta x_1 = x_1 - a$, $\Delta x_2 = x_2 - x_1$, $\ldots$ $\Delta x_n = b - x_{n-1}$. The *norm* of the subdivision is the largest interval length.

The lower Riemann sum L and the upper sum U can be formed, and the more finely divided that $[a, b]$ is the closer the values of the lower and upper sum U and L. The upper and lower sums may be written as:

$$L = \min_1 \Delta x_1 + \min_2 \Delta x_2 + \cdots + \min_n \Delta x_n$$

$$U = \max_1 \Delta x_1 + \max_2 \Delta x_2 + \cdots + \max_n \Delta x_n$$

$$\lim_{\text{norm}\to 0} U - L = 0 \left(\text{i.e., } \lim_{\text{norm}\to 0} U = \lim_{\text{norm}\to 0} L \right)$$

Further, if S = Σ $f(c_k)$ Δx_k (where c_k is any point in the subinterval and $\min_k \leq f$ $(c_k) \leq \max_k$ and we have

$$L \leq S \leq U$$

$$\lim_{\text{norm} \to 0} U = \lim_{\text{norm} \to 0} S = \lim_{\text{norm} \to 0} L$$

Integral Existence Theorem (Riemann Integral)

If f is continuous on $[a, b]$ then

$$\int_a^b f(x)dx = \lim_{\text{norm} \to 0} \sum f(c_k)\Delta x_k$$

exists and is the same number for any choice of the numbers c_k.

Properties of Definite Integrals

The following are some algebraic properties of the definite integral.

(i) $\int_a^a f(x)dx = 0$

(ii) $\int_a^a f(x)dx = - \int_a^b f(x)dx$

(iii) $\int_a^a kf(x)dx = k \int_a^b f(x)dx$

(iv) $\int_a^b f(x)dx \geq 0$ if $f(x) \geq 0$ on $[a, b]$

(v) $\int_a^b f(x)dx \leq \int_b^b g(x)dx$ if $f(x) \leq g(x)$ on $[a, b]$

(vi) $\int_a^b f(x)dx + \int_b^c f(x)dx = \int_a^c f(x)dx$

(vii) $\int_a^b |f(x) + g(x)dx| = \int_a^b f(x)dx + \int_a^b g(x)dx$

(viii) $\int_a^b |f(x) - g(x)dx| = \int_a^b f(x)dx - \int_a^b g(x)dx$

27.3.2 Fundamental Theorems of Integral Calculus

We present two fundamental theorems of integral calculus.

First Fundamental Theorem: (Existence of Anti-Derivative)

If f is continuous on $[a, b]$ then $F(x)$ is differentiable at every point x in $[a, b]$ where $F(x)$ is given by

$$F(x) = \int_a^x f(t)dt$$

Further,

$$\frac{dF}{dx} = \frac{d}{dx}\int_a^x f(t)dt = f(x)$$

That is, if f is continuous on $[a, b]$ then there exists a function $F(x)$ whose derivative on $[a, b]$ is f.

Second Fundamental Theorem: (Integral Evaluation Theorem)

If f is continuous on $[a, b]$ and F is any anti-derivative of f on $[a, b]$ then:

$$\int_a^b f(x)dx = F(b) - F(a)$$

That is, the procedure to calculate the definite integral of f over $[a, b]$ involves just two steps:

(i) Find an anti-derivative F of f
(ii) Calculate $F(b)–F(a)$

For a more detailed account of integral and differential calculus, the reader is referred to Finney (1988).

27.4 Numerical Analysis

Numerical analysis is concerned with devising methods for approximating solutions to mathematical problems. Often an exact formula is not available for solving a particular equation $f(x) = 0$, and numerical analysis provides techniques to approximate the solution in an efficient manner.

An algorithm is devised to provide the approximate solution, and it consists of a sequence of steps to produce the solution as efficiently as possible within defined accuracy limits. The maximum error due to the application of the numerical methods needs to be determined. The algorithm is implemented in a programming language such as Fortran.

There are several numerical techniques to determine the root of an equation $f(x) = 0$. These include techniques such as the bisection method, which has been used since ancient times, and the Newton–Raphson method developed by Sir Isaac Newton.

The bisection method is employed to find a solution to $f(x) = 0$ for the continuous function f on $[a, b]$ where $f(a)$ and $f(b)$ have opposite signs (Fig. 27.10). The method involves a repeated halving of subintervals of $[a, b]$, with each step locating the half that contains the root. The inputs to the algorithm are the endpoints a and b, the tolerance (TOL), and the maximum number of iterations N. The steps are as follows:

    1.   Initialise i to 1
    2.   while $i \leq N$
           i.   Compute midpoint p
                        $(p \rightarrow a + (b - a) / 2)$
          ii.   If $f(p) = 0$ or $(b - a) / 2 < $ TOL
                        Output p and stop
         iii.   If $f(a)f(p) > 0$
                        Set endpoint $a \rightarrow p$
                        Otherwise set $b \rightarrow p$
         iv. $i \rightarrow i + 1$

The Newton–Raphson method is a well-known technique to determine the roots of a function. It uses tangent lines to approximate the graph of $y = f(x)$ near the points where f is zero. The procedure for Newton's method is:

Newton's Method

(i)   Guess a first approximation to the root of the equation $f(x) = 0$.
(ii)   Use the first approximation to get a second, third and so on.
(iii) To go from the n-th approximation x_n to the next approximation x_{n+1} then use the formula:

Fig. 27.10 Bisection method

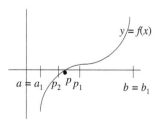

$$x_{n+1} = x_n - \frac{f(x_n)}{f'(x_n)}$$

where $f'(x_n)$ is the derivative of f at x_n.

Newton's method is very efficient for calculating roots as it converges very quickly. However, the method may converge to a different root than expected if the starting value is not close enough to the root sought.

The method involves computing the tangent line at $(x_n, f(x_n))$, and the approximation x_{n+1} is the point where the tangent intersects the x-axis.

Fixed-Point Iteration and Algorithm

A solution to the equation $g(x) = x$ is said to be a fixed point of the function g. There are sufficient conditions for the existence and uniqueness of a fixed point of a function. If $g \in C[a, b]$ and $g(x) \in [a, b]$ for all $x \in [a, b]$. Then g has a fixed point. Further, if g' exists on (a, b) and $0 < k < 1$ exists such that $|g'(x)| \leq k < 1$ for all $x \in (a, b)$ then g has a unique fixed-point p in $[a, b]$.

The approximation of the fixed point of a function g involves choosing an initial approximation p_0 and generating the sequence $\{p_n\}_1^\infty$ by letting $p_n = g(p_{n-1})$ for each $n \geq 1$. If the sequence converges to p and g is continuous,[3] then

$$p = \lim_{n\to\infty} p_n = \lim_{n\to\infty} g(p_{n-1}) = g\left(\lim_{n\to\infty} p_{n-1}\right) = g(p)$$

This technique is called fixed-point iteration and the algorithm is as follows:

(i)   Make an initial approximation p_0 to the fixed-point p.
(ii)   Use the first approximation to get a second, third and so on by computing $p_i = g(p_{i-1})$.
(iii) Continue for a fixed number of iterations or until$| p_k{-}p_{k-1}| < $ TOL.
(iv)   The approximation to the fixed point is p_k if$|p_k{-}p_{k-1}| < $ TOL.

27.5 Fourier Series

Fourier series is named after Joseph Fourier, a nineteenth-century French mathematician, and is used to solve practical problems in physics. A Fourier series consists of the sum of a possibly infinite set of sine and cosine functions. The Fourier series for f on the interval $0 \leq x \leq l$ defines a function f whose value at each point is the sum of the series for that value of x.

---

[3] For a function that is continuous at x_0 then for any sequence $\{x_n\}$ converging on x_0 then $\lim_{n\to\infty} f(x_n) = f(x_0)$.

$$f(x) = \frac{a_0}{2} + \sum_{m=1}^{\infty} \left[a_m \cos \frac{m\pi x}{l} + b_m \sin \frac{m\pi x}{l} \right]$$

The sine and cosine functions are periodic functions.

Note 1: (Period of Function)

A function f is periodic with period $T > 0$ if $f(x + T) = f(x)$ for every value of x. The sine and cosine functions are periodic with period 2π: i.e. $\sin(x + 2\pi) = \sin(x)$ and $\cos(x + 2\pi) = \cos(x)$. The functions $\sin m\pi x/l$ and $\cos m\pi x/l$ have period $T = 2\ l/m$.

Note 2: (Orthogonality)

Two functions f and g are said to be orthogonal on $a \le x \le b$ if:

$$\int_a^b f(x)g(x)dx = 0$$

A set of functions is said to be mutually orthogonal if each distinct pair in the set is orthogonal. The functions $\sin m\pi x/l$ and $\cos m\pi x/l$ where $m = 1, 2,...$ form a mutually orthogonal set of functions on the interval $-l \le x \le l$ and they satisfy the following orthogonal relations:

$$\int_{-l}^{l} \cos \frac{m\pi x}{l} \sin \frac{n\pi x}{l} dx = 0 \quad \text{all } m, n$$

$$\int_{-l}^{l} \cos \frac{m\pi x}{l} \cos \frac{n\pi x}{l} dx = \begin{cases} 0 & m \ne n \\ l & m = n \end{cases}$$

$$\int_{-l}^{l} \sin \frac{m\pi x}{l} \sin \frac{n\pi x}{l} dx = \begin{cases} 0 & m \ne n \\ l & m = n \end{cases}$$

The orthogonality property of the set of sine and cosine functions allows the coefficients of the Fourier series to be determined. Thus, the coefficients a_n, b_n for the convergent Fourier series $f(x)$ are given by

$$a_n = \frac{1}{l} \int_{-l}^{l} f(x) \cos \frac{n\pi x}{l} dx \quad n = 0, 1, 2, \ldots$$

$$b_n = \frac{1}{l} \int_{-l}^{l} f(x) \sin \frac{n\pi x}{l} dx \quad n = 1, 2, \ldots$$

The values of the coefficients a_n and b_n are determined from the integrals and the ease of computation depends on the particular function f involved.

$$f(x) = \frac{a_0}{2} + \sum_{m=1}^{\infty} \left[a_m \cos \frac{m\pi x}{l} + b_m \sin \frac{m\pi x}{l} \right]$$

The values of a_n and b_n depends only on the value of $f(x)$ in the interval $-l \leq x \leq l$. The terms in the Fourier series are periodic with period $2l$ and the function converges for all x whenever it converges on $-l \leq x \leq l$. Further, its sum is a periodic function with period $2l$ and therefore $f(x)$ is determined for all x by its values in the interval $-l \leq x \leq l$.

27.6 The Laplace Transform

An integral transform takes a function f and transforms it to another function F by means of an integral. Often, the objective is to transform a problem for f into a simpler problem, and then to recover the desired function from its transform F. Integral transforms are useful in solving differential equations, and an integral transform is a relation of the form:

$$F(s) = \int_{\alpha}^{\beta} K(s,t) f(t) dt$$

The function F is said to be the transform of f, and the function K is the kernel of the transformation.

The Laplace transform is named after the well-known eighteenth-century French mathematician and astronomer, Pierre Laplace. The Laplace transform of f (denoted by $\mathscr{L}\{f(t)\}$ or $F(s)$) is given by

$$\mathscr{L}\{f(t)\} = F(s) = \int_{0}^{\infty} e^{-st} f(t) dt$$

The kernel $K(s,t)$ of the transformation is e^{-st} and the Laplace transform is defined over an integral from zero to infinity. This is defined as a limit of integrals over finite intervals as follows:

$$\int_{a}^{\infty} f(t) dt = \lim_{A \to \infty} \int_{a}^{A} f(t) dt$$

Theorem (Sufficient Condition for Existence of Laplace Transform)

Suppose that f is a piecewise continuous function on the interval $0 \leq x \leq A$ for any positive A and $|f(t)| \leq Ke^{at}$ when $t \geq M$ where a, K, M are constants and K, $M > 0$ then the Laplace transform $\mathcal{L}\{f(t)\} = F(s)$ exists for $s > a$.

The following examples are Laplace transforms of some well-known elementary functions:

$$
\begin{aligned}
\mathcal{L}\{1\} &= \int_0^{\infty} e^{-st} dt = \tfrac{1}{s}, & s > 0 \\
\mathcal{L}\{e^{at}\} &= \int_0^{\infty} e^{-st} e^{at} dt = \tfrac{1}{s-a} & s > a \\
\mathcal{L}\{\sin at\} &= \int_0^{\infty} e^{-st} \sin at \, dt = \tfrac{a}{s^2+a^2} & s > 0
\end{aligned}
$$

27.7 Differential Equations

Many important problems in engineering and physics involve determining a solution to an equation that contains one or more derivatives of the unknown function. Such an equation is termed a differential equation, and the study of these equations began with the development of the calculus by Newton and Leibniz.

Differential equations are classified as ordinary or partial based on whether the unknown function depends on a single independent variable or on several independent variables. In the first case, only ordinary derivatives appear in the differential equation, and it is said to be an *ordinary differential equation*. In the second case, the derivatives are partial, and the equation is termed *a partial differential equation*.

For example, Newton's second law of motion $(F = ma)$ expresses the relationship between the force exerted on an object of mass m and the acceleration of the object. The force vector is in the same direction as the acceleration vector. It is given by the ordinary differential equation:

$$
m \frac{d^2 x(t)}{dt^2} = F(x(t))
$$

The next example is that of a second-order partial differential equation. It is the wave equation and is used for the description of waves (e.g. sound, light and water waves) as they occur in physics. It is given by

$$
a^2 \frac{\partial^2 u(x,t)}{\partial x^2} = \frac{\partial^2 u(x,t)}{\partial t^2}
$$

There are several fundamental questions with respect to a given differential equation. First, there is the question as to the existence of a solution to the differential equation. Second, if it does have a solution then is this solution unique.

A third question is to how to determine a solution to a particular differential equation.

Differential equations are classified as to whether they are linear or nonlinear. The ordinary differential equation $F(x, y, y', \ldots, y^{(n)}) = 0$ is said to be *linear* if F is a linear function of the variables $y, y', \ldots, y^{(n)}$. The general ordinary differential equation is of the form:

$$a_0(x)y^{(n)} + a_1(x)y^{(n-1)} + , \ldots, a_n(x)y = g(x)$$

A similar definition applies to partial differential equations and an equation is *nonlinear* if it is not linear.

27.8 Review Questions

1. Explain the concept of the limit and what it means for the limit of the function f at a to be l.
2. Explain the concept of continuity.
3. Explain the difference between average velocity and instantaneous velocity, and explain the concept of the derivative of a function.
4. Determine the following:

   (a) $\lim\limits_{x \to 0} \sin x$

   (b) $\lim\limits_{x \to 0} x \cos x$

   (c) $\lim\limits_{x \to \infty} |x|$

5. Determine the derivative of the following functions:

   (a) $y = x^3 + 2x + 1$
   (b) $y = x^2 + 1, x = (t+1)^2$
   (c) $y = \cos x^2$

6. Determine the integral of the following functions:

   (a) $\int (x^2 - 6x)dx$.
   (b) $\int \sqrt{(x-6)}dx$
   (c) $\int (x^2 - 4)^2 3x^3 dx$

7. State and explain the significance of the first and second fundamental theorems of the calculus.
8. What is a periodic function and give examples?
9. Describe applications of Fourier series, Laplace transforms and differential equations.

27.9   Summary

This chapter provided a brief introduction to the calculus including differentiation, integration, numerical analysis, Fourier series, Laplace transforms and differential equations.

Newton and Leibniz developed the calculus independently in the late seventeenth century. It plays a key role in describing how rapidly things change and may be employed to calculate areas of regions under curves, volumes of figures and in finding tangents to curves.

In calculus, the term Δx means a small change in x and Δy means the corresponding change in y. The derivative of f at x is the instantaneous rate of change of f and is defined as

$$\frac{dy}{dx} = \lim_{\Delta x \to 0} \frac{\Delta y}{\Delta x} = \lim_{\Delta x \to 0} \frac{f(x + \Delta x) - f(x)}{\Delta x}$$

Integration is the inverse operation of differentiation and involves determining the original function from the known derivative. The integral of a function consists of all those functions that have the function as a derivative.

Integration is applicable to problems involving area and volume, and it allows the area of a region with curved boundaries to be determined.

Numerical analysis is concerned with devising methods for approximating solutions to mathematical problems. Often an exact formula is not available for solving a particular problem, and numerical analysis provides techniques to approximate the solution in an efficient manner.

A Fourier series consists of the sum of a possibly infinite set of sine and cosine functions. A differential equation is an equation that contains one or more derivatives of the unknown function.

This chapter has sketched some important results in the calculus, and the reader is referred to Finney (1988) for more detailed information.

Reference

Finney T (1988) Calculus analytic and geometry, 7th edn. Addison Wesley

Epilogue

28

We embarked on a long journey in this book to give readers a flavour of the mathematics used in the computing field. The first chapter introduced analog and digital computers, and the fundamental architecture underlying a digital computer. Chapter 2 discussed the foundations of computing, including the binary number system and the step reckoner calculating machine that were invented by Leibniz. Babbage designed the difference engine as a machine to evaluate polynomials, and his analytic engine provided the vision of a modern computer. Boole's symbolic logic is the foundation for digital computing.

Chapter 3 provided an introduction to fundamental building blocks in mathematics including sets, relations and functions. A set is a collection of well-defined objects and it may be finite or infinite. A relation between two sets A and B indicates a relationship between members of the two sets and is a subset of the Cartesian product of the two sets. A function is a special type of relation such that for each element in A there is at most one element in the codomain B. Functions may be partial or total and injective, surjective or bijective.

Chapter 4 presented a short introduction to algorithms, where an algorithm is a well-defined procedure for solving a problem. It consists of a sequence of steps that takes a set of values as input and produces a set of values as output. A computer program implements the solution algorithm in some programming language.

Chapter 5 presented the fundamentals of number theory and discussed prime number theory and the greatest common divisor and least common multiple of two numbers.

Chapter 6 discussed algebra including simple and simultaneous equations and the method of elimination and substitution that are used to solve simultaneous equations. We showed how quadratic equations may be solved by factorization, completing the square or using the quadratic formula. We presented the laws of logarithms and indices and discussed monoids, groups, rings, integral domains, fields and vector space that are used in abstract algebra.

© Springer Nature Switzerland AG 2020
G. O'Regan, *Mathematics in Computing*, Undergraduate Topics
in Computer Science, https://doi.org/10.1007/978-3-030-34209-8_28

Chapter 7 discussed sequences and series, and permutations and combinations. Arithmetic and geometric sequences and series were discussed as well as their applications to the calculation of compound interest and annuities.

Chapter 8 discussed mathematical induction and recursion. There are two parts to a proof by induction (the base case and the inductive step). We discussed strong and weak induction, and how recursion is used to define sets, sequences and functions. This led to structural induction, which is used to prove properties of recursively defined structures.

Chapter 9 discussed graph theory where a graph $G = (V, E)$ consists of vertices and edges. It deals with the arrangements of vertices and edges between them, and has been applied to practical problems such as the modelling of computer networks, determining the shortest driving route between two cities and the travelling salesman problem.

Chapter 10 discussed cryptography, which is an important application of number theory. The codebreaking work done at Bletchley Park in England during the Second World War was discussed, as well as private and public key cryptosystems.

Chapter 11 presented coding theory which is concerned with error detection and error correction codes. The underlying mathematics was discussed, which included abstract mathematics such as vector spaces.

Chapter 12 discussed language theory including grammars, parse trees and derivations from a grammar. The important area of programming language semantics was discussed, including axiomatic, denotational and operational semantics.

Chapter 13 discussed computability and decidability. The Church–Turing thesis states that anything that is computable is computable by a Turing machine. Church and Turing showed that mathematics is not decidable. In other words, there is no mechanical procedure (i.e. algorithm) to determine whether an arbitrary mathematical proposition is true or false, and so the only way is to determine the truth or falsity of a statement is try to solve the problem.

Chapter 14 discussed matrices including 2×2 and general $n \times m$ matrices. Various operations such as the addition and multiplication of matrices were discussed, as well as the calculation of the determinant and inverse of a matrix was discussed.

Chapter 15 presented a short history of logic, and we discussed the Greek contributions to syllogistic logic, stoic logic, fallacies and paradoxes. Boole's symbolic logic and its application to digital computing were discussed, and we considered Frege's work on predicate logic.

Chapter 16 provided an introduction to propositional and predicate logic. Propositional logic may be used to encode simple arguments that are expressed in natural language, and to determine their validity. The nature of mathematical proof was discussed, and we presented proof by truth tables, semantic tableaux and natural deduction. Predicate logic allows complex facts about the world to be represented, and new facts may be determined via deductive reasoning. Predicate calculus includes predicates, variables and quantifiers, and a predicate is a characteristic or property that the subject of a statement can have.

Chapter 17 presented some advanced topics in logic including fuzzy logic, temporal logic, intuitionistic logic, undefined values, theorem provers and the applications of logic to AI. Fuzzy logic is an extension of classical logic that acts as a mathematical model for vagueness. Temporal logic is concerned with the expression of properties that have time dependencies, and it allows temporal properties about the past, present and future to be expressed. Intuitionism was a controversial theory on the foundations of mathematics based on a rejection of the law of the excluded middle, and an insistence on constructive existence. We discuss three approaches to deal with undefined values, including the logic of partial functions; Dijkstra's approach with his cand and cor operators; and Parnas's approach which preserves a classical two-valued logic.

Chapter 18 discussed the nature of proof and theorem proving, including automated and interactive theorem provers. We discussed the nature of mathematical proof and formal mathematical proof. Chapter 19 discussed software engineering and the mathematics to support software engineering.

Chapter 20 discussed software reliability and dependability, including topics such as software reliability and software reliability models; the cleanroom methodology, system availability, safety and security-critical systems, and dependability engineering.

Chapter 21 discussed formal methods, which consist of a set of techniques to rigorously specify and derive a program from its specification. Formal methods may be employed to rigorously state the requirements of the proposed system; they may be employed to derive a program from its mathematical specification; and they may provide a rigorous proof that the implemented program satisfies its specification. They have been mainly applied to the safety-critical field.

Chapter 22 presented the Z specification language, which was developed at Oxford University in the U.K. Chapter 23 discussed automata theory, including finite-state machines, pushdown automata and Turing machines. Finite-state machines are abstract machines that are in only one state at a time, and the input symbol causes a transition from the current state to the next state. Pushdown automata have greater computational power than finite-state machines, and they contain extra memory in the form of a stack from which symbols may be pushed or popped. The Turing machine is the most powerful model for computation, and this theoretical machine is equivalent to an actual computer in the sense that it can compute exactly the same set of functions.

Chapter 24 discussed model checking which is an automated technique such that given a finite-state model of a system and a formal property, then it systematically checks whether the property is true or false in a given state in the model. It is an effective technique to identify potential design errors, and it increases the confidence in the correctness of the system design.

Chapter 25 discussed probability and statistics and included a discussion on discrete and continuous random variables; probability distributions; sample spaces; sampling; the abuse of statistics; variance and standard deviation; and hypothesis testing. The application of probability to the software reliability field was discussed.

Chapter 26 discussed complex numbers and quaternions. Complex numbers are of the form $a + bi$ where a and b are real numbers, and $i^2 = -1$. Quaternions are a generalization of complex numbers to quadruples that satisfy the quaternion formula $i^2 = j^2 = k^2 = -1$.

Chapter 27 gave a very short introduction to calculus and provided a high-level overview of limits, continuity, differentiation, integration, numerical analysis, Fourier series, Laplace transforms and differential equations.

Chapter 28 is the concluding chapter in which we summarized the journey that we travelled in this book.

Glossary

ABC Atanasoff–Berry Computer

ACL A Computational Logic

ACM Association for Computing Machinery

AECL Atomic Energy Canada Ltd.

AES Advanced Encryption Standard

AMN Abstract Machine Notation

APL A Programming Language

ATP Automated Theorem Proving

BCH Bose, Chaudhuri and Hocquenghem

BNF Backus–Naur Form

CCS Calculus Communicating Systems

CICS Customer Information Control System

CMM Capability Maturity Model

CMMI® Capability Maturity Model Integration

COBOL Common Business-Oriented Language

CP/M Control Program for Microcomputers

CPU Central Processing Unit

CSP Computational Tree Logic

DES Data Encryption Standard

DPDA Deterministic PDA

DSA Digital Signature Algorithm

DSS Digital Signature Standard

© Springer Nature Switzerland AG 2020
G. O'Regan, *Mathematics in Computing*, Undergraduate Topics
in Computer Science, https://doi.org/10.1007/978-3-030-34209-8

DVD Digital Versatile Disc

EDVAC Electronic Discrete Variable Automatic Computer

ENIAC Electronic Numerical Integrator and Computer

FSM Finite-State Machine

GCHQ General Communications Headquarters

GSM Global System Mobile

HR Human Resources

HOL Higher Order Logic

IBM International Business Machines

IEC International Electrotechnical Commission

IEEE Institute of Electrical and Electronics Engineers

ISO International Standards Organization

ITP Interactive Theorem Proving

LEM Law of Excluded Middle

LISP List Processing

LT Logic Theorist

LTL Linear Time Logic

LPF Logic of Partial Functions

MIT Massachusetts Institute of Technology

MOD Ministry of Defence

MTBF Mean Time Between Failure

MTTF Mean Time to Failure

NATO North Atlantic Treaty Organization

NBS National Bureau of Standards

NIST National Institute of Standards & Technology

NQTHM New Quantified Theorem Prover,

OTTER Organized Techniques for Theorem proving and Effective Research

PDA Pushdown Automata

PL/1 Programming Language 1

PMP Project Management Professional

PVS Prototype Verification System

RDBMS Relational Database Management System

RSA Rivest, Shamir and Adleman

SAM Semi-automated Mathematics

SCAMPI Standard CMM Appraisal Method for Process Improvement

SECD Stack, Environment, Code, Dump

SEI Software Engineering Institute

SRI Stanford Research Institute

TM Turing Machine

UML Unified Modelling Language

UMTS Universal Mobile Telecommunications System

VDM Vienna Development Method

VDM✦ Irish School of VDM

Y2K Year 2000

Bibliography

O' Regan G (2012) A brief history of computing, 2nd edn. Springer

© Springer Nature Switzerland AG 2020
G. O'Regan, *Mathematics in Computing*, Undergraduate Topics
in Computer Science, https://doi.org/10.1007/978-3-030-34209-8

Index

A
Abstract algebra, 108
Abuse of statistics, 400
Ada Lovelace, 21, 22
Agile development, 309
Algebra, 99
Algorithm, 215
Alonzo Church, 53, 215
Alphabets and words, 186
Analog computers, 2
Analytic engine, 19
Annuity, 123
Application of functions, 53
Applications of relations, 47
Arithmetic sequence, 119
Arithmetic series, 120
Artificial intelligence, 286
Automata theory, 32, 57, 373
Automath system, 298
Axiomatic approach, 341
Axiomatic semantics, 193

B
Babbage, 17
Backus–Naur form, 190
Bags, 363
Bandera, 390
Bertrand Russell, 210
Bijective, 51
Binary numbers, 15
Binary relation, 32, 41, 60
Binary system, 95
Binary trees, 149
Binomial distribution, 399
Block codes, 174
Bombe, 73, 158, 161, 166
Boole, 22
Boole's symbolic logic, 241

Boyer–Moore theorem prover, 299

C
Caesar cipher, 156
Calculus Communicating System (CCS), 348
Capability Maturity Model Integration (CMMI), 316
Cayley–Hamilton theorem, 230
Central limit theorem, 405
Chomsky hierarchy, 188
Church–Turing thesis, 215
Classical engineers, 306
Classical mathematics, 314
Claude Shannon, 25, 242
Cleanroom, 320
Cleanroom methodology, 323
Coding theory, 171
Combination, 127
Communicating Sequential Process (CSP), 348
Commuting diagram property, 368
Competence set, 47
Completeness, 214
Complete partial orders, 203
Complex numbers, 412
Compound interest, 117
Computability, 215
Computability and decidability, 209
Computable function, 54, 198
Computational tree logic, 389
Computer algorithms, 61
Computer representation of numbers, 95
Computer representation of sets, 40
Computer security, 329
Concurrency, 387
Conditional probability, 396
Correlation, 399
Covariance, 398
Cramer's rule, 229

© Springer Nature Switzerland AG 2020
G. O'Regan, *Mathematics in Computing*, Undergraduate Topics
in Computer Science, https://doi.org/10.1007/978-3-030-34209-8

CPSIA information can be obtained
at www.ICGtesting.com
Printed in the USA
LVHW050102070220
646091LV00002B/40

9 783030 342081